REF
QH
455
.E67
2003
HWLCSC
Chicago Public Library
R0403373553
D0165300

Chicago Public Library
C
P
L
REFERENCE
Form 178 rev. 11-00

Geographical Genetics

MONOGRAPHS IN POPULATION BIOLOGY

EDITED BY SIMON A. LEVIN AND HENRY S. HORN

Titles available in the series (by monograph number)

1. *The Theory of Island Biogeography,* by Robert H. MacArthur and Edward O. Wilson
2. *Evolution in Changing Environments: Some Theoretical Explorations,* by Richard Levins
3. *Adaptive Geometry of Trees,* by Henry S. Horn
4. *Theoretical Aspects of Population Genetics,* by Motoo Kimura and Tomoko Ohta
5. *Populations in a Seasonal Environment,* by Steven D. Fretwell
6. *Stability and Complexity in Model Ecosystems,* by Robert M. May
7. *Competition and the Structure of Bird Communities,* by Martin L. Cody
8. *Sex and Evolution,* by George C. Williams
9. *Group Selection in Predator-Prey Communities,* by Michael E. Gilpin
10. *Geographic Variation, Speciation, and Clines,* by John A. Endler
11. *Food Webs and Niche Space,* by Joel F. Cohen
12. *Caste and Ecology in the Social Insects,* by George F. Oster and Edward O. Wilson
13. *The Dynamics of Arthropod Predator-Prey Systems,* by Michael P. Hassel
14. *Some Adaptations of Marsh-Nesting Blackbirds,* by Gordon H. Orians
15. *Evolutionary Biology of Parasites,* by Peter W. Price
16. *Cultural Transmission and Evolution: A Quantitative Approach,* by L. L. Cavalli-Sforza and M. W. Feldman
17. *Resource Competition and Community Structure,* by David Tilman
18. *The Theory of Sex Allocation,* by Eric L. Charnov
19. *Mate Choice in Plants: Tactics, Mechanisms, and Consequences,* by Mary F. Wilson and Nancy Burley
20. *The Florida Scrub Jay: Demography of a Cooperative-Breeding Bird,* by Glen E. Woolfenden and John W. Fitzpatrick
21. *National Selection in the Wild,* by John A. Endler

Geographical Genetics

BRYAN K. EPPERSON

PRINCETON UNIVERSITY PRESS
Princeton and Oxford

Published by Princeton University Press, 41 William Street,
Princeton, New Jersey 08540
In the United Kingdom: Princeton University Press, 3 Market Place, Woodstock,
Oxfordshire OX20 1SY

Library of Congress Cataloging-in-Publication Data

Epperson, Bryan K., 1957–
Geographical genetics / Bryan K. Epperson.
p. cm. — (Monographs in population biology ; 38)
Includes bibliographical references (p.).
ISBN 0-691-08668-0 (cl : alk. paper) — ISBN 0-691-08669-9 (pb : alk. paper)
1. Population genetics. 2. Medical geography. 3. Population geography.
I. Title. II. Series.

QH455 .E67 2003
577.8′8—dc21 2002190855

British Library Cataloging-in-Publication Data is available

This book has been composed in Times Roman

Printed on acid-free paper. ∞

www.pupress.princeton.edu

Printed in the United States of America

10 9 8 7 6 5 4 3 2 1

Dedicated to the Memory of

Gustave Malécot

Contents

Preface xi

1. Space–Time Population Genetics 1

2. Geographical Patterns Observed in Nature 11
   - Standard Patterns of Genetic Isolation by Distance 14
   - Directional Patterns 25
   - Complex Paths of Migration 31
   - Other Complexities of Migration Processes 34
   - Effects of Temporal Changes on Spatial Patterns 36
   - Spatial Patterns Caused by Selection 37
   - Partition of Genetic Variation 40

3. Ancient Events in Spatial–Temporal Processes 43
   - Out of Africa 45
   - Patterns Caused by More Recent Expansions of Human Populations 63
   - Patterns in Other Species 64

4. Spatial and Space–Time Statistics 67
   - Spatial Autocorrelation Statistics 68
   - Moran's *I* Statistic 71
   - Correlograms 77
   - Mantel Tests for Correlations of Genetic Distance with Geographic Distance 79
   - Measures of Geographic Distances and Special Weighting Schemes 81
   - Directional Autocorrelation and Special Weighting Schemes 84
   - Spatial Statistics for Localized and Systemic Spatial Nonstationarity 87
   - Multiple Genetic Characters 92

Spatial Models and Statistical Analyses of Spatial Patterns 95
Space–Time Analysis 99
Measures of Kinship 106
Measures of Overall Degree of Genetic Differentiation of a System 107
Other Estimators 110

5. Theory of Genetics as Stochastic Spatial–Temporal Processes 114
Spatial Probabilities of Identity by Descent 116
Spatial and Space–Time Genetic Correlations 124
Stochastic Migration 136
Negatively Shared Stochastic Migration Effects 147
Migration Matrices 148
F Statistics and Wright's Island Model 149
Space–Time Probabilities of Identity by Descent 152
Space–Time Coalescence Probabilities 158
Heterozygosity 160
Mutation Models 160
Clines 164
Population Expansions 167
Multilocus Processes 168
Theoretical Dispersal Curves 170

6. Synthesis: Tying Spatial Patterns among Populations to Space–Time Processes 172

7. Spatial Patterns Observed within Populations 183
Standard Forms of Genetic Isolation by Distance 185
Spatial Autocorrelation Observed within Populations—Moran's *I* Statistics 192
Observed Join-Count Statistics 209
Values for Other Measures 217
Biparental Inbreeding Caused by Spatial Structure and Its Interaction with Inbreeding Depression and Selection 220

The Roles of Demographic Factors in Spatial Structure 223
Using Spatial Patterns to Detect Natural Selection 233

8. Statistical Methods for Spatial Structure within Populations 244
Distribution Theory for Join Counts 245
Moran's *I* Statistics in Terms of Join Counts 252
Statistical Properties of Join-Count Statistics 255
Statistical Properties of Moran's *I* Statistics for Quadrats 267
Statistical Properties of Moran's *I* Statistics for Individual Genotypes 268
Covariances of Moran's *I* for Converted Individual Genotypes of Multiallelic Loci 272
Multiple-Locus Estimators 279
F Statistics 280
Measures of Kinship 284

9. Theory of Spatial Structure within Populations 288
Genealogies and Identity by Descent 288
Kinship and A Priori Covariance 299
Relationship of Various Measures of Observable Spatial Covariances to the Theoretic Values 301
Wright's View 305
Mechanistic Models of Dispersal 308

10. Emerging Study 317
Emerged Conclusions 317
Future Studies 323

Literature Cited 329

Index 353

Preface

This monograph considers the spatial distribution of genetic variation as an integral part of population genetic processes. Its central thesis is the connection of spatial or geographic patterns to the space–time processes that produce them. The monograph reviews genetic survey studies and uses spatial statistics to relate these to theoretical processes. The intended audience includes faculty and other researchers, and graduate students in advanced course subjects. Their areas of specialization may include population genetics, evolutionary biology, ecological genetics, conservation genetics, spatial statistics, genetic epidemiology, genomic diversity, human genetics, and physical anthropology.

The monograph has two major sections: one for spatial or geographic patterns of genetic variation among populations (chapters 2–6), the other for spatial structure of genetic variation within populations (chapters 7–9). Although there are significant similarities between the two, there are many more important differences. Inside each major section, the chapters follow the sequence that scientists typically study the field of geographical genetics. Initially, they want to know how spatial statistics have been applied to experimental systems and what it is that they measure. Second, because their specific systems are often complicated, it may be necessary to tailor spatial statistical methods, and this requires knowledge of the distribution theory of the statistics. Third, to make detailed inferences about the space–time processes of interest it is frequently necessary to tie the distributions of statistics to theoretical models that are usually stochastic processes. The precise assumptions made in the available models can be critical, and sometimes these models must be modified to fit a particular experimental system. In the monograph the first level reviews the types of spatial patterns observed in nature and how they have been measured with spatial statistics. The second goes more deeply into the mathematical statistics of spatial measures and issues of statistical power, and develops the connection of the measures to stochastic space–time processes. The third presents a number of theoretical models of stochastic space–time processes. The monograph contains reviews of a wide range of works by others, as well as published works and new results from the author.

Many friends, family, and colleagues have discussed with me various ideas in the monograph. However, I especially thank Don Dickmann, Alan Fix, and Rosemarie Walter for many helpful comments on the manuscript at various stages, as well as the two readers of the manuscript. Special thanks go to Robert Sokal, for his very thoughtful and thorough comments on chapters 2, 3, and 4, as well as for his work, which inspired me to focus on spatial statistics some twenty years ago.

Geographical Genetics

CHAPTER 1

Space–Time Population Genetics

> I invoke the first law of geography: everything is related to everything else, but near things are more related than distant things.
>
> —Waldo Tobler (1970)

Spatial patterns of genetic variation have three facets. First, spatial structure is interdependent with many population genetic processes, including microenvironmental selection, clinal selection, biparental inbreeding, and inbreeding depression. In many cases, valid models of genetic processes must include spatial context. This is true within a single population as well as in the context of a system of populations. The integral role of spatial structure of genetic variation in evolutionary processes was recognized in the formation of the neo-Darwinian synthesis. For example, when Sewall Wright (e.g., 1931) took his studies of genetics and breeding systems to the population level, his models of adaptation and evolution incorporated spatial distributions, e.g., in the isolation by distance and shifting balance theories. The second facet is that, because spatial patterns capture the cumulative effects of forces acting over many generations, it may be possible to use them to study the presence and operation of these forces. Statistical analyses of spatial structure can have high levels of statistical power for detecting a number of forces operating in population genetic processes. In contrast, detection of selective forces is notoriously difficult when based on deviations from Hardy–Weinberg proportions or short-term temporal changes in gene frequencies (e.g., Lewontin and Krakauer 1973). Third, many populations exhibit strong spatial patterns of genetic variation, usually in the form of positive spatial autocorrelations, and this violates the assumption of independent, identically distributed elements in samples, which is made in the formulation of standard statistical methods.

At the heart of spatial and space–time analysis of population genetics is the connection between observed spatial patterns and the space–time processes that generate them. Advances in the understanding of this connection have accelerated in recent years. While much of the theoretical basis for spatial genetic distributions has been in place for many decades (e.g., Wright 1943; Malécot 1948), until fairly recently the connection has been relatively imprecise, because of a number of difficulties, including (1) singularities and other difficulties in obtaining analytic results on theoretical values for a number of very important processes; most notably Nagylaki (1978) proved that there are analytical singularities for diffusion approximations of the spatial structure at short distances in systems with two spatial dimensions, for genes experiencing genetic isolation by distance; (2) difficulties in developing unbiased estimators of theoretic parameters (Morton 1973b); and (3) lack of sophisticated spatial statistics. The first problem has diminished recently for several reasons, most significantly advances in mathematical studies and the ready availability of fast computers for simulating large numbers of complex spatial–temporal processes. The second has been improved through recent developments in statistics and large numbers of molecular genetic markers. The third actually has been improved for some time now, because spatial statistics were well developed by the mid-1970s (Cliff and Ord 1973) and introduced into population genetics in 1978 by Sokal and Oden (1978a,b).

In general, unless migration is unimportant, spatial patterns are highly dependent on the status of an underlying space–time processes. Among exceptions may be included the combination of strong clinal or other forms of environmental selection with low migration at selection equilibrium. Such a process creates a static spatial pattern, and because migration is limited enough the genetic values at different locations scarcely affect one another directly; rather, they are determined by the environment. In contrast, if there is nonnegligible immigration into a population from any other populations, then its genetic dynamics depend on the spatial pattern of genetic variation among nearby populations and many other aspects of spatial–temporal processes.

Most evolutionary questions addressed by genetic survey data traditionally have involved some sort of averaging over space. One important set of questions centers on the partitioning of genetic variation; how much of the variance on average is contained within populations rather than as differences among populations (Wright 1951). These questions do not explicitly involve spatial *patterns* of genetic variation per se. However, if gene flow occurs over limited distances, even the partitioning of variance may be affected by spatial structure (e.g., Wright 1946). Other questions are explicitly spatial. For example, on average how do the rates of migration and genetic differentiation among pairs of populations depend on the geographic distance separating them? How far do seed and pollen disperse within large populations and how closely related are individuals neighboring one another? How much does limited dispersal within populations contribute to inbreeding and the loss of genetic variation? What do spatial patterns reveal about natural selection or the effects of environmental factors on diseases partially determined by genetics?

One of the key concepts of spatial–temporal processes is genetic isolation by distance. The most common and proximal cause of spatial structure is limited dispersal of either (diploid) individuals, for most animals, or seed and pollen. Limited dispersal causes genetic isolation by distance to build over generations. For example, within a population, even if initially the spatial distribution of individual genotypes were random, limited dispersal means that individual genotypes in the next generation will not be. Relatives will exhibit some degree of spatial proximity, and spatial neighbors will exhibit some average excess of relatedness. Then, when dispersal is limited during the mating cycle of the next generation, there is an excess of matings among spatially proximal individuals who are related beyond average. The progeny of a third generation will be inbred, as well as more intensely spatially structured than the second. In principle, the same occurs among populations. Usually as this process proceeds the degree of relatedness via spatial proximity reaches a plateau. Under standard conditions the amount of genetic isolation by distance for selectively neutral genes depends solely and inversely on the rates and distances of dispersal, but there can be other contributing factors, such as direction of dispersal.

In studies that seek evidence for environmental factors, standard forms of genetic isolation by distance often provide the proper *null hypothesis*. This null hypothesis is robust, but exact statistical tests for deviations from it may be complicated by several forms of statistical interactions. In contrast, using a null hypothesis that genotypes are randomly distributed in space will usually be invalid. The occurrence of patterns that seem to fit a particular selection model (e.g., clinal selection) does not necessarily mean that selection is acting, because isolation by distance for neutral genes can often produce similar patterns at some spatial scales. Strategies for using spatial patterns to infer population genetic processes often can incorporate various types of ancillary information. Examples include known geographic distributions of environmental factors, such as measures of mean annual temperature, and independent information on migration rates.

A number of ecological genetic processes, which typically operate on shorter time scales than do evolutionary processes, may also be addressed using averaged spatial patterns, and space–time models. With respect to spatial context, ecologists are particularly interested in using genes as markers in tracing the movements and history of successful establishment of populations. Spatial patterns of genetic diversity can also be critical to conservation genetics, because they show how to maximize genetic diversity in sampled ex situ or conserved in situ populations. Similarly, spatial distributions inform expectations about habitat fragmentation, both in terms of the containment of genetic diversity within versus among fragments and the spatial structures and levels of biparental inbreeding within fragments.

Observed spatial patterns may also be used to infer ancient events, such as the geographical origination of genetic variants, population contractions, and species range expansions, as well as recent events such as admixture. Ancient events may leave signatures in spatial or spatial–temporal patterns of genetic variation. These signatures may be long-lasting, but ultimately they are transient, in contrast to the stable patterns produced by selection, genetic drift and migration averaged over long periods. Their study requires different (nonequilibrium) models, experimental designs, and statistical methods. Most empirical studies of ancient events aim, in effect, to parse off particular features of the past from the spatial–temporal processes and contexts in which they exist. What are the conditions under which time or space may be ignored at various steps in the inferential procedures? Success usually requires large data sets. Modern molecular data, such as DNA sequences, may be uniquely suitable when they incorporate temporal information in the form of multiple mutations, which provides a certain "temporal depth" (Templeton 1998). Recently, some tailored statistical methods have been developed, but further statistical developments are called for. Theoretical models that are dynamic and spatially explicit also need further development. For example, we lack theoretical results on the geographic distribution of haplotypes following a range expansion (Cann et al. 1987; Templeton et al. 1995). Studies of signature effects of ancient events versus studies of stable patterns are at opposite ends of the spectrum with regard to how they utilize fluctuations of processes and the readiness of these to "average out" temporally and spatially.

Studies of spatial distributions of genetic variation may also concern the spatial association of genetic traits with one or more other variables representing environmental factors. Such cross-correlations are particularly important in studies that seek to identify genetic and environmental factors of cancer (e.g., Sokal et al. 2000) and other partially genetic diseases in humans. However, simple correlation coefficients, as measures of spatial association of two variables, may be nonzero even if there is no causal relationship between them. For example, if the spatial pattern for isolation by distance for neutral genetic variation is randomly overlain with a determined pattern of a purported environmental factor, generally some sort of correlation will result at least at some spatial scales. Moreover, spatial structuring of either variable violates the assumptions of the standard coefficient of correlation. Spatial statistics have been designed for properly estimating cross-correlations (e.g., Cliff and Ord 1981), but only the Mantel statistic, which actually measures the association of two sets of distance measures (Mantel 1967; Smouse et al. 1986), has yet been widely used in genetic studies. As for spatial autocorrelation of genetic variables, appropriate null hypotheses for spatial cross-correlations should incorporate spatial patterns for selectively neutral genetic variation.

Connecting spatial patterns to an underlying space–time process is complicated when the process includes direct interactions among the genetic values at different locations due to spatial proximity (usually via migration) and stochastic inputs (e.g., genetic drift). Spatial *statistical* interactions, e.g., in the form of spatial correlations, do not usually *directly* correspond, either in terms of spatial scale or magnitude, to *process* interactions via spatial proximity. Perhaps the clearest example of the difference is the contrast between spatial autoregressive and space–time autoregressive models, which is discussed in chapter 4.

Advances in statistical analysis of spatial patterns, particularly the development of spatial statistics, were led largely by British statistical geographers (e.g., Cliff and Ord 1973; Haining 1977; Bennett 1979). Most notably, the works of Cliff and Ord (1973, 1981) developed much of the probability theory for spatial autocorrelation statistics. Arguably, the single most important development for inferring stochastic space–time processes was the derivation (Cliff and Ord 1973) of the distribution theory for autocorrelation under the null hypothesis of randomization. Under this null hypothesis the expected value of the variable (at all locations) need not be known or estimated. If a pattern is subject to stochastic processes, it is generally impossible to know the true mean of the process, and in many cases it is impossible to obtain unbiased estimates of it no matter how much data are available. For example, the observed mean may depend on the spatial scale over which the sample is taken (e.g., Epperson 1993a). In population genetics, this development provided the first proper statistical tests for the presence of spatial patterns per se (not only tests for differentiation among populations), and it further avoided problems of bias in estimators of spatial genetic correlation or kinship.

After spatial autocorrelation statistics were introduced into the fields of population biology and population genetics, largely through the papers of Sokal and Oden (1978a,b; see also Jumars et al. 1977), they quickly became widely used (e.g, Sokal et al. 1989a; Waser and Elliott 1991; see reviews by Heywood 1991 and Epperson 1993a). Methods of spatial time series analysis have also been developed (e.g., Bennett 1979; Cliff and Ord 1981; Upton and Fingleton 1985), and these are particularly useful for space–time data. In contrast, their application has been limited, largely because until recently (Lee and Epperson, n.d.) no sophisticated statistical computing tools were available. There have been many other important statistical advances, notably methods for utilizing temporal information contained in haplotype data, such as the nested cladistic methods of Templeton and colleagues (e.g., Templeton 1998).

Experimental design, particularly the spatial distribution of sample points, either individuals (usually for within-population studies) or populations, strongly affects the spatial patterns observed. A recurring theme in this monograph is the importance of having somewhat regular spacing of sample points wherever possible. Regular spacing generally improves statistical power by increasing the effective sample size in terms of numbers of pairs of locations that have any given degree of spatial proximity. It promotes the degree of spatial replication in data, and this can improve the ability to tie the statistical results to stochastic space–time processes.

Choices of spatial *scales* for sampling and statistical analyses have particularly critical effects on the measures obtained. For example, in genetic isolation by distance the dominant features are spatial genetic "patches," or regions of concentrations of allele frequencies (Rohlf and Schnell 1971; Turner et al. 1982), that are repeated, spatially scaled units of structure. They result in spatial correlations that generally decrease with distance, hence observed correlations will depend on the distance among sample locations. Indeed, the "diameter" of patches corresponds to the distance at which Moran's spatial autocorrelation statistics (Moran 1950) become negative (Sokal and Wartenberg 1983). Moreover, comparisons of physical scales of structures for different species must consider standardized dispersal distances. Within populations, the physical scale of spatial genetic structure depends primarily on measures of dispersal distances *relative* to density of individuals (Wright 1943; Slatkin and Barton 1989). Similarly, the spatial scale of geographic patterns of genetic variation among populations depends on distances of migration among populations, relative to the density of populations per unit area. Moreover, spatial scale affects data-based contrasts between patterns for selectively neutral loci and other types, such as clinal selection.

Two types of spatial genetic distributions traditionally have been distinguished. One is the spatial distribution of individual genotypes *within* a large, more or less continuously distributed population, and the other of gene frequencies *among* discrete, well-delineated populations. Although there are important similarities between the *among* and *within*, there are much more important differences at all levels:

1. The nature of the study data differs dramatically. For example, *within*-population studies tend to have hundreds of spatially located data points (typically individual genotypes), whereas most *among*

studies have 20 or fewer populations. As a result, pairwise measures such as spatial autocorrelation typically can utilize a degree of averaging over space that is two orders of magnitude greater for *within* compared to *among*. Thus idiosyncracies in the processes or data are much more important in *among* studies.

2. The ranges of statistical methods used are also considerably different. Moreover, statistics used for both (e.g., Moran's *I* statistics) typically have differing stochastic and sampling properties.

3. The connections between statistics and stochastic processes are diametrically opposed in the two cases. For *among* populations the most precise connection is the space–time autoregressive (STAR) model, and for *within* populations the best existing connection is very extensive computer simulations. STAR models will not work for *within*, and, perhaps surprisingly, there are few published simulation results for *among*.

4. There are important differences in the theoretical stochastic processes. One key difference is that theoretical models designed for *among* typically do not allow correlated gene movements, which is a common feature of gene flow *within* populations. Diffusion approximations sometimes bridge both (Nagylaki 1986), but they fail in systems with two spatial dimension, the case of most interest. In addition, *within*-population dispersal is intimately connected with the mating system (e.g., Allard 1975; VanStaaden et al. 1996), whereas for the *among*-population processes gene flow among populations may usually be separated from the mating system within populations.

For these reasons, separate treatment for *among* (chapters 2–6) and *within* (chapters 7–9) avoids having to make nearly constant distinctions and conditioning statements.

The distinction of within and between may not be obvious in nature. There are many cases where individuals are distributed more or less discontinuously over some spatial scales, yet discrete populations cannot be well delineated. Studies should be dealt with on a case by case basis, with careful consideration of fit to either the continuous or discrete population models, or recognition that neither is appropriate. For example, the discrete population model may be appropriate if populations can be constructed such that the distances among them are great enough that any spatial structure within populations does not affect the genetic makeup of emigrants from it–i.e., the location of an individual genotype within a population does not substantially affect its probability of emigrating, relative to others in the same population, to any other given population.

Quantity of genetic data can strongly influence the inferences that can be made. For any given gene, the spatial patterns, spatial correlations, and degrees of genetic differentiation that are produced by a particular space–time stochastic process can vary substantially at several levels. First, there can be variation in the process parameters; for example, migration rates may vary over time. Second, predicted values of measures of spatial patterns are based on at least two things: the controlling parameters (e.g., rates of mutation and migration) of the stochastic process; and either an initial distribution or an equilibrium (stationary) distribution. In other words, theoretical values are based on "a priori" (Malécot 1948) expected values of stochastic variables. In fact, as a system evolves stochastic changes occur from generation to generation, and the distribution is only one realization of a stochastic process. Generally, different predicted values obtain for each generation n when they are based on knowledge, i.e., "conditioned on" or a posteriori of the status of a system at say generation $n - 1$. In addition, there is a variance associated with such a posteriori measures, and on top of this there may be sampling error. In many cases there is simply too much variation in spatial patterns for individual genetic loci to precisely estimate model parameters. (It should be noted that single-locus data are often sufficient for tests of the null hypothesis that the spatial distribution of genetic variation is random.) Multiple replications of the process are needed, hence multiple alleles and/or loci, but fortunately the numbers are within a practical range.

Qualities of genetic data can also be important. One of the key distinctions is exemplified in true multilocus data versus multiple locus data. In general terms, data in the form of multilocus genotypes contain more information whenever there is linkage disequilibrium. It appears that in most cases there will be little linkage disequilibrium among different neutral genetic loci under isolation by distance, at equilibrium. Thus spatial measures that are averaged over loci will contain nearly all of the information. The criticality of linkage disequilibrium also holds for DNA-sequence data or haplotype data.

There is little or no recombination among polymorphic sites along a sequence, and there can be very high amounts of (or even complete) linkage disequilibrium. In such data there can be substantial additional information in the form of multiply mutated haplotypes, and analyses can gainfully employ the infinite sites mutation model (e.g., Ewens 1974), rather than the infinite alleles mutation model (e.g., Crow 1989). However, because the additional information requires multiple segregating polymorphic sites, it depends on the population size being large enough that drift does not overpower the rate of production of mutations. Ewens (1974) showed that for a single population of most species (applicable to the *within* case) there should be almost no additional information in the infinite sites mutation model. A system of populations typically contains a larger total number of individuals, hence additional information is somewhat more likely.

Looking toward the future, it is worthwhile to consider the potential of space–time genetic data, i.e., genetic data collected over different time periods as well as over space. Arguably, the default of genetic surveys should be to collect space–time data, except when this is infeasible or inefficient. It seems likely that space–time data are efficient in terms of overall sample size. Compromises in spatial extent or per-location sampling effort, or possibly in the numbers of loci, may be warranted. Space–time data can resolve a number of problems of inferences based on genetic surveys. Statistical interactions can be directly related to process parameters primarily for drift, migration, and selection, as is shown in chapter 4. For example, data-based and direct identification of the maximum spatial distance over which there are interactions (e.g., migrations) among locations requires space–time data in spatial time series methods (e.g., Pfeifer and Deutsch 1980a). However, sampling over time may involve other difficulties, or may be infeasible if not impossible. In part, feasibility will depend on the length of time of the generation, being very different say for viruses compared to humans. Space–time data may also be inordinately valuable for studies of ancient events. It may be anticipated that as high-throughput genotyping becomes increasingly efficient, and as ability to obtain usable DNA from ancient samples improves, the collection of space–time genetic data may become a more attractive alternative.

CHAPTER 2

Geographical Patterns Observed in Nature

> Even the most sanguine "selectionist" will not claim that selection coefficients in excess of a few percent are the rule. But the power to detect selection of this magnitude by observing changes in gene frequency, or differences in components of fitness between genotypes, is very low for samples sizes within the range of practicality. . . . The alternative is then to use information about the spatio-temporal distribution of allele frequencies. . . . The difficulty with this procedure is that various parameters enter in a confounded way into the determination of the observations.
>
> —Richard Lewontin and Jesse Krakauer (1973)

What can spatial patterns reveal about population genetic processes? Variation of a species across its native range has always been central to the study of evolution, both in terms of how it is *partitioned* within versus among populations and its *spatial pattern*. Both depend strongly on the amount of migration, and natural selection wherever it may be acting. Any geographic survey of genetic variation, whether for purposes of studying evolution, ecological genetics, genotype by environmental interactions, or genetic disease, generally involves, either explicitly or implicitly, some aspects of space–time processes. Spatial patterns acquire the accumulated effects of migration and selection, averaged over space and long periods of time. This chapter reviews studies of patterns of genetic variation in nature, and shows how spatial statistics have been used to measure them. The chapter also shows how inferences have been made on the averaged effects of migration, selection, and other factors, in the context of the full complexity of space–time processes.

The starting point for spatial inference is the pattern expected for selectively neutral genes, in the setting of data for multiple markers. The intensity and scale of spatial patterns of genetic variation, as measured by various spatial statistics, foremost reflects the rates and distances of migration among populations in a manner that is fairly standardizable. Nonstandard spatial features, measured by specialized spatial statistics, may reveal prevailing directions of migrations, or paths of migration, and physical or other barriers to gene flow. Still other spatial statistical methods may be used to study factors such as linguistic groups or other social barriers to gene flow.

Populations should be considered in their full spatial context, because spatial or geographic patterns of genetic variation are generally produced by complex space–time processes. Patterns that are observed in a subset of populations may be connected to patterns and processes in the larger system of populations in which they are embedded. For example, a pair of populations, one small and one large, may be both somewhat isolated geographically from other populations. If a high degree of similarity is observed among the two populations, it may be tempting to conclude that the smaller population was recently founded from the larger one. However, perhaps instead all or most of the small population is derived from immigrants from other unstudied populations, which may also be highly similar. Moreover, the gene frequencies in populations may drift substantially over time, and thus similarity does not always mean ancestry. Two populations may be similar simply by chance.

Spatial patterns for neutral markers form the appropriate *null hypothesis* for inferences about natural selection. Dating from the original works of Wright (1951), theoretical studies have shown that geographic distributions of genetic variation should differ strongly from random or uniform distributions, even for neutral loci. Generally, the null hypothesis should not be that genetic variation is without pattern, i.e., that the spatial distribution is "random." The presence of a pattern does not necessarily mean that natural selection is acting. To distinguish selection from genetic drift often requires detailed knowledge of genetic isolation by distance for selectively neutral loci, which, for example, can exhibit clines at some spatial scales. Such knowledge can be gained through direct measures of migration. However, the use of multiple genetic loci is a particularly important strategy for understanding the evolutionary forces acting on each (Lewontin and Krakauer 1973; Sokal and Wartenberg 1983; Sokal 1986; Sokal et al. 1989b; Epperson 1990a), because loci sharing the same influences should have similar geographic patterns. Generally it may be assumed that all neutral loci are subjected to the same degree of genetic drift and migration, at least in expectation. Loci under natural selection can also be subjected to the same level of migration, but their effective migration can also differ when fitness values vary among populations. Spatial contrasts may exist between selected versus neutral loci, but quantification and interpretation of contrasts still often requires detailed understanding of the processes for neutral theory and the stochastic and statistical variation of spatial measures of genetic variation under neutrality (Sokal 1986; Slatkin and Arter 1991a,b; Sokal and Oden 1991). In addition, the presence of spatial autocorrelations of genetic variation among populations complicates such modeling efforts and can cause substantial problems for more standard statistical methods (Epperson 1990b, 1993a).

Studies of the geographic variation of genes and traits that affect fitness may utilize prior information on "candidate" genes and environmental factors. In some cases, system-wide geographic patterns are very strong and clear, particularly if migration rates are very small and selection is strong. Such cases may be understood without particularly detailed statistical analyses or modeling of the spatial–temporal processes. In contrast, if migration is relatively strong then patterns cannot be understood without relating them to some sort of spatial–temporal model. This is true not only for spatial correlations of genetic variation, but also for the spatial "cross-correlations" of genetic traits with environmental factors, for example, the association of a spatial distribution of a genetic disease with geographic distributions of causal environmental factors.

Selection generally interacts with migration, and models of processes should incorporate both. This also creates statistical interactions between the spatial effects of genetic isolation by distance with those of selection. This is true in the multilocus setting as well as for single loci. As an example, the spatial distribution of multilocus genotypes plays a key role in the first step of shifting balance theories, even for loci that may be neutral at some states. Because of population differentiation in gene frequencies, even alleles that are in low frequency system-wide may occur in relatively high frequencies within some populations. This increases the likelihood of there being formed multilocus genotypes with multiple low-frequency alleles at different loci. If the array of multilocus genotypes is changed by population structure, then loci that were effectively neutral can become subjected to selection, because some of the new multilocus combinations may have higher fitnesses. They may spread within a population, but whether it is likely that they subsequently spread throughout a system of populations remains a controversial aspect (the third phase) of Wright's shifting balance theory (Coyne et al. 2000). Wade and Goodnight (1998) have argued that metapopulation dynamics may be important. Peck and colleagues (1998) suggest that if the spatial pattern is considered, rather than simply differentiation among populations, it becomes more likely that locally formed traits can spread. The partitioning of genetic variation within and among populations is one of the most-studied aspects of geographic surveys of genetic variation, but it is only sufficient under Wright's island model (Wright 1951), where it is presumed that all populations exchange migrants at the same rate. This assumption is substantially violated in most systems. When migration and gene flow are restricted, spatial–temporal models are required, as well as more complex spatial or spatial temporal statistical analyses.

The study of gene flow with or without selection in the spatially explicit context is by its nature complicated, but in many situations it is necessary. Similarly, teasing out effects of selection from isolation by distance is worthwhile but requires sophisticated spatial or space–time analyses.

STANDARD PATTERNS OF GENETIC ISOLATION BY DISTANCE

Simple limits to the distances and rates that populations exchange migrants causes strikingly nonrandom spatial patterns of genetic variation. Over time, limited migration produces large genetic correlations among spatially proximal populations, and such correlations usually drop off smoothly as the distances among populations increase. Generically termed genetic isolation by distance (Wright 1943), the spatial patterns produced take on a fairly standard form for selectively neutral loci under conditions that are widely met, if only approximately. When the rates of migration among populations depend primarily on the distances separating populations, spatial patterns display symmetry both within and among spatial dimensions, in systems of populations that exist essentially in one, two, or three dimensions.

Standard spatial patterns of isolation by distance are best studied using pairwise measures of genetic similarity or correlation among populations, although Wright (1943) originally used hierarchical measures. Malécot (1948) showed that pairwise measures rather fully characterize spatial patterns in theoretical models of isolation by distance. An example using spatial correlations for a system in two spatial dimensions is shown in figure 2.1. Correlations of gene frequencies are largest between pairs of populations separated by the shortest distances, and the correlations decrease as distance increases. Values of correlations at short distances as well as the form of the decrease with distance depend on the number of spatial dimensions, as well as the rates and distances of dispersal. For example, the decrease is exponential only in one-dimensional systems, such as may occur for a riparian or riverine system of populations, despite the frequent claim that all systems are exponential (e.g., Morton 1973b, 1982).

Moran's I statistic (1950) is a pairwise measure that can be applied to genetic data in a wide range of natural systems of populations. In its standard form it can be calculated for the spatial distribution of the frequencies of a particular gene among populations. In most cases it is calculated for separate distance classes, which normally are mutually exclusive and exhaustive. In other words, all pairs of populations are placed into one of a set of mutually exclusive classes each of which represents a range of distances of separation. For each distance class k, the Moran statistic is calculated by

$$I(k) = \frac{n \sum_i \sum_j w_{ij}(k) Z_i Z_j}{W_k \sum_i Z_i^2} \tag{2.1}$$

Each Z_i is the mean-adjusted value of some genetic trait, $X_i - X$, where X_i is the value (e.g., frequency of an allele) of the trait, in a sample taken from population i, and X is the mean value in all n

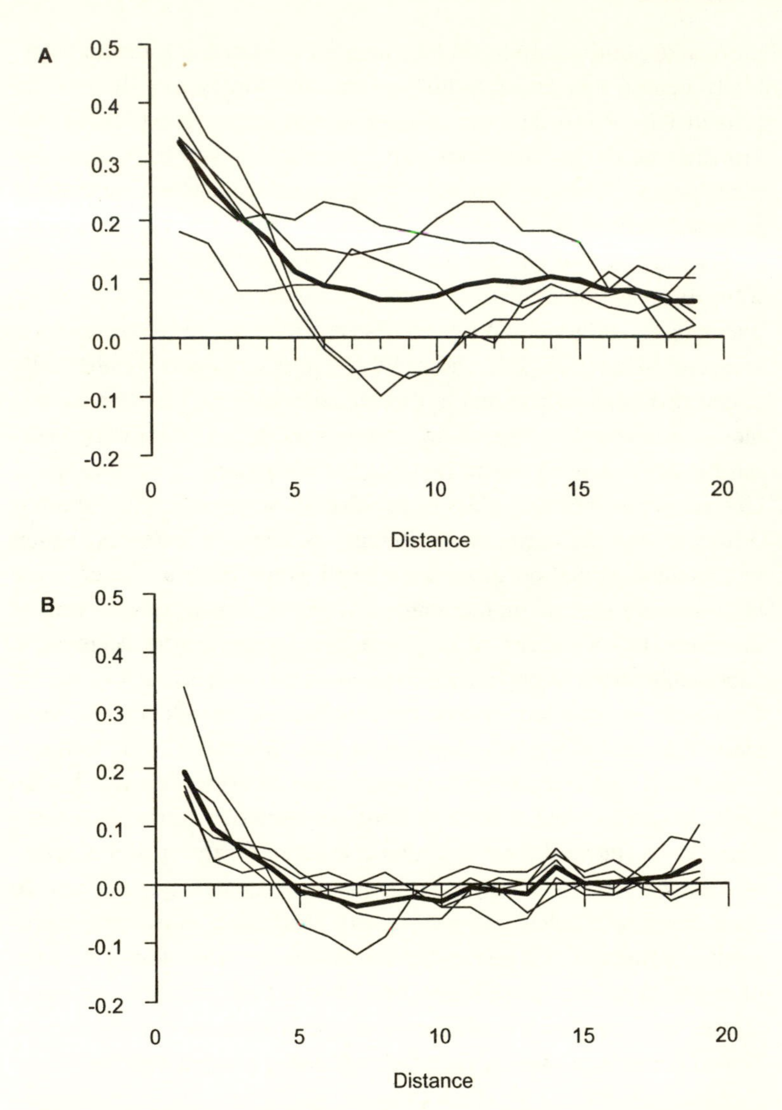

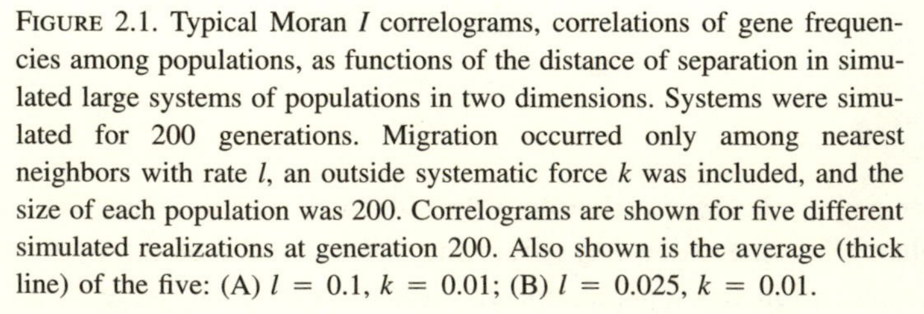

FIGURE 2.1. Typical Moran I correlograms, correlations of gene frequencies among populations, as functions of the distance of separation in simulated large systems of populations in two dimensions. Systems were simulated for 200 generations. Migration occurred only among nearest neighbors with rate l, an outside systematic force k was included, and the size of each population was 200. Correlograms are shown for five different simulated realizations at generation 200. Also shown is the average (thick line) of the five: (A) $l = 0.1$, $k = 0.01$; (B) $l = 0.025$, $k = 0.01$.

populations in the system. Spatial distribution of populations is specified by the values of the $w_{ij}(k)$, in the present case binary (0, 1) variables specifying the inclusion (1) or exclusion (0) of the pair of populations i and j in class k, and W_k is the sum of the weights or twice the number of pairs of populations placed into class k (Sokal and Oden 1978a; Cliff and Ord 1981). Moran's I statistic is an average correlation among populations separated by a range of distances specified by distance class k. Various extensions of Moran's statistic are possible. The set of values of I statistics for all distance classes is termed a correlogram.

Moran's spatial autocorrelation statistics generally have very similar values as the spatial correlations in a theoretical model. The most notable difference is that I correlograms typically become slightly negative at long distances, whereas theoretical correlations may not (Epperson 1993b). The distance at which an I correlogram becomes negative also provides an estimate of the spatial scale of autocorrelation. Simulation studies have shown that the dominant feature of spatial distributions of genetic variation in two-dimensional systems of standard genetic isolation by distance is the development of patches or regions of concentration of allele frequencies (figure 2.2). Analogous genetic patches are produced *within* large continuous populations (Rohlf and Schnell 1971; Turner et al. 1982; Sokal and Wartenberg 1983). This is an important result that is not necessarily obvious from the analytical theories of Wright (1943) and Malécot (1948), because the latter are based on expected correlations, and not on stochasticity in correlation values. Perhaps the best way of intuiting genetic patches is that they represent regional balances between stochastic fluctuations (i.e., local genetic drift) and the tendency for migration to smooth out such regional fluctuations over larger geographic scales, and possibly over the entire system of populations. In systems where the sizes of populations are small, stochastic fluctuations are marked, and genetic patches are stronger both in terms of the degree of localized differentiation and in the sharpness of the boundaries of genetically similar groups of populations. When migration rates and/or distances are greater, patches are less differentiated and have smoother boundaries. For example, for a locus with two alleles, A and a, patches with high frequencies of A alleles are produced and patches of low frequencies of A (high frequencies of allele

FIGURE 2.2. A two-dimensional gene frequency surface of a simulation where each population exchanged migrants with its four nearest neighbors with rate 0.1. The outside systematic force had recall coefficient 0.01 (e.g., mutation rate). The effective size of each population was 200 and the surface is at generation 200. Populations with larger circles had greater values of the gene frequency. To emphasize the pattern, values less than the grand mean are indicated in gray and values greater than the mean (ca. 0.5) are in black. Note that there are large patches of relatively homogeneous areas. Note also that in this simulation the uppermost patch (indicated by small gray dots) is more elliptic than circular, having greater diameter from left to right. This is somewhat unusual, but reflects the stochasticity in patches.

a) are similarly produced. In systems with large numbers of populations, spatial genetic patches are repeated spatial units of structure. Patch size increases with the rates and distances of migration (Sokal and Wartenberg 1983). The distance at which *I* correlograms become negative provides an operational estimate of the diameter of genetic patches (e.g., figure 2.3).

Examples of standard genetic isolation by distance are seen in some of the genetic variables spatially distributed among 15 tribes in Kenya (Sokal and Winkler 1987). Overall, the 19 head and body measures and 5 pigmentation variables (all quantitative traits with

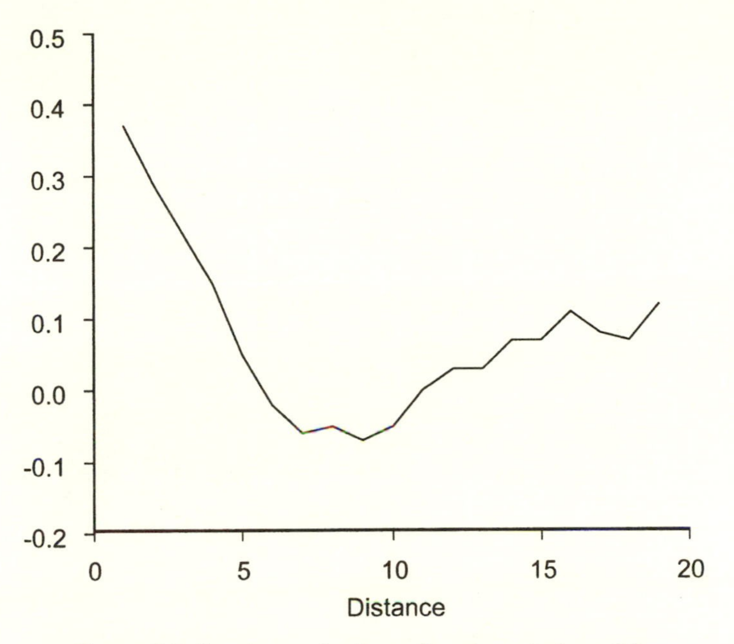

FIGURE 2.3. Correlogram for the surface shown in figure 2.2.

considerable genetic components) were far from randomly distributed. Quantitative traits have spatial correlations that are similar to those for gene frequencies, under processes of isolation by distance (e.g., Rogers and Harpending 1983; Lande 1991; Nagylaki 1994). Moran's *I* statistics calculated for each variable and for four distance classes averaged 0.26, −0.31, −0.13, and −0.13 for female individuals, respectively for distance classes 0–125 kilometers, 125–250 kilometers, 250–500 kilometers, and 550–1000 kilometers (figure 2.4A). The corresponding average values for males were very similar: 0.27, −0.24, −0.09, and −0.22. The dominant spatial pattern corresponds closely with the shape of autocorrelation correlograms produced by theoretical models of genetic isolation by distance: there are quite large positive values at short distances, negative correlations at intermediate distances, and, importantly, no tendency for average values to continue to decrease (become more negative) further as distances increase to the limits of the sample (Sokal and Winkler 1987). In the spatial distributions of morphological traits in Kenyan tribes, genetic patches appear to have diameters of about 125 kilometers.

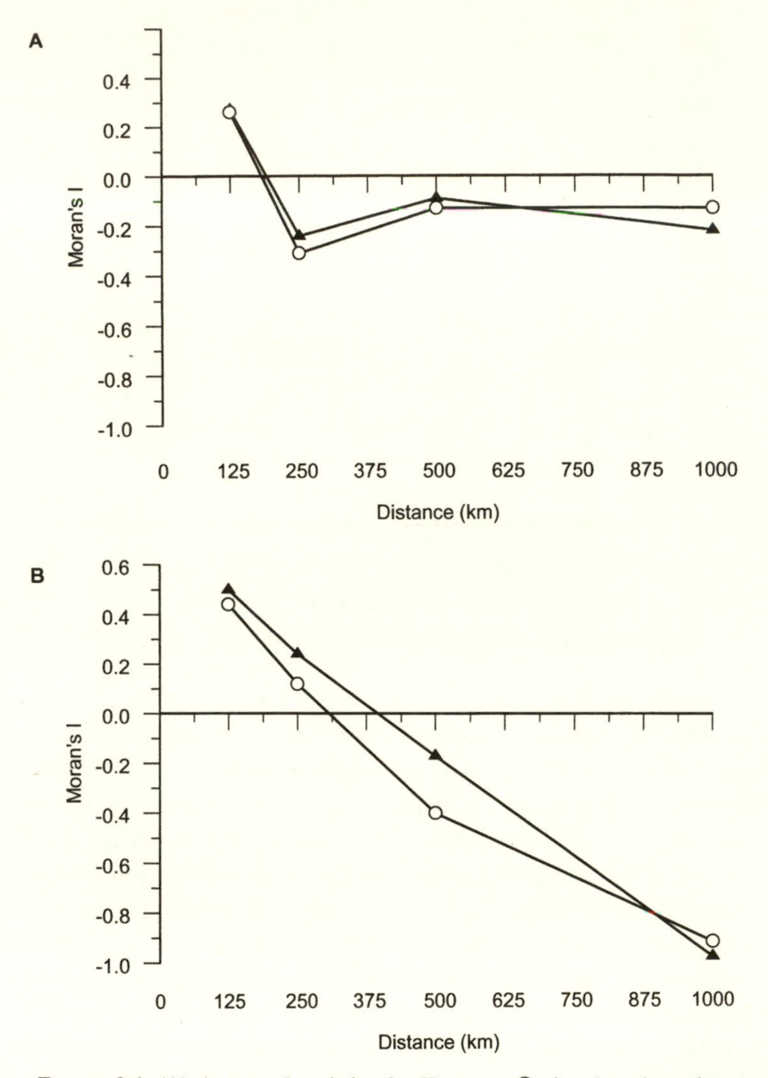

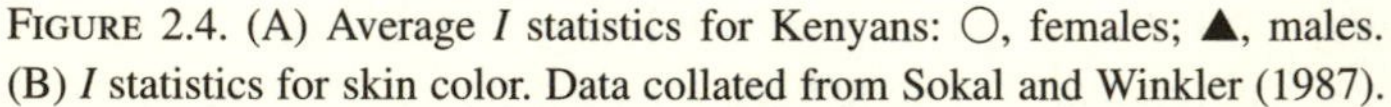

FIGURE 2.4. (A) Average I statistics for Kenyans: ○, females; ▲, males. (B) I statistics for skin color. Data collated from Sokal and Winkler (1987).

Precise comparisons of the actual values of I for short distances to theoretical values in models require knowledge of the numbers of unsampled populations, because correlations depend on distance *relative* to density of populations (number of populations per unit area). Thus precise estimation of the migration rates is not possible. Typically, in studies of patterns of variation among populations the density of populations is unknown or at least unreported.

Importantly, some of the traits in Kenyan tribes exhibited a different type of correlogram; in particular, that for skin color (figure 2.4B) steadily decreased with distances well beyond the genetic patch size, having large negative values at 1000 kilometers. Such declines are generally interpreted as evidence of a cline (Sokal and Winkler 1987). Another important feature was evidence that the migration patterns have been rather complex since the tribes moved into the region (less than two to three hundred years ago). However, for many variables such complexities appeared to have averaged out, and this indicates how quickly complex migration processes average out to produce a fairly standard form of genetic isolation by distance.

Genetic isolation by distance in humans with differing spatial scale is observed in villages in the central part of Bougainville Island of the Solomon Islands (Sokal and Friedlaender 1982). A total of 121 variables, including blood group polymorphisms, anthropometric variables, dermatoglyphic variables, and dentometric variables, were analyzed for 18 villages. Many of the variables did not exhibit much autocorrelation, but those that did yielded patterns that were largely consistent with genetic isolation by distance. Most of the autocorrelated variables exhibited high positive correlations only at the shortest distance classes (0–15 kilometers and 15–30 kilometers), correlations were generally negative at 30 to 45 kilometers, and most correlograms showed little tendency to continue to decrease as distances increased further. The spatial scale of genetic patchiness is much smaller than in Kenya, and this is probably due to the fact that the Bougainville tribes live on a relatively small island, are traditionally gardening peoples, and villages typically move less than a few kilometers during a life span. Most married people have migrated less than 10 kilometers from their birthplaces (Friedlaender 1975). Here, too, the process appears to be more complicated, since some variables did exhibit clinal features, and Sokal and Friedlaender (1982) suggest that there were two or more major immigrations, each originally from different external sources.

Similar results, but on a much smaller spatial scale, were obtained in a study of two geographic regions (actually two different city blocks) containing multiple colonies of *Helix aspera*, the European garden snail (Selander and Kaufman 1975). The genetic variables in this case were allozyme gene frequencies. One of the city blocks exhibited the standard genetic isolation by distance for nearly all allozyme loci. The other block was quite different, exhibiting nearly random gene frequency distributions, which might have resulted from repeated colonizations from disparate external sources (Sokal and Oden 1978a). In contrast, positive autocorrelation extended up to 6–50 kilometers in the European banded snail, *Cepaea nemoralis* (Jones et al. 1980; Sokal and Wartenberg 1981). The large scale of autocorrelation may be attributed to a combination of limited dispersal and environmental selection.

In one of the largest studies of spatial patterns of genetic variation in nonhuman species, Sokal et al. (1980) applied spatial autocorrelation analysis to 33 morphological traits in 118 populations of the aphid, *Pemphigus populicaulis*, throughout eastern North America. They were able to demonstrate the genetic component of many of these traits, in part because individuals within galls are genetically identical. In maps of the mean values of traits, the dominant feature is large patch areas of homogeneity. Patch areas sometimes corresponded among traits, but this could be partially due to common genetic basis. Multiple discriminant functions of the 33 traits also had similar spatial distributions. Correlograms of Moran's I statistics, calculated for distance classes, indicated highly nonrandom spatial distributions. The statistics for the first distance class (200 kilometers) were significant for all 33 traits, and had large positive values ranging from 0.13 to 0.50 (Sokal et al. 1980). Most correlograms approached zero at about 600 kilometers. At larger distances there appear to be some differences in the shapes of correlograms: some had small positive values at 1000 and 1200 kilometers, and others had fairly small but sometimes significant negative values at the maximum distances, which ranged from 2600 to 3000 kilometers among the variables. There was no evidence of strongly clinal patterns, and overall the patterns fit well the patchy distributions expected for neutral characters in genetic isolation by distance. Spatially located values for the five most important discriminant functions were also subjected to spatial autocorrelation analysis. All five had significant values (ranging from 0.14 to 0.79) at distance class one, 200 km, and most of the five correlograms had approached zero by about 600 kilometers.

Bird et al. (1981) conducted spatial analyses of the coefficients of variation for a subset of the same data. Every character showed significant variation in variability among sample locations. Based on multivariate variability profiles, the locations clustered into three groups. Locations near each other tended to have the same type of variability profiles, and the types occurred in patches, which varied in size from about 100 kilometers (for members of group 1), 200 kilometers (group 3), to 400–600 kilometers (group 2).

One of the most extensive spatial datasets studied is human populations in Europe and the Middle East. Sokal and Menozzi (1982) and Sokal and Wartenberg (1981) report spatial analyses of 21 allele frequencies for HLA blood group loci in 58 populations. Distance classes were formed approximately as multiples of 700 kilometers. Although the correlograms vary, the general trend is large autocorrelations at the shortest distance class (~700 kilometers), lesser positive values for distance class two (~1400 kilometers), and negative values for distance class three (~2100 kilometers). For many alleles, the statistics were near zero for the remaining distance classes (~2800 kilometers to 4900 kilometers), and these fit the form for standard genetic isolation by distance. However, the values for some alleles continued to decrease, exhibiting clinal features. The average values across all alleles were affected by the negative values. The averages are 0.213, 0.036, −0.070, −0.131, −0.258, −0.344, and −0.473, respectively for 0–700, 700–1400, 1400–2100, 2100–2800, 2800–3500, 3500–4200, and 4200–4900 kilometers. The negative values observed for some alleles generally follow a single, strong trend that is oriented northwest–southeast. The trend is believed to reflect major migrations out of the southeast, coincidental with the spread of agriculture. Thus, there is some directionality to the spatial patterns for some alleles, and in a narrow sense the patterns for some alleles do not fit the standard genetic isolation by distance model. For these alleles, apparently the regionalized autocorrelation caused by stochastic effects in standard genetic isolation by distance has superimposed upon it directional migration effects, which have been characterized in theoretical models (Sokal et al. 1989b; Epperson 1993a,b).

Many studies of spatial patterns of geographically distributed data have used the Mantel (1967) statistic, either as an alternative to or in conjunction with Moran statistics. The Mantel statistic can be used to measure the association of one n by n matrix **Y** of measures of genetic distance (for example, Nei's [1972]) among pairs of n populations with a second n by n matrix **X** of measures of geographical distance among populations (Sokal 1979b). The Mantel statistic M is calculated as

$$M = \sum_{i \neq j} X_{ij} Y_{ij} \tag{2.2}$$

Generally, aspects of spatial patterns that are revealed by spatial autocorrelation statistics such as Moran's I statistics are also reflected in the Mantel statistics. In fact, Mantel tests can be configured to test spatial correlations in distance classes.

Prior to the development of spatial autocorrelation statistics, many studies of geographic variation and isolation by distance used similar measures, particularly the genetic covariance and other so-called *kinship* measures. Most notably, N. E. Morton has conducted many analyses, particularly on human populations, for a variety of genetic markers. Usually, a smoothly decreasing function on distance d is observed, which is in a general sense consistent with genetic isolation by distance models. In many cases, the data fit the function $R(d) = (1 - L)\, ae^{-bd} + L$, where a, b, and L are constants fitted to the data. This exponential function is not appropriate for two-dimensional systems (which generally are not exponentially decreasing). Nonetheless, the fact that data could be fitted to it does mean that there is a steadily decreasing function on distance. For example, Imaizumi et al. (1970) found that independent measures of migration and immigration for Oxfordshire parishes produced a predicted function of kinship on distance that was well-matched by measures based on pedigree information from marriage records. Morton (1969) reviewed various studies of inbreeding, homozygosity, and pedigree analyses for many human populations. Morton (1982) compared isolation by distance from 28 different sets of data on humans, including studies on wide-ranging ethnic groups, continents, social structures, etc., and concluded that generally the fit of data to the standard isolation by distance models was good.

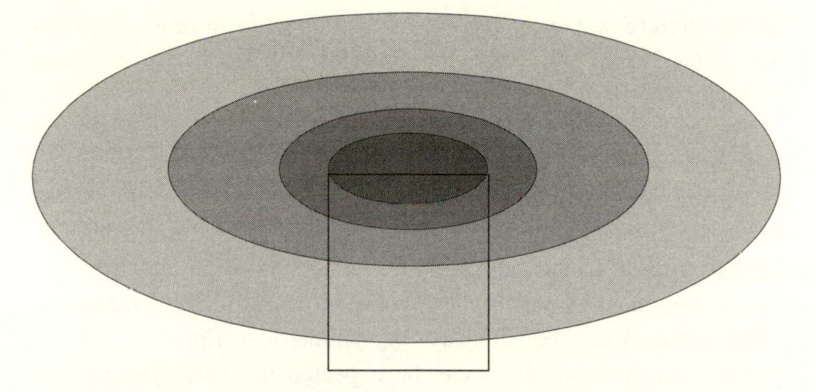

FIGURE 2.5. Example of an elliptical mound. Darker shaded areas represent regions where populations would have high values for the frequency of a gene, and lighter areas as well as the white background would contain populations with lower gene frequencies. Note that if populations were sampled only throughout the area of the small rectangle, the surface would appear very much like a cline. There would be large negative values of correlations between populations at the northern and southern extremes, and large positive correlations between populations at the same latitude but east and west. If the entire area were sampled, then there would be still be some directionality but of a different nature. Among the relatively short-distance distance/direction classes, the correlations would remain larger for longer distances for east–west, than for north–south. This surface could be approximated by a purely neutral genetic isolation by distance process.

DIRECTIONAL PATTERNS

Substantial directionality to spatial patterns may occur in systems of populations with multiple dimensions. For example, in the hypothetical gene frequency distribution in figure 2.5 correlations are positive for longer distances in the horizontal dimension. To some degree asymmetrical correlations deviate from assumptions in standard forms of genetic isolation by distance processes. Moreover, directionality can be used to study the operation of causal factors other than simple distance limits to migration, by using specialized measures. It has been demonstrated in theoretical studies that directionalities of various sorts may be caused by clinal selection, some configurations of immigration (Sokal et al. 1989b), or directionality in migration rates (Epperson 1993a), although such complex processes have not been fully characterized.

The most basic cause of spatial directionality is directionality in migration rates. Correlations may be twice as large or more in one dimension compared to another (Epperson 1993a). In real systems where the distribution of populations is highly nonuniform it may be difficult to form a consistent definition of system-wide directionality. This is important in part because correlations in different directions within a dimension should always be symmetrical (Epperson 1993a). Nonetheless, directionalities can be expected in many systems. For example, in species where migration is passive transport by wind, migration may be more frequent along the direction of prevailing winds. Correlations should be measured differently for distances along an axis aligned with appropriate systemic factors, such as prevailing wind direction, than for distances perpendicular to that axis.

Most of the theoretical models of genetic isolation by distance have been based on "isotropic" migration models, where, for example, in the two-dimensional case isotropy means that the migration rates are the same in all four directions. However, it is possible to incorporate directional migration into theoretical models, and arguably such processes could be considered as within the definition of standard genetic isolation. Because of stochasticity, even spatial patterns produced by isotropic migration will display some degree of directionality (e.g., figure 2.2). Moreover, precisely how a system is sampled may affect the degree of directionality (figure 2.5). Careful consideration must be given to sampling, because, for example, pure genetic isolation by distance can cause what appears to be a cline at some geographic scales of sampling. Observed clines are usually attributed to selection.

Directionality of various types may be detected by tailored Moran's I statistics, which classify pairs of populations according to direction as well as distance. The method of Oden and Sokal (1986) classifies all pairs of populations by a range of compass directions from one to another, as well as distances. Moran's I statistics are calculated using equation 2.1 for each distance/direction class k. Similarly, specialized Mantel statistics can be constructed for distance/direction classes. Two-dimensional correlograms can be constructed, and they may be displayed as in figure 2.6. The graph consists of sectored concentric rings. Each ring represents a distance class, and each sector a range of compass directions or bearings. Each "strip," defined by ring and sector, represents a distance/direction class. Furthermore, the magnitudes of the *I* statistics in classes may be indicated by degrees of shading. Note that the graph is symmetrical about both the 45–225° and 135–315° axes.

Oden and Sokal (1986) examined the properties of distance/direction Moran statistics in artificially generated data surfaces in two dimensions. Strong clines, represented in an artificial surface of a north–south gradient of values, where values are identical east–west, show the strongest directionalities. In this case, for large distances there are large negative values northerly–southerly and large positive correlations easterly–westerly. A single (circular) genetic patch (circular mound) shows no meaningful directionality and resembles isotropic isolation by distance. An intermediate pattern, an elliptical mound, shows an intermediate degree of directionality (Oden and Sokal 1986). The latter case corresponds fairly closely to the important consideration that a certain amount of directionality could be produced in samples taken from isotropic genetic isolation by distance processes, depending on how much stochastic variation there is in patch shape and how the area is sampled relative to the spatial scale of genetic patches. For example, one could (unknowingly) collect from a rectangular total sample area that contained a large genetic patch for much of its center area, but whose corners are outside the patch. A genetic patch tends to be circular, but axial symmetry is limited, and it could appear at times elliptic. If it were, the usual correlograms, as well as the distance/direction correlograms, could resemble those for an elliptic mound.

A large survey of 57 range-wide populations of *Drosophila buzzatii* in Australia illustrates several points on inferences based on directionality rather than distance alone (Sokal et al. 1987). Standard *I* correlograms revealed substantial spatial autocorrelations, and for 6 of the 12 allozyme allele frequencies the correlograms were statistically significant. None of the correlograms exhibited a strong tendency to decrease at large distances, a diagnostic for clines (see, e.g., figure 2.4B), although some did exhibit a weak tendency. In contrast, a summary Mantel statistic analysis, using geographic distance–di-

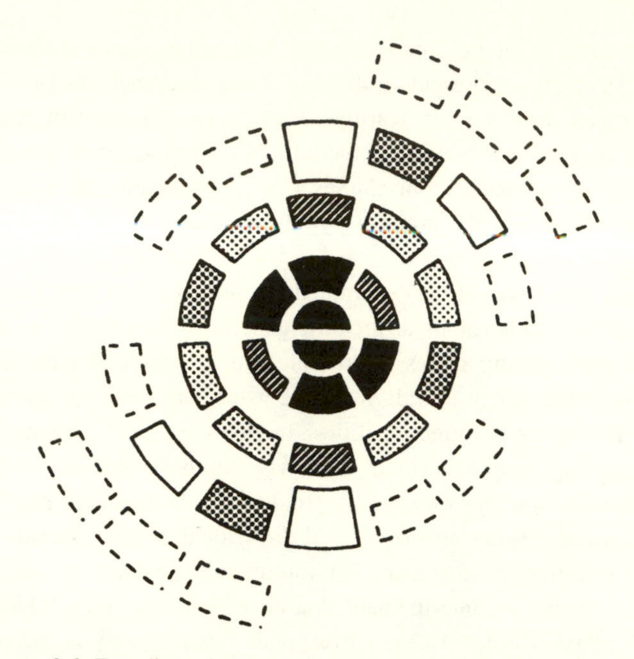

FIGURE 2.6. Two-dimensional correlogram summarizing *Drosophila buzzatii* data for Australia. Each sector represents a distance/direction class. Including the center half-circles there are five rings representing five distance classes with upper bounds 197, 526, 987, 1580, and 2305 km, respectively. The number of sectors, each representing a range of compass directions in polar coordinates on the Cartesian plane, is larger for intermediate distance rings, because there are more pairs of populations in those distance classes. Missing sectors are distance–direction classes for which there were no pairs of populations, and sectors with dashed outlines had insufficient numbers of pairs for appropriate statistical power. In this figure are represented quintiles of Mantel statistics based on Nei genetic distance measures based on many allele frequencies for allozyme loci, although similar figures could be drawn for Moran's I statistics for individual alleles. Sectors with no shading had normalized Mantel statistics with values ranging from −0.1764 to −0.1201; those with small dots from −0.0360 to −0.0067; large dots from 0.0089 to 0.0131; stripes from 0.0196 to 0.0704; and black from 0.1098 to 0.1264. Figure from Sokal et al. (1987).

rection classes and Nei's genetic distance for all loci combined, showed a fairly strong north–south cline and a lesser trend northeast–southwest (figure 2.6). For example, the north–south cline in figure 2.6 is indicated by the large negative value in the strip located at the top of the graph. Thus, although the patterns support that there are some limits to gene flow and some genetic drift, i.e., isolation by distance, there are also some large-scale trends, particularly the north–south trend, and these could be also influenced by selection (Sokal et al. 1987).

Further evidence of clines were exhibited in the distance/direction Moran statistics for each allele. In general terms, the results of the distance/direction correlograms were consistent with those for the distance correlograms, in that the six alleles that showed little autocorrelation solely by distance also showed little distance/direction autocorrelation. However, additional information was gained for the other six alleles, which did show significant distance autocorrelation. One allele (*Est1-a*) showed a large negative correlation at long distances west–east, but also a positive correlation at long distances north–south. Another allele (*Est-2c*) showed the opposite, a north–south cline. A third allele (*Est-2c*) showed two clinal trends, one north–south, the other northeast–southwest, and the latter was also found for *Adh-1b*. Finally, *Pgm-a* showed an east–west difference, and the remaining two alleles showed no strong directional trends in the two-dimensional correlograms.

The interpretation of the causes of directionalities in the *D. buzzatii* data also illustrate the importance of using multiple genetic markers. Gene frequency clines can be caused by repeated immigrations from populations outside the study system (and having different gene frequencies), to only a part of a system of populations, for example, the southern part of Australia (Sokal et al. 1989b). This should cause high correlations in the spatial distributions of frequencies for different loci. In contrast, it appears that under the standard genetic isolation by distance process, if the system is large, there should be little or no correlations for different loci, even if they are closely linked (Epperson 1995b). *D. buzzatii* was introduced into Australia recently, most likely in the late 1920s to early 1930s (Barker and Mulley 1976), and it quickly spread naturally as well as inadvertently with biological control programs for invasive species of *Opuntia*. In the subsequent 300 generations or so, *D. buzzatii* has developed substantial spatial structure over its range in eastern Australia. It appears that repeated immigrations did not occur in *D. buzzatii* because there are almost no correlations among the gene frequency surfaces for different alleles (Sokal et al. 1987). In other systems, such as the previously mentioned *Helix aspera* and humans in Europe, high correlations among loci are observed and interpreted as supportive of immigration.

Complex migrational processes may be highly idiosyncratic and stochastic, and highly directional during short periods, but over certain scales of space and time the system may average out and produce generically the same spatial patterns as the standard genetic isolation by distance model. A clear example of this is the Yanomama Indians (Sokal et al. 1986). Twelve of fifteen alleles had significant spatial distance correlograms, but correlations among the allele surfaces were low, among 50 villages in southern Venezuela and Brazil (Sokal et al. 1987). As a set, the distance/direction correlograms did not indicate strongly clinal patterns (Sokal et al. 1987). However, allele *C* of the *Rh-Cc* system did exhibit a general east to west trend. Nonetheless, the dominant pattern is one of fairly standard genetic isolation (drift-migration) processes, since the founding of the region about ten generations earlier. The maximum distance among villages, 579 kilometers, is about 50 times larger than the average distance between the birthplaces of parents and their offspring (Sokal et al. 1987). Moreover, the geographic structure is highly concordant with the verbal history of the tribes. Recent population dynamics of the Yanomama have been relatively undisturbed by western intrusion. A number of studies (e.g., Neel 1967) have shown that populations exist in rather isolated villages, each of which contains a few but extended lineages. With fairly regular frequency villages experience "fissions," or splits along lineages, subsequent migrational movements, and frequently fusions with other villages. Interestingly, single locus genotypic frequencies within villages are near Hardy–Weinberg values (Neel and Ward 1972), but the multilocus genotypic frequencies within villages exhibit considerable gametic phase linkage disequilibrium (suggesting admixture), which nonetheless tends to average out across the total region. The fusion-fission process is the main determinant of the spatial pattern of genetic variation. Smouse and colleagues (1986) showed, using a multivariate extension of the Mantel statistic test, that genetic distance is highly correlated with geographic distance ($r = 0.415$). Genetic distance is also highly correlated with distances based on linguistic affinity and fission/fusion history ($r = 0.365$). However, fission history distance is very highly correlated with geographic distance ($r =$

0.773). The highly irregular migration history, documented by the Mantel analysis, nonetheless produced spatial patterns that in some respects are similar to those produced by the standard genetic isolation by distance model.

COMPLEX PATHS OF MIGRATION

The rates of migration between populations may be affected by factors other than distance and direction. The spatial arrangement of populations itself may affect which populations exchange migrants. Moreover, physical or other barriers may prevent migration among populations that are close geographically. These and other complex features of migration affect spatial patterns of genetic variation. Thus spatial patterns might be used to reveal complex paths of migrations, using modified spatial statistical analyses. Alternatively, complex migration paths can be accommodated in spatial analysis of modified forms of isolation by distance.

Perhaps the most fundamental modifications account for the mobility and "behavior" of organisms in the context of irregularly distributed populations and geographic features. These can be incorporated into spatial analyses by the construction of *connectedness* schemes, which reflect likely migration paths among populations. For example, migrations for some species may occur only through mountain passes, not over peaks, and migrations from valley populations on either side must go through those in the passes. Hence, migrational distance should be measured along the paths through the passes, not simply the physical or Euclidean distances. Connectedness among populations in different valleys should not be direct, rather such populations should be indirectly connected through pass populations, and distances should be measured along the connectedness paths. Spatial distance predictive of correlations should also be measured along connectedness paths.

In less extreme cases, where there may not be obvious or strict barriers, a priori criteria may be needed to represent paths of migrations. This is particularly important for species that in some sense actively seek to migrate to "neighbor" populations. The difficulty is in defining neighbor populations in a way that is meaningful for a given system. However, without prior knowledge of precisely how migrations occur a more objective criteria for connectedness is needed. Then other considerations are paramount, such as geometric properties. Among the many possible forms of connectedness, the Gabriel connectedness criteria has the most favorable properties (Gabriel and Sokal 1969; Matula and Sokal 1980). It ensures that all populations are included in a network of migration paths, and it has favorable conditions for allowing definition of not only "nearest neighbor" but also second nearest neighbors and higher order neighbors. This is important for realizing the spatial replication needed to statistically analyze genetic correlations as functions of degree of spatial proximity of populations. Two populations are Gabriel connected if and only if the distance between them is less than the sum of their (two) distances to any other population, and this can be formatted into computation algorisms, such as in D. Wartenberg's SAAP program. Connectedness networks can also be modified to incorporate obvious barriers, typically by removing some connections (Sokal 1986). Gabriel connectedness and the practice of measuring distances along connectedness paths have been applied in many studies. Moran's *I* statistics and the Mantel statistic can be formulated in the same way, but with distances classes based on distances measured along the connectivity network. Moreover, directional measures of distance along the network can be incorporated into the geographic distance matrix used in the Mantel statistic.

Gabriel and modified Gabriel connectedness have been used in a number of experimental studies, including many of the studies discussed above. In all of the above studies the systems had fairly regular spacing of populations, and the use of Gabriel connectedness generally did not substantially change the conclusions about genetic isolation by distance. However, in other studies use of connectedness schemes has been critical, particularly when there are strong barriers to migration. Populations on either side of the barriers can have much lower genetic correlations than expected based on the short distance between them.

Another way of analyzing the effects of barriers is based on the idea that the degree of genetic differentiation may be much higher among populations on either side than expected based on the short distances separating them. When the gene frequencies in populations are viewed as a "surface," the slope of the surface is very steep in the zone separating populations on opposite sides of a barrier. A statistical technique called wombling seeks to identify unusually steep, localized slopes caused by barriers, using a fairly complicated criteria (Barbujani et al. 1989). Although steep slopes for single loci can occasionally occur in standard genetic isolation by distance models, when they geographically coincide for several genetic loci, barriers are indicated. Such geographic locations can be checked a posteriori for possible reasons that barriers exist.

Wombling revealed more details of the spatial patterns in the Australian *Drosophila buzzatii* (Barbujani et al. 1989). There is a large relatively homogeneous area in the center of the range, characterized by the fact that the slopes of genetic variation are not greater than those expected for standard genetic isolation by distance. This area corresponds to the main infestation of the primary host of *D. buzzatii*, invasive *Opuntia* species. At the periphery of this area are several localized areas where the degree of genetic differentiation is particularly high. Three of the areas of steep change correspond to zones between the main area and isolated peripheral populations of *D. buzzatii* and isolated patches of *Opuntia*. In addition, there are areas with large slopes that identified four small clusters of populations, and again each of these are peripheral areas where *Opuntia* is rare, occurring in isolated patches. Overall, the pattern suggests that there are several peripheral populations that may have experienced considerable founder effects and possibly selection (Barbujani et al. 1989), in addition to the genetic-drift-migration (i.e., genetic isolation by distance) processes occurring throughout.

Wombling analysis of the Yanomama data revealed many areas where there is strong localized differentiation (Barbujani et al. 1989). Unlike the surfaces for *D. buzzatii* there was no large central region lacking such features. Many parallel steep slopes separate the easternmost range, where three villages of the Ninam dialect-speaking peoples occur. A second group, just to the west of the Ninam, corresponds closely with the Yanomam dialect. A third distinct group is located in the northwest, and this corresponds exactly with the Sanema dialect. The remaining populations all exist in the southwest corner, in a region where there is high differentiation nearly throughout, and these villages form the Yanomame dialect. The peoples of the Yanomame dialect occupy the largest number of villages and they are more diverse than those of the other dialects. They contain at least five recognizable subgroups, which may have differentiated as this group expanded and underwent repeated fissions (Neel 1978). The correspondence of geographic genetic differentiation and linguistics, also revealed in the Mantel tests, is striking. Similar results are observed in Europe. Localized areas of spatially abrupt changes in gene frequencies are numerous, some 33 in one analysis (Barbujani and Sokal 1990). The connection to linguistics is again strong; out of the 33 localized areas all but two correspond to language boundaries, and many of them also correspond to obvious montane or marine physical barriers (Barbujani et al. 1989).

OTHER COMPLEXITIES OF MIGRATION PROCESSES

Other aspects of migration processes may change spatial patterns of genetic variation and affect inferences about amounts of migration and genetic drift. Most prominent are effects under the general heading of "stochastic migration." Stochastic migration obtains whenever there are stochastic changes in gene frequencies during the migration process (Latter and Sved 1981; Fix 1975, 1978, 1993; Rogers 1987). There are several forms of stochastic migration. Measures such as F_{ST} of average differentiation are affected by all forms. The predicted inverse relationship of F_{ST} to migration rates m and population size N, generically $F_{ST} = 1/(1 + 4Nm)$, is fundamentally altered when there is stochastic migration. When the stochastic changes in gene frequencies during migration dominate those associated with standard genetic drift during population size regulation within populations, F_{ST} *increases* with increasing rates of migration. Stochastic migration is not important with respect to F_{ST} when very large absolute numbers of gametes migrate (and do not migrate), relative to the size of mature populations after population size regulation (Epperson 1993b). It is moderately important in species where postdrift, say adult diploid, individuals constitute migrants. It is very important when there are correlations among the individuals in each specific migrant group that migrates from one population to another. For example, a migrant group may consist largely of a single family, clan, or similar group, in which there are high kinship and genetic correlations relative to the population from which they arose (Fix 1978; Rogers 1987). The stochasticity in sampling a small but highly correlated group causes its gene frequencies to be very different and the gene frequency in the population receiving the group will be stochastically changed to a disproportionate degree. Greater migration results in greater stochastic changes, analogous to *lesser* population size in genetic drift.

The simplest form of stochastic migration does not change the spatial correlations predicted for isolation by distance (Malécot 1948; Rogers 1987; Epperson 1994), although it may increase the role of chance in any realized outcome of isolation by distance. This form occurs whenever the stochastic changes during migration are not shared among multiple populations receiving migrants from the same source population (Epperson 1994). When they are shared the predicted spatial genetic correlations are affected. Shared stochastic migration effects can be categorized, according to their effects on spatial correlations, into *positively shared* effects versus *negatively shared* effects (Epperson 1994). Positively shared stochastic migration effects are probably much more common than negatively shared ones. They may occur whenever only a portion of each population as a source is in condition to contribute migrants to multiple recipient populations. Positively shared stochastic migration effects appear to generally increase the spatial correlations, especially for relatively short distances (Epperson 1994, 2000a). The effects will usually be small if migrants are nearly random samples of a population, but they can be dominating when there is a high degree of kin structure or other forms of intraclass correlations to the group of emigrants from each source population (Epperson 1994, 2000a). Many real systems of migration will have a mixture of shared and unshared stochastic migration effects as well as drift during population size regulation, and the spatial correlations will be a weighted average of the processes (Epperson 1994).

Negatively shared stochastic migration effects may be rare, but they can cause spatial patterns and correlations to differ dramatically from the standard genetic drift-migration process of isolation by distance. Correlations may obtain large *negative* values even for short distances (Epperson 1994). A good example may be the fusion/fissions of Yanomama villages. Because fissions, say of two groups, are along clan lines, there should be a large negative genetic correlation between the two groups, *relative* to the entire village before fission. There must also be substantial genetic drift within villages as well as interclan marriages that occur between fusion/fission events. The spatial correlations should be a weighted average of those for the pure negatively shared process and the standard genetic isolation process (Epperson 1994). Nonetheless, it appears that the fusion/fission aspect of Yanomama life should cause the autocorrelation function to decrease more rapidly and become negative at shorter distances compared to the standard isolation by distance model. In fact, in the Yanomama, negative correlations do occur at relatively short distances, and the spatial aspect of fusion/fission is consistent with the observed patterns of genetic variation. Moreover, the wombling analysis revealed many regions where the slope of genetic differentiation is higher than expected for the standard genetic isolation by distance model. This is consistent with negative autocorrelations for short distances. Finally, note that this should be the case, because both geographical distances and fusion/fission history (themselves highly correlated) show a tight fit with genetic distances, in multivariate Mantel statistics (Smouse et al. 1986).

Unfortunately, the effects of stochastic migration have not been examined explicitly in natural populations, even though there is a firm theoretical basis for expecting some aspects of the spatial pattern to be affected. Stochastic migration distinguishes different stages of life cycle according to the timing of population size regulation and migration. Levels of inbreeding, variance in gene frequencies among populations, and spatial correlations may differ for each stage of multistage life cycles (Rogers 1988). Hence, in principle, the various stochastic migration effects could be detected by stratifying samples according to life stage, and, for example, measuring spatial correlations in adults separately from juveniles.

EFFECTS OF TEMPORAL CHANGES ON SPATIAL PATTERNS

Theoretical models of standard genetic isolation by distance usually assume that the sizes of populations remain constant (Malécot 1948; Kimura and Weiss 1964). The genetic effects of changes in size is well understood for single populations, and in many circumstances the main effects can be captured using an "effective" population size, which is usually closely related to the harmonic mean (Wright 1938) averaged over relatively long time periods. The short-term effects of a single-population size bottleneck have also been characterized extensively in the population genetics literature. Little theoretical work has been done in the context of geographical genetics. However, Bodmer and Cavalli-Sforza (1968) suggest that spatial variation in population size has little effect on the spatial structuring of genetic variation, and it appears likely that the same is true for variation in population sizes over time. More work is needed on this subject. Intuitively, it seems that at least over sufficiently long timescales population size fluctuations should average out, and in the long run the standard form of isolation by distance should be produced. Unusually small populations should create a temporary localized increase in stochasticity, which eventually should be smoothed throughout the system of populations.

There are similar issues in the metapopulation context (e.g., Levins 1969; Hedrick and Gilpin 1997), where there are extinctions and recolonizations (Wright 1940). However, to date metapopulation theory generally has not resolved spatially explicit systems and does not offer much insight on spatial patterns. Most metapopulation models have no spatial context, and are based on Wright's island model, where migration among populations, sources of propagules for newly founded populations, and locations of populations going extinct do not depend on spatial proximity (Whitlock and McCauley 1990). The primary concerns are in terms of the total amount of genetic variation maintained and its partitioning within versus among populations. Recently, Nunney (1999) extended metapopulation models to hierarchical systems. Whitlock and Barton (1997) examined a variety of models, and their study included derivation of the effective size of a system of populations with limited migration, using migration matrix approaches. Other recent trends in metapopulation studies include selection and the evolution of dispersal rates (e.g., Olivieri et al. 1995).

SPATIAL PATTERNS CAUSED BY SELECTION

Strong evidence for selection can be sought in the frequent observation of clinal patterns for gene frequencies or the genetic component of quantitative traits. Clines imply that there is an axis to a multidimensional spatial pattern, and in some cases the axis may be aligned with an identified environmental factor. Clines have been a staple of the theory of natural selection, and they have been extensive reviewed by Endler (1977). A particularly common axis for environmental gradients is latitude, to the extent that there are recognizable "rules" that hold for many species with large geographic ranges (e.g., Allee et al. 1949; Clarke 1966; Ricklefs 1972). For example, in many animals the ratio of body volume to surface area tends to increase and lengths of extremities decrease for populations tending northward (in the northern hemisphere) along latitudinal gradients. In plants, flowering time is often genetically variable with latitude, because of day length and/or temperature trends.

Strong clinal patterns are produced only when selection is very strong and/or migration is limited relative to the spatial scale of the environmental gradient. Thus absence of a cline does not necessarily mean that clinal selection does not act. However, if migration rates and distances are small compared to the strength and spatial scale of operative selective influences, then the genotypic frequencies in each population are determined by its environment, independent of other populations. Gene frequency surfaces precisely follow the pattern of the environmental factors, in a self-explanatory way, which need not be discussed here in more detail. The latter is not limited to clinal environmental factors, but may be quite idiosyncratic. Thus almost any pattern of genetic variation and values of spatial statistics could result from environmental selection (Sokal and Oden 1978a; Sokal 1979a). Spatial distributions of genetic variation may depend further on the mode of genic action (Endler 1977).

If selection is weak or if there is considerable migration a clinal pattern may exist, but it may be fairly flat and difficult to infer. Clines are frequently observed for critical quantitative traits, but the detection of clinal selection for specific genetic loci is more difficult. The latter may often require more extensive data, as well as sensitive spatial statistical analyses, and comparisons to complex theoretical models combining natural selection and migration. In such cases, there are complex interactions between selection and migration, and there are equally complex statistical interactions. Spatial patterns of selected genetic variation are no longer determined simply by the spatial distribution of fitness values. This is probably a common situation, particularly for individual genetic loci (as opposed to quantitative traits), where selection is not strong enough that migration has no effect. Moreover, we may not know a priori which genes are those that are responding to either known or unknown gradients of environmental factors. Here the starting point is to outline how observed spatial patterns may be used to infer selection on individual loci without such a priori information. In doing so investigators need to consider that the pattern effects produced by selection may be confounded with the "noise" that occurs from stochastic effects combined with gene flow. The logic that there should be correlations between an environmental factor and responsive gene frequencies, if selection is acting (as opposed to spatial structures in the gene frequency surface per se), must also take into account that such correlations can also be caused by purely stochastic processes for neutral loci. Spatial patterns caused by purely stochastic processes will generally show correlations with spatial patterns of environments, at least at some spatial scales.

A few experimental studies have conducted sophisticated spatial statistical methods that attempt to separate pattern influences of gene flow from those of selection, and have convincingly demonstrated clinal variation whether due to selection or directional migration. One of the best examples is the study by Bocquet-Appel and Sokal (1989), who used trend surface analysis to separate clinal variation from smaller scale autocorrelations caused by limited gene flow for cranial measurement data from three different time periods and over 100 samples in Europe. In population systems with two spatial dimensions, the genetic variable (usually a gene frequency) q_{xy} for populations (x, y denote Cartesian coordinates) is modeled as a polynomial with a stochastic and/or statistical error term ε:

$$q_{xy} = \sum_{i=0}^{s} \sum_{j=0}^{v} \beta_{ij} x^i y^j + \varepsilon \tag{2.3}$$

Bocquet-Appel and Sokal (1989) found that the long-distance trends represented by either a linear (both s and v are either zero or one, but both are not zero) or a quadratic ($s = v = 2$) explain a good portion of the autocorrelation observed. Moreover, the de-trended residuals showed positive autocorrelation at short distances consistent with that caused by limited gene flow in the standard isolation by distance process. In other words, it appears that they were able to demonstrate the presence of both isolation by distance and larger scale autocorrelation caused by other factors. In an extreme counter-example, Sokal et al. (1989b), showed, using simulations, that the spatial scale of autocorrelations caused by selection can be nearly the same as for neutral loci, if the sizes of the environmental patches are similar to the genetic patches expected for neutral loci. It is also known that simple selection that does not vary with environment generally reduces the degree of autocorrelation and spatial patterning (Malécot 1948; Epperson 1990a).

PARTITION OF GENETIC VARIATION

Many genetic surveys report the level of genetic variation within *versus* among sampled populations. Although the partitioning per se as opposed to spatial patterns of genetic variation is important in itself for several reasons, here the focus is on how summaries of surveys, many of which do not report detailed spatial distributions, yield insights into biological factors that influence migration rates. In many surveys, the number of populations is too small to examine spatial patterns. Typically, measures of variation among populations involve the variance of allele frequencies, or some function of it such as the standard formula for F_{ST} (equation 4.22) and Nei's (1973a,b) genetic diversity statistic, G_{ST}. In effect, these statistics measure the relative proportion of the total variation that exists among populations rather than within. Statistics for genetic diversity within populations include average effective heterozygosity and average actual heterozygosity. The general expectation is that as population sizes N or migration rates m decrease, F_{ST} and similar measures increase in value. The total variation in the complete system (i.e., variation within together with that which is among) decreases with decreasing size of populations. This view of population genetics usually implicitly or explicitly assumes Wright's island model, which is not spatially explicit and assumes all populations in the system exchange migrants at the same rate m. It should be pointed out that the general expectation (e.g., that F_{ST} increases as the product Nm decreases) also depends on details of the timing of migrations in the life cycle.

For example, Fix (1994) has examined models where migration occurs at the adult stage, after population size regulation (and genetic drift). If migrations are substantially kin structured, then stochastic migration effects contribute substantially to the variance among populations, and the usual expectation that even low rates of migration are sufficient to retard genetic differentiation is incorrect. Nagylaki (1979) has shown that variation in migration rates also increases the variance among populations.

In a most remarkable compilation, Hamrick and Godt (1990) reviewed some 653 isozyme studies on 449 plant species. Differences in G_{ST} values occurred for several life-history factors. Taxonomic classes differed in the degree of relative differentiation among populations. Dicots had the largest values (average $G_{ST} = 0.273$), followed by monocots (0.231) and gymnosperms (0.068). Life form was also important, with annual plants having the highest values of G_{ST} (0.357), followed by short-lived herbaceous perennials (0.233) and long-lived herbaceous perennials (0.213), short-lived woody perennials (0.088), and finally long-lived woody perennials (0.076). Larger geographic ranges corresponded with lower degrees of differentiation among populations, with average values of 0.248, 0.242, 0.216, and 0.210 for endemic, narrowly distributed, regionally distributed, and widespread ranges, respectively. Breeding system was particularly important: primarily self-fertilizing species had the highest values (0.510), presumably caused in part by the fact that there is little chance for migration via pollen. Animal-pollinated species also had relatively high values (0.216 and 0.197 for mixed selfing/outcrossing and outcrossing species, respectively), again reflecting limited rates of pollen migration among populations. Wind-pollinated species had lower differentiation (0.100 and 0.099 for mixed mating and primarily outcrossing, respectively), again consistent with the generally longer distances of pollen movements when dispersed by wind. Similar results were observed corresponding to general features reflecting the likelihoods or rates of seed migrations among populations. Finally, successional status was important. Colonizing or early succession species have the highest value (0.289), followed by mid (0.259) and late (0.101) successional species, presumably reflecting the greater role of founder effects in early successional species.

Major reviews of the partitioning of genetic diversity in animal species include those by Nevo and colleagues (Nevo 1978; Nevo et al. 1984). Various life-history traits are correlated with both the total amount of diversity and the degree of differentiation among populations. Theoretical expectations are widely confirmed. For example, in a review of 333 species of vertebrates and invertebrates, representing 27 animal groups, Bohonak (1999) found a large negative correlation (-0.72) between the level of genetic differentiation among populations and the dispersal ability of species.

Generally, the features observed in distributions of allele frequencies are also seen for other types of data, such as DNA sequence data, other haplotype data, and microsatellite data. For the first two, any differences may be attributed to the underlying mutation model, usually the infinite sites mutation model rather than the infinite alleles mutation model used for allele frequencies. Differences in the types of data obtained center on the fact that haplotype or sequence data can include the number of segregating site differences among types as well as their frequencies. Sequences may contain information beyond allelic status and, although substantial additional information in single populations is apparently rare (Ewens 1974), that for systems of populations has yet to be characterized. Nonetheless, we expect greater sequence differentiation under more or less the same conditions (i.e., when the product Nm is smaller) as those for greater differentiation in gene frequencies or lesser probabilities of identity by descent. Moreover, spatial patterns of haplotypes also are similar to those for alleles. For example, the spatial distribution of pairwise coalescences within populations has been studied (Barton and Wilson 1995), and is controlled by dispersal distances similarly to the spatial distributions of pairs of genotypes. Microsatellite variants are produced by a largely stepwise mutation process, and the differences in sizes of fragments reflect differentiation in addition to population frequencies of fragments (alleles). A number of DNA studies are discussed in chapter 3, and Bachmann (2001) reviews many of these studies and the general relationship of various markers used in evolutionary studies.

CHAPTER 3

Ancient Events in Spatial–Temporal Processes

> The occurrence of new alleles each produced by a unique mutation or recombination [causes] . . . the "gametic pool" of Sewall Wright to be extended, as time progresses, to an indefinitely increasing number of new alleles; now called infinitallelism, which might have been the germ of the molecular clock had not . . . the (falsely Mendelian) consensus about fixed genes . . . paralyzed Post-Darwinism imagination.
>
> —Gustave Malécot (1998 personal communication)

Major events in the distant past may leave transient signatures in spatial and spatial–temporal patterns of genetic variation. Important ancient events include the time and place of genetic "innovations," refugia, range expansions, colonizations or major immigration events, and fragmentation. Transient effects of ancient events contrast with stable patterns that can be produced by selection, genetic drift, and migration averaged over long periods. The study of transient effects of major events in the distant past calls for a somewhat different emphasis in the context of spatial–temporal processes. For example, in the theoretical works of Malécot (e.g., 1948) the focus was on deriving spatial distributions produced over long periods or at equilibrium, and they provide an entirely appropriate basis for empirical studies of stable geographical patterns of genetic variation. Equilibrium results do not show the transient effects of major events. Moreover, we should distinguish long-lasting yet transient effects of ancient events from more recent or shorter term transient effects. The latter may be analyzed using the migration matrix approach or short-term space–time correlations.

The signatures left by most nonrecurring ancient events are delible and continued gene flow will eventually erase them. Many empirical studies aim to detect the trace of an ancient event, in effect to parse off particularly important features of the past from the spatial–temporal context. Frequent goals are to infer the geographic origins of new genetic variants or by extension the geographic origins of species themselves. In this chapter examples are used to illustrate the primary issues of the general conditions required for valid separation of particular temporary spatial or spatial–temporal patterns from the space–time process in which they are embedded. Typically, such studies do the following: (1) use population differentiation per se of molecular variation at the present time; (2) conduct phylogenetic reconstruction to infer the gene genealogies, and the ancestral genotype (generally without including information on spatial proximity, structure, and migration in the probability models); and then (3) use present spatial patterns of types that are most like the inferred "ancestral" type, and hence infer the past location of ancestral type. Examination of such studies also illustrates some of the key distinctions of modern molecular data from gene frequency data. Phylogenetic studies can sometimes take advantage of a unique kind of temporal "depth" (Templeton 1998) to spatially distributed molecular data. Steps 1 and 2 may not encounter serious problems, particularly if the timescale on which mutations accumulate is much slower than that on which migrations occur, and when coalescences within populations are much more recent than those among populations (Nordborg 1997). The rationale for the third step has received the least attention, and the step often appears to be made subconsciously. While it may seem to be a safe assumption that a given gene sampled at the present time is most closely related to the past or ancient genes in the same population, the gene may also be very closely related to ancient genes in other populations, perhaps in some cases quite distant geographically. This is a recurring theme in the inference of major events in spatial–temporal processes.

The study of ancient events in population genetics generally requires extensive datasets. Most survey data are contemporary, but as ancient DNA samples become more available, they may become disproportionately important and informative. There are still few species for which sufficient, even purely spatial (i.e., not space–time) datasets exist. There can be no doubt that in the near future sample sizes orders of magnitudes larger will become available for some species, including humans. But because data requirements have been yet rarely met in surveys, much of this chapter is devoted to a specific example, the origin of anatomically modern humans. Other examples of ancient population genetics follow, for humans and other species.

OUT OF AFRICA

Studies of human genetics typically have especially large numbers of widely separated study populations, with large sample sizes and numbers of genetic markers, and this makes them well suited for inferring the location of an ancestral population from the distant past. We will examine issues around the so-called out-of-Africa hypothesis or theory of the origins of modern humans. I wish to make it very clear that this examination is not necessarily intended to challenge this theory. It very well may be true, and there are many considerations not addressed here that support the theory.

The paramount features of the out-of-Africa hypothesis surround the geographic location and isolation of the first *anatomically* modern humans. By some half-million years ago so-called archaic hominids had spread throughout much of the Old World. The simplest form of the out-of-Africa hypothesis is that the first anatomically modern humans evolved in a small population of probably less than 10,000 individuals, in complete reproductive isolation, somewhere in Africa, about 200,000 years ago. Once evolved, this population began to increase in size and spread geographically. One scenario has it that humans spread along the coastal zones of Africa, and later into Asia and Europe. As they spread further to various regions of the Old World, they must have come into contact with the archaics already present, but did not interbreed at all. All genes today are descended from those in the isolated original population in Africa. This theory has become widely accepted in the last decade or so, while enthusiasm for a contrasting theory, the "multiregional hypothesis" (e.g., Wolpoff 1989), waned. The multiregional hypothesis states that the gene pool of anatomically modern humans, us, contains substantial contributions from many prior regionalized and differentiated archaic populations. Much of the cited support for the out-of-Africa theory is genetic, in particular the pattern of genetic differentiation among geographically separated groups, especially ethnic populations that have not undergone recent global migrations. There is also some physical evidence for the theory (e.g., Klein 1995; Lahr 1996; Sokal et al. 1997b).

The out-of-Africa theory became widely supported following phylogenetic studies of mitochondrial DNA, mtDNA. Mitochondrial DNA is strictly maternally inherited (e.g., Stoneking and Soodyall 1996), hence the genealogy of mtDNA is matrilineal. In a highly influential and widely publicized paper, Cann et al. (1987) studied samples of mtDNA from global populations and inferred a "mitochondrial Eve," the woman who carried the most recent common ancestor (MRCA) of all mitochondria today. In other words, all mitochondria lineages trace back to or "coalesce" in the mitochondrial Eve. Using polymorphic sites in the data and the molecular clock, Cann et al. (1987) estimated that "Eve" lived about 200,000 years ago. While this feature captured headlines and public imagination, the existence of a mitochondrial Eve is a necessary outcome of life, because *any set of genes must trace to a common ancestor at some time in the past.* Moreover, because mtDNA does not recombine, the entire mtDNA genome must all be descended from a single woman. The estimate that the time back to the most recent common ancestor, or TMRCA, was 200,000 years ago is interesting in part because it coincides with fossil evidence of the appearance of anatomically modern humans. However, this coincidence in itself also does not mean much, because there is no a priori reason to expect the TMRCA of a set of haplotypes or DNA sequences to necessarily coincide with the event of origination.

If there is selectively neutrality, the TMRCA should depend primarily on a function of the overall effective population size, and estimates of the TMRCA depend also on the mutation rates. For example, the TMRCA could in principle occur long after the origin of modern humans, in particular if human population size bottlenecks occurred after the origin. The TMRCA could also have been much earlier than the origination, if populations remained large prior to the formation and throughout the existence of modern humans. There could have been polymorphism within the theorized isolated single original population under the out-of-Africa hypothesis, depending on what its size was and for how long it was isolated. Based on the estimated value of TMRCA, the human effective population size, which is usually a function of the harmonic mean, since the time Eve existed has been estimated at around 10,000. A number of studies have argued that this is too small to fit the multiregional hypothesis, because, they maintain, the Old World population of archaics must have been much larger simply to have been sustained (e.g., Harpending et al. 1998).

Effective population size N_e is an important concept in population genetics, and it is worth noting some of its general properties. It simplifies the extension of theoretical models, originally constructed for an "ideal" population, to many other situations. Typically, the ideal population is constant in size, with monoecious random mating, and certain constraints on the variance of numbers of progeny produced per parent (e.g., see discussion about the process Equation 5.1 in chapter 5). Generally, an N_e is a function of the actual population size N and other, modifying factors, and it can simply be substituted for N in the model equations formulated for the ideal population. Equations are variously expressed in terms of probabilities of identity by descent, coalescences, or gene frequency covariances or variances, hence N_e is "effective" with respect to these measures. The most common are the "inbreeding effective number" and the "variance effective number" (e.g., see review by Crow and Denniston 1988). In some cases, but not others, they are equivalent. In general, population biological factors that substantially affect N_e include unequal numbers of females and males, large variance in numbers of offspring, and various forms of systemic inbreeding (e.g., Caballero and Hill 1992). However, when examining out-of-Africa the increase in population size is far more important. N_e is strongly disproportionately affected by any small sizes, as, for example, when N_e is a function of the harmonic mean (e.g., Crow and Denniston 1988).

While the TMRCA of mtDNA appears to be on fairly solid footing, it is much more difficult to determine the sequence and timing of how population size may have expanded and contracted during the last 200,000 years or so. The TMRCA for mtDNA is not so satisfying because it represents only the mitochondria. It does not mean that all other (e.g., autosomal) genes also coalesced at the same time, nor even that all genes came from the same population as the mitochondrial Eve. More recent analyses of larger datasets of mtDNA polymorphisms (e.g., Stoneking and Soodyall 1996) generally support the TMRCA reported by Cann et al. (1987). Together with the TMRCA, the genotype of the most recent common ancestor is also inferred from the gene genealogy, based solely on probabilities of mutations. It should be noted that the actual probabilities of common ancestry are also functions of geographic structure and migration rates, as well as mutation rates, and, although the former two factors may not make much difference, they were not considered in the calculations. However, this may not matter much if mutation effects occur on a much slower timescale than migration effects (e.g., Nordborg 1997; Fu 1997).

The geographic location of a most recent common ancestor is even more difficult to assess. Cann et al. (1987) found that among sampled populations, those in Africa were most similar to the inferred ancestral mtDNA sequence. Many studies have stated that (provided the data and gene genealogical conclusions are reliable) this means the ancestral sequence and the mitochondrial Eve existed in an African population. However, usually such statements have been made without further comment and are unsupported by any population genetic arguments or models. Recent theoretical developments have shown that the restricted presence of the inferred ancestral gene in specific present-day populations may or may not be closely related to the likelihoods of those populations having originated the ancestral gene, depending on the rates of mutation and migration (Epperson 1999a; 2002). The inferred ancestral gene (if selectively neutral) is in essence a randomly chosen representative of a population, hence relevant theoretical results can be expressed as space–time probabilities of identity by descent, a space–time extension of Malécot's definition of spatial probabilities of identity by descent. Under the conditions of the models (Epperson 1999a; 2002; and chapter 5), the relative values of these probabilities equal the probabilities of origination, because of the simple fact that one of the populations must have contained the ancestor of any given gene at present. Particularly important is how the probabilities of descent depend on the geographic distance between a potential ancestral (origination) geographically located population and the location of a present population. For example, consider Africa and Asia as alternative potential locations of origination, containing the ancestral mtDNA (mitochondrial Eve), and that now only African populations contain the ancestral type mtDNA. What are the *relative* likelihoods that Africa rather than Asia was the location? The two likelihoods depend on the migration rates and mutation rates (Epperson 1999a; 2002), and they should be nearly equal when the amount of migration is high and the rate of mutation is high. In such cases, the location of a gene today (e.g., ancestral type mtDNA in some African populations today) has almost nothing to do with where its ancestor was a long time ago. In other words, the spatial or geographic pattern of genetic variation today contains almost no information about where the origination was, whether one uses haplotype frequencies or gene genealogies based on phylogenetic reconstruction and degrees of differentiation (e.g., among DNA sequences).

For example, in a system with one spatial dimension (appropriate if populations were mostly coastal), and with migration only between nearest neighbor populations with rate l_1 equal to 10 percent, and a per-sequence mutation rate k of 10^{-6}, an ancient (10,000 generations ago) gene from a population located 100 populations away is about 82 percent as likely to share identity as if from the same population itself (figure 3.1). With $l_1 = 0.10$ and $k = 10^{-4}$, a population located 100 populations away has probability 23 percent as large, in contrast to the purely spatial probabilities of identity by descent, which are only 4 percent as large. Moreover, when mutation rates are high the probabilities become very small, suggesting that they would be difficult to estimate from genetic data. The function of relative likelihoods of origination on distance is not quite so flat when migration rates are lower, for example, at 1 percent (figure 3.1), indicating that the amount of migration is critical within the range of 1–10 percent. We do not know what the actual migration rates were among ancient humans. If the system of populations was primarily two dimensional, then even greater flatness may be expected, as occurs for the space–time correlations of gene frequencies (Epperson 1993b).

The above models assume that migration is isotropic, i.e., the rates are the same in both directions. If migration were anisotropic, flowing more out of Africa than into it, it may be that migration causes distant populations to be *more* likely the ancestral source of present

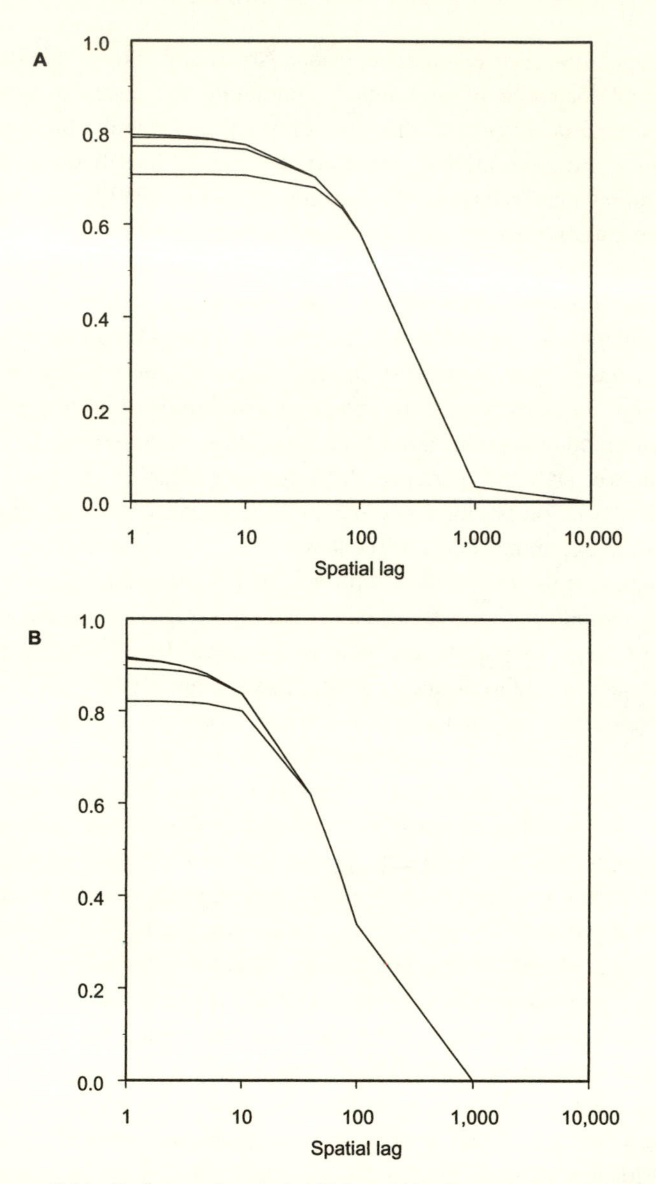

FIGURE 3.1A–D. Space–time probabilities of identity by descent in migration models with different parameter values. All cases are migration models with nonzero isotropic ($l_1 = l_{-1}$) migration only between adjacent populations in a system with one spatial dimension. For each graph, the X axis is spatial distance of separation on a log scale, and there are four curves representing, from top to bottom, time lags of zero (purely spatial

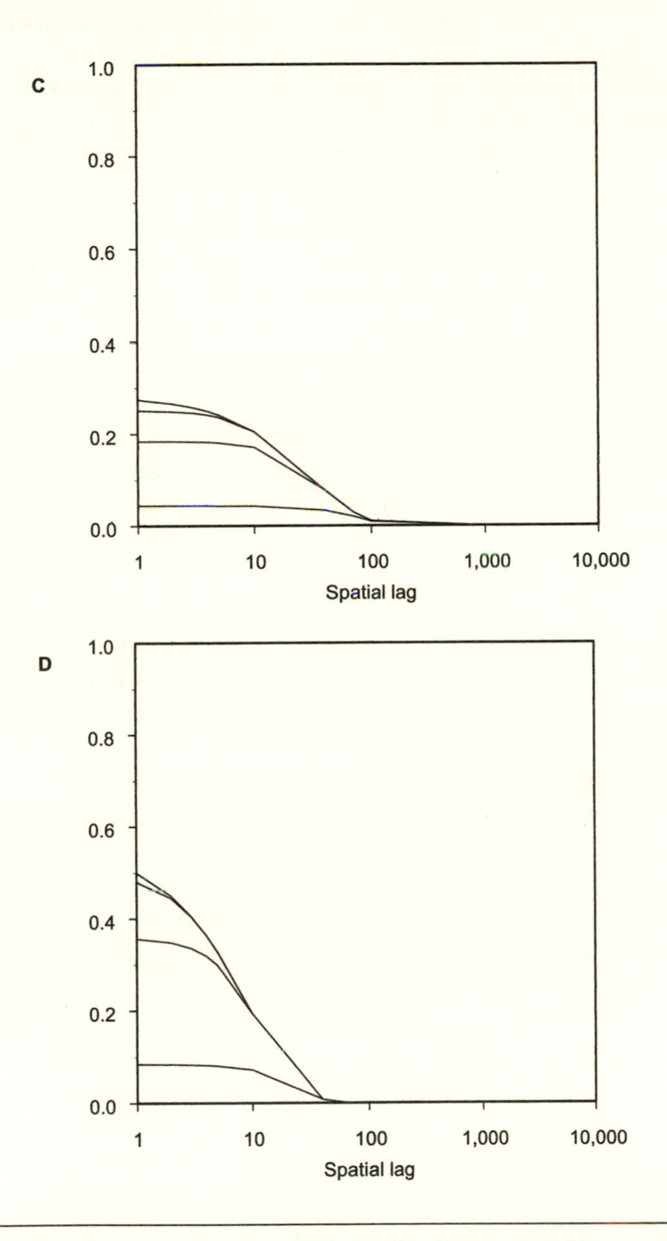

FIGURE 3.1 (*continued*) probabilities), 100, 1000, and 10,000 generations, respectively. The parameters for the four models are (A) $k = 10^{-6}$, $l_1 = 0.1$; (B) $k = 10^{-6}$, $l_1 = 0.01$; (C) $k = 10^{-4}$, $l_1 = 0.1$; and (D) $k = 10^{-4}$, $l_1 = 0.01$. Note that the "flattening" effects of large time lags on the spatial curves are greater in graphs C and D.

variants than the same (or very nearby) populations (Epperson 1993b). In other words, the mtDNA data, particularly the similarity of present day African populations to the mtDNA Eve, could indicate that Africa is *less* likely than other populations (perhaps even Asia) to be the ancestral location of Eve! The anisotropic model has not been examined for space–time probabilities of identity by descent, but there may be reasons to expect them to be similar to the space–time correlations, which can show this effect of anisotropy (Epperson 1993b). However, both models also assume that populations have always existed at a fixed size *N*. Thus their application to the out-of-Africa hypothesis, which implies that populations exist only as they become founded, is clouded. I want to make very clear that I am not arguing here that humans originated in Asia. There are as yet no completely satisfactory theoretical results, but the results do illustrate how difficult inference of origination can be. Such considerations also point to the potential importance of "space–time" genetic data, in particular from ancient DNA samples, and the space–time probabilities of identity by descent or space–time coalescence may provide the probability theory needed to properly utilize such data.

Other evidence for the mtDNA Eve and out-of-Africa theory has been assigned explicitly in terms of details of the geographic distribution of haplotypes together with the degree of differences (reflecting numbers of mutations) among haplotypes and the inferred ages of haplotypes. Haplotypes that have the largest number of inferred mutational changes from the inferred ancestral type are assumed to be more recent. This temporal dimension could in principle contain critical information in an explicitly nonequilibrium context (e.g., Templeton 1998). The geographic distribution may contrast under isolation by distance versus range expansion. If Old World human populations had been stable and exchanging migrations, in a restricted manner, throughout existence, then old haplotypes should be widespread and younger ones less so. Cann et al. (1987) argue that a range expansion, such as is implicit with out-of-Africa, would have caused some ancient haplotypes to remain in the origination area and not spread globally. Having been lost during colonization events and by mutations, they are restricted to the origination areas today. Some of the more recently mutated haplotypes can become widespread if they were not lost during colonizations associated with expansion.

This is, in fact, the pattern observed in the mtDNA data, where Africa is the origination area (Cann et al. 1987). However, there is no basis for this idea from theoretical models, either analytic or computer simulated (Templeton et al. 1995). Theoretical studies of range expansions are in their infancy. Generalities may not be tractable because the pattern may depend on many details about population bottlenecks in the ancestral area, losses of haplotypes in colonization events, intervening migration among established populations, and other details about the sequence of events as populations spread geographically.

The best-developed statistical method tailored to geographic distributions of phylogenies is the nested cladistic analysis of Templeton (1993). The method first constructs a phylogenetic tree for the haplotypes. Such "gene genealogies" may be based on coalescence theory, or other methods such as parsimony. As was noted earlier, this ignores the geographical structure of genetic variation, although this is probably acceptable under many conditions. Next, the unrooted haplotype tree is partitioned into a hierarchical structure, according to the nested cladistic criteria (e.g., see figure 3.2). The most recently mutated haplotypes occur at the "tips" of the tree, the next most recent ones occur at the first nodes in from the tips, and the nodes positioned more toward the interior of the tree represent still older halotypes (Templeton et al. 1995). Among various possible procedures, nested cladistic analysis then uses the tree to first form groups of tip haplotypes and the nearest interior node, i.e., haplotypes that differ by one polymorphic site (one mutation), because each such group has a greater degree of shared ancestry. Then these "one-step clades" are pruned off the tree, and the procedure is repeated for tip haplotypes of the pruned tree. The pruning and forming of one-step clades is repeated until all haplotypes have been classified into one-step clades. The one-step clades are usually still connected by a tree, and the next step is to use them in the same way to classify all one-step clades into two-step clades. This is repeated until the entire tree has been hierarchically classified. Sometimes special rules are needed if there are ambiguities (reticulation) in the tree or symmetries (Templeton et al. 1995). This is a rigorous method, but it may involve some loss of information and it relies on the topology of the tree being true.

The frequencies of the nested clades (sets of related haplotypes) are tested for significant differences among geographic localities. Nested clades can be used to test for the presence of specific geographic patterns. The same could be done for the frequencies of haplotypes themselves, but the use of nested clades may be helpful because it combines data of like type, e.g., haplotypes that share recent ancestry. Templeton et al. (1995) developed a nested contingency table analysis, where geographic location is one of the categorical variables, and the others are clade types nested in hierarchical levels. Exact tests can be constructed, and statistical significance implies that the clade types are not evenly distributed across the geographic locations of populations. However, this inference in itself does not describe the features of the geographic pattern that might be used to distinguish which types of processes have occurred. One way to proceed further is to calculate the geographic distances over which different clades are distributed. Various measures of distance could be used, such as the Euclidean or Great Circle distance measures. For each clade X, the average distance from where each representative of it is found to the geographic center of its range can be calculated, $D_c(X)$ (Templeton et al. 1995). Hence $D_c(X)$ is a measure of how widespread clade X is. This value may be contrasted with the average distance of clade X types from the geographical center of the next higher level clade group of which X is a member, $D_n(X)$. The contrasts may distinguish different processes, including standard genetic isolation by distance versus range expansion. The former could be understood from examination of the space–time probabilities of identity by descent, because it includes information on the relative likelihoods of descendants of an ancestral haplotype being found at varying distances from its origin, in the standard isolation by distance model.

A translation of the predictions of Cann et al. (1987) for haplotypes in expansions into a form for nested clades in expansions was also developed by Templeton and colleagues. Templeton et al. (1995) suggest that a fairly recent range expansion should show large values of $D_c(X)$ and $D_n(X)$ for some but not all tip clades (i.e., wide range expansion for haplotypes that are "young," formed during the range expansion). It should also show small values for some interior nodes, because, according to Cann et al. (1987), some old or ancestral types persist only in a limited area at or near the origin. There may be other details that differentiate contiguous (short-distance) range expansion versus long-distance colonizations, in the expansion process. These patterns contrast those expected for both standard isolation by distance and population fragmentation.

Templeton (1998) demonstrated several advantages of a rigorous statistical method like nested cladistic analysis over graphical methods (e.g., Avise 1994). For example, such methods can be used to study the effects of sampling (Templeton 1998). However, more powerful methods, based on haplotype distributions in explicit space–time models, might be developed in the future. For example, violations in the assumptions of the three steps outlined at the beginning of this chapter could be substantial, and further study is needed to understand the conditions under which they are. Moreover, there may be loss of power. For example, Smouse (1998) has weighed some pros and cons of using phylogenetic trees in biogeographical studies.

Templeton (1993, 1997) reexamined the mtDNA datasets for the out-of-Africa problem using nested cladistic analysis. The results indicated that restricted gene flow, as in standard isolation by distance among long-standing populations existing throughout the Old World, was the prevailing factor. No intercontinental range expansions were evident among Old World populations. In contrast, there was evidence of a much more recent expansion within Europe as well as one associated with colonization of the New World (Templeton 1997).

Many studies argued that the fact that there is greater mtDNA variation in Africa (Vigilant et al. 1991) indicates origination in Africa. The general thinking is that genetic variation was lost through repeated founder effects as new populations were founded and humans spread throughout the rest of the Old World. However, loss of variation in populations from founder effects is temporary if there is gene flow among populations afterward. Theoretical results on the "flattening" of space–time probabilities of identity by descent and space–time correlations (Epperson 1993b), as time goes on, provide an indirect framework for how long differences in diversity among populations should persist. Moreover, the scenario of repeated founder effects is just one of many nonequilibrium processes that can explain the higher level of diversity within African populations. For example, a quite opposite scenario is that the diversity reflects the temporary effects of a "recent" admixture event, such as a major migration back into Africa, as has been suggested based on nested cladistic analysis of Y-chromosome data (Hammer et al. 1998). Finally, it is worth noting that a few samples of DNA from Neanderthal remains (Krings et al. 1997) may provide some support for an expansion throughout Europe during that time period. Space–time probabilities and space–time coalescence probabilities could provide the probability theory needed to more precisely interpret these samples, which are likely to become increasingly important.

Other methods of using modern molecular data for detecting population expansions are based on the frequency distribution of the numbers of pairs of haplotypes that have any given number i of mutational differences. Theoretical models of single, unstructured, populations show that if the population has undergone a fairly recent expansion, the frequency distribution of pairwise differences will have a shape (Rogers and Harpending 1992) that differs from the one expected for a population that has had stationary size (Watterson 1975). Rogers and Harpending (1992) showed how the distribution appears immediately after an expansion, and how it changes over the succeeding generations. Let $x_i(\tau)$ be the frequency of i pairwise differences, at time τ, measured in units of $\tau = 2\mu t$ (t = number of generations; μ = rate of mutation), expired since an instantaneous population expansion event. This distribution differs markedly from the equilibrium distribution (for any time t), where $x_i(t)$ is geometrically decreasing with i. Actual distributions for any given gene, DNA sequence, or haplotype may differ substantially because of stochasticity (Slatkin and Hudson 1991). The distribution has a sizable mode at intermediate values of i following a sudden increase in population size. The properties of the mode resemble those of a wave, because it tends to move toward the right (increasing values of i) as time proceeds further (Rogers and Harpending 1992). Contractions in population size can also cause modes in the distributions (Rogers and Harpending 1992), although the distribution tends to be more "ragged," with many minor peaks, or less smoothly varying with i, compared to that following expansion (Harpending 1994). The value at which the mode occurs depends on the values of θ, pre- and postexpansion (or contraction), where θ is $2N\mu$, and N is either the population size (for haplotype data such as mitochondrial haplotypes), or the number of autosomal genes (for diploid genotypes).

Rogers and Harpending (1992) showed that the distribution of pairwise difference among the worldwide mtDNA samples of Cann et al. (1987), with 147 sampled individuals, exhibits the proper shape. Indeed, it could be closely fitted by sudden expansion, with initial (preexpansion) $\theta = 2.44$, secondary (post-sudden expansion) $\theta = 410.69$, and with a value of $\tau = 7.18$. Using available information on mutation rates, this corresponds to an expansion from about 800–1600 females to 137,000–274,000 about 60,000–120,000 years ago (Rogers and Harpending 1992). These results are reassuring, in part because some sort of expansion to the present size of some six billion should be reflected in the data. Moreover, when mtDNA data on regional populations (Di Rienzo and Wilson 1991), are analyzed separately, the distributions also exhibit modes, including datasets from Sardinia, Japan, the Middle East, and American Indians, all of which have undergone recent expansions (Rogers and Harpending 1992). The overall shape of the worldwide distribution has been argued to support the out-of-Africa theory, again largely because it suggests that the population size during the era in which modern humans evolved was too small to have been maintained across the Old World. Hence widespread archaics could not have contributed to our gene pool (Rogers and Harpending 1992), using the same logic mentioned earlier in this chapter. Note that this argument alone does not specify the geographic location of origination; it is based on a single population model, and hence is not spatially or geographically explicit.

In a geographic setting, the definition of "population" may be confounded. This is a good example of the difficulties in trying to parse out a purely temporal effect (sudden expansion of the total population size of the modern human species) from the more complex spatial–temporal process and context in which it occurred. Such analyses must distinguish overall effective population size (averaged over long periods), from sudden expansions and contractions, as well as the role of migrations and range expansions. Moreover, analyses must consider the complications that may arise from migration among populations and even the definition of population itself. For example, Graven et al. (1995) analyzed a large data set collected from Senegalese Mandenka, and maintained that the distribution of pairwise differences did not have the wave typical of sudden expansion. However, Eller and Harpending (1996) conducted extensive computer simulations of the dataset and found that the data did not reject the possibility of demographic expansion nor did it reject stationarity. They also found that at least the postexpansion population sizes were rather high, and noted that the definition of population from which the sample was taken should be considered to include more than just the Mandenka. They point out that the Mandenka presumably have been exchanging migrants and genes with other regional populations, and hence that West Africa populations as a whole should be considered as the reference population for the samples. Moreover, even if one population, such the Mandenka, has not experienced an expansion, this does not mean that the same is true for other African populations (Eller and Harpending 1996).

Genetic data of a second type, Y-chromosome haplotypes, provides added information on the origin of modern humans, in terms of demographics of males (Hammer and Zegura 1996; Hammer et al. 1997; Underhill et al. 1997). The nonrecombining part of the Y chromosome is the patrilineal counterpart to mtDNA. Harpending et al. (1998) analyzed sequence differences among approximately 20,000 sites from Y chromosomes of 718 sampled individuals (Underhill et al. 1997). They concluded that the distribution of frequencies of polymorphism was consistent with population expansion, as well as inconsistent with that expected from a selective sweep.

Hammer et al. (1998) conducted a nested cladistic analysis of a large data set of Y chromosomes, from 1544 individuals representing 35 populations in total. By obtaining nine polymorphic (diallelic) sites, ten different haplotypes were found. In the first step of their cladistic analysis, the authors used parsimony to reconstruct an unambiguous tree rooted with outgroups of four species of great apes. For the ten Y-chromosome haplotypes, there were five one-step clades, and three two-step clades (figure 3.2). The null hypothesis was strongly rejected for the entire cladogram, as well as for the majority of one-step and two-step clades (Hammer et al. 1998), indicating that there is a spatial pattern of shared ancestry. The most common pattern for Y-chromosome clades was isolation by distance, but there were also three evident range expansions (Hammer et al. 1998). Most importantly, there appeared to have been a range expansion based on some of the oldest clades, because they had small D_c values, and this supported out-of-Africa. Nonetheless, as has been noted earlier, this re-

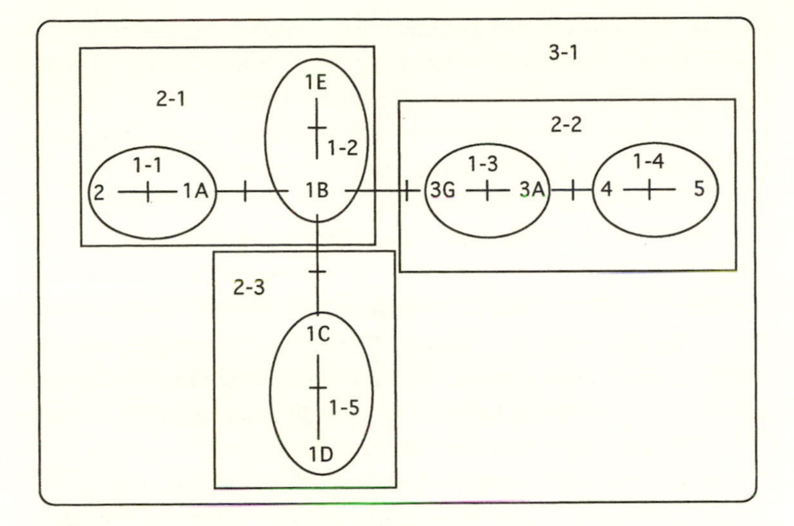

FIGURE 3.2. Nested cladogram for *Y*-chromosome haplotype data. The structure of the cladogram is indicated by lines connecting ten haplotypes, 1A, 1B, 1C, 1D, 1E, 2, 3A, 3G, 4, and 5. There are five one-step clades contained in ovals and denoted 1–1 through 1–5, three two-step clades contained in rectangles and denoted 2–1 through 2–3, and a single three-step clade that contains all ten haplotypes. Original figure provided by M. Hammer (1998).

lies on the unproven presumption that some old haplotypes are likely to persist in the original population in which they arose. Another range expansion was global, as it should be. Perhaps most interestingly, the third expansion was out of Asia and *into* Africa, but one in which Y chromosomes were not completely replaced. This would have caused significant admixture in the African populations, and could help to explain the greater levels of genetic diversity in Africa today. Interestingly, the difference in patterns for Y chromosomes and mtDNA suggests differences in the demographics of males and females.

The limitations of the mtDNA and Y-chromosome data also indicate that conclusive evidence is more likely to come from the nuclear genome. Most of the reported mtDNA variation is limited to a single "hypervariable" segment of about 1 kilobase in size (Brown 1985). The mtDNA pattern and the Y-chromosome pattern are each just one outcome of a process that presumably has been subjected to a large degree of stochasticity. Moreover, both the mitochondrial genome and Y chromosome are, in principle, highly subject to natural selection, and it may be difficult to separate out the effects of any "selective sweeps," for example, large changes in mtDNA haplotype frequencies, perhaps even fixation and losses, due to natural selection (e.g., Knight et al. 1996; Jorde et al. 1997). On the other hand, the population genetics of nuclear genes are in some cases more difficult to model, especially if there is recombination. Statistical methods are also sometimes more complicated. Whereas for haplotypes such as mtDNA a single type is observed per sampled individual, for autosomal loci two genes are sampled from an individual, and they may not be independent (e.g., if there is inbreeding).

Data on isozyme loci and nuclear restriction fragment length polymorphism (RFLP) do not particularly support out-of-Africa. These markers are not more diverse in African populations (e.g., Bowcock et al. 1991; Nei and Roychoudhury 1993), in contrast to mtDNA. However, it has been argued (Jorde et al. 1997) that this may reflect an ascertainment bias, because many of the markers were developed as part of a search for polymorphic markers among European populations.

One of the first extensive nuclear DNA sequence datasets collected was for the HLA major histocompatibility complex. In comparison to most of the other sequence datasets, HLA has very high levels of polymorphism, and the inferred phylogenetic tree is deep. It appears that many alleles are very old, and that the coalescences of some alleles occurred on the order of millions of years ago (Ayala 1995; Erlich et al. 1996), and hence many alleles were segregating in the founding population(s) of modern humans. Correspondingly, the inferred long-term effective population size is large, on the order of 100,000 (Takahata 1993). Sherry et al. (1997) argue that because the time frame includes a long period before the formation of modern humans, when population sizes may have been large relative to the founding population of modern humans in the out-of-Africa scenario, this is not inconsistent with out-of-Africa. However, Ayala (1995) argued that the data suggest that the population size was around 100,000 during the period in which modern humans arose. This was countered by an argument that a smaller effective population size (10,000), together with a certain level of diversifying selection, could also explain the data (Erlich et al. 1996). Nonetheless, the possibly confounding role of diversifying selection makes conclusions from the HLA data difficult.

Extensive nuclear data are available in the form of polymorphisms for *Alu* retroposable elements. More than 500,000 *Alu* exist in families of up to 500–2000 mostly noncoding (Deininger and Batzer 1993) and presumably selectively neutral elements. Some elements have inserted recently enough that not all humans carry an element at a given insertion site, and other inserted elements are fixed, although they vary in sequence because of mutations (Batzer et al. 1996, 1997). Knight et al. (1996) surveyed 29–60 worldwide individuals for three insertions that had varying levels of nucleotide diversity. One insertion had a diversity value that, by comparison to nucleotide divergence of humans from apes, suggested an average age of sequence divergence in the range of 30,000 to 55,000 years ago. Knight et al. (1996) found that the data fit more closely a one-population model, which reflects the out-of-Africa hypothesis, than they fit a model of two isolated populations that diverged 1.5 million years ago. However, there are many possible forms of the multiregional model, and migration could affect the conclusions.

Sherry et al. (1997) examined 13 *Alu* elements from 122 individuals. The ratio of dimorphic (not present in all individuals) elements to elements fixed in humans but absent from apes, can be used to estimate effective population size, under the assumption that an element is never lost once it becomes inserted in a new chromosomal location. However, it is not clear that this assumption is consistent with loss through genetic drift. Sherry et al. (1997) estimated that the population size is about 18,000, effective over the past one to two million years, and they showed that the data are consistent with a population bottleneck and subsequent expansion. However, other scenarios are possible. There does not seem to be any reason to assume that the insertions occurred simultaneously with the population isolation (and contraction in size) associated with the speciation in Africa (Stanley 1997). Again, the strongest argument is simply that 18,000 is too small to include the archaics that had spread throughout the

Old World (Harpending et al. 1998). Finally, in these studies there was no particular evidence that the founder population was in Africa.

A recent study of diversity of microsatellites does suggest that African populations have slightly greater nuclear genomic diversity (Jorde et al. 1997). Among 60 unlinked microsatellite loci from 255 individuals sampled from all over the Old World, average heterozygosity was slightly greater in Africans ($H = 0.76$) than in Asians (0.70) or Europeans (0.73), but the difference was not statistically significant. When allele size was taken into account, reflecting multiple mutations in a stepwise fashion, Africans had statistically significant greater variances. Moreover, the contrasts were greater for those microsatellite loci that had overall smaller variances in allele sizes. This suggests that loci with higher variance also had higher mutation rates, and thus the pattern created by expansion has been erased to a greater degree for the more variable loci. While high mutation rate tends to cause greater spatial patterning, in terms of the slope of the isolation by distance function, it also causes all spatial correlations to be small, making detection of genetic patterns, especially from the distant past, more difficult (Epperson 2002). Other explanations, such as recent admixture in Africa, are also possible.

The above discussions do not mean that the out-of-Africa hypothesis is incorrect. Perhaps there are just too many coincidences in the datasets, each of which points in some way toward out-of-Africa. This data-rich example illustrates the statistical and stochastic issues in inferring events from the distant past in a spatial–temporal context. It is also important to note that this subject, like many subjects in human genetics, is of inherent interest to many people. There can be little doubt that much more data, much of it sequence data from the nuclear genome, will be generated, and to some degree this may offset the stochasticity inherent in such processes. Much larger datasets may detect even the faintest of signals, which we may conclude become very faint indeed over such lengths of time. This also means that we need precise models of the processes, under out-of-Africa and competing theories, together with optimal points of comparison. Finally, the importance of the role of ancient DNA samples is yet to be understood, particularly, how many of such samples are required to give representations of past populations at a point in time and space.

PATTERNS CAUSED BY MORE RECENT EXPANSIONS OF HUMAN POPULATIONS

Spatial correlation and trend surface analysis indicate that there are both isolation by distance and clines for some genetic markers in Europe (Sokal and Wartenberg 1981; Bocquet-Appel and Sokal 1989), and, as was discussed in chapter 2, particularly the latter may have been caused by a population expansion from the southeast. Sokal (1988) analyzed the spatial patterns for 107 genetic variables, grouped into 27 systems, at over 3000 locations in Europe. He used a multivariate extension of the Mantel test statistic for the association of one matrix of genetic distances with several other matrices of distances (Smouse et al. 1986). In Europe, the matrix of genetic distances among samples, GEN, is highly correlated with geographic distance, GEO, as expected for any kind of genetic isolation by distance. However, GEN was also highly correlated with linguistic differences, LAN, and, most importantly, there remains a positive partial correlation of GEN and LAN, conditioned on GEO. This shows that language, which tends to have a branching structure often with known polarity, is a major factor in traces of genetic heritage, and that "language barriers" among populations can reduce gene flow between them (Sokal 1988).

Sokal et al. (1992) examined a similar dataset further, to test some specific theories about the nature of the expansion, in particular whether it was associated with three waves of Kurgan migrations or with the immigration of Indo-Europeans together with agriculture. Using Mantel procedures, they constructed two additional distance matrices that specified the relevant features of the two hypotheses. Both hypotheses failed to cause the multiple regressions to fit significantly better, and hence neither help to explain the spatial patterns seen today. Nonetheless, this analysis is a good example of how various hypotheses as well as external information about specific space–time processes may be included and tested in proper statistical models, without constructing phylogenetic trees (Smouse 1998). The results suggest that expansion or demic diffusion of Indo-Europeans (Sokal et al. 1991) is not necessarily closely tied to the spread of agriculture. Sokal et al. (2000) showed that cancer incidences are also associated with genetics and ethnohistory in Europe.

Recently, spatial autocorrelation analysis called AIDAs (autocorrelation indices for DNA analysis [Bertorelle and Barbujani 1995]) was conducted on a large sample (>2600) of mtDNA sequences in Europe, the Near East, and the Caucasus (Simoni et al. 2000). It was concluded that Paleolithic expansion and a Neolithic diffusion of agricultural communities may be responsible for the north–south clines in genetic differentiation of mtDNA diversity, although the pattern for nuclear genes reflects a somewhat more complex process. This appears to fit with Templeton's (1997) nested cladistic analysis of Old World mtDNA, in which the only range expansion detected was a relatively recent (compared to the scale of out-of-Africa) expansion within Europe. In contrast, the Y-chromosome data for Europe was dominated by isolation by distance (Hammer et al. 1998), suggesting that demographic diffusion of women was more important than that of men in the expansion in Europe, although one Y-chromosome haplotype does seem to reflect demic diffusion (Ammerman and Cavalli-Sforza 1984) and expansion in Europe between 5000 and 10,000 years ago (Hammer et al. 1997).

Similarly, other fairly recent range expansions have been detected, using nested cladistic analyses, in the settlement of remote islands in the Pacific (Sykes et al. 1995), and of Siberia and North America (Torroni et al. 1993a,b). In all of these cases there was strong independent evidence of expansions, and thus these genetic studies provided a good test of the ability to detect range expansions with genetic data. The success, in considerable contrast to the out-of-Africa hypothesis, may be related to the fact that the posited events are much more recent. They should be more detectable, according to general considerations in the theory based on space–time probabilities of identity by descent and coalescences (Epperson 1999a; 2002). As time proceeds the signature left in spatial patterns declines, especially if migration rates are relatively high (Epperson 1999a).

PATTERNS IN OTHER SPECIES

The ability to detect the temporary effects of ancient events in a spatial–temporal context is well documented for many species other than human. For example, Templeton (1998) reviewed studies on mtDNA in 11 animal species or subspecies. In every case there was prior knowledge that the species had fairly recently expanded their ranges, and in all cases but one expansion was indicated in the present geographic pattern of genetic variation, as examined with nested cladistic analyses (Templeton 1998). Total sample sizes ranged from 34 to 613 individuals and with 4–58 distinct haplotypes. In three cases, two subspecies of darter, *Etheostoma blennoides blennoides* and *E. b. pholidotum* (Wilson 1997), and the gopher, *Geomys bursarius*, the species are North American with current ranges that include large areas that were under glaciers during the Pleistocene. They expanded ranges northward since the Wisconsinian glaciation, which reached glacial maximum about 18,000 years ago. Similarly, two North American subspecies of salamander, *Ambystoma tigrinum tigrinum* and *A. t. mavortium* (Templeton et al. 1995), have expanded into areas that while not glaciated were nonetheless uninhabitable during the time of glaciation. Range expansions were detectable in all five cases, fragmentation was also detected for the darters, and a regional colonization event was detected for the gopher. Similarly, the lichen grasshopper, *Trimerotropis sáxatalis*, expanded into the Ozarks from the southwest and this expansion was followed by a fragmentation event (Gerber 1994). Range expansion and fragmentation were also detected in the fish *Galaxias truttaceus* in colonizing lakes created by melting Pleistocene glaciers in Tasmania (Templeton 1998).

Expansions that are very recent were also detected (Templeton 1998). Both contiguous (gradual) expansion and (long-distance) colonization of *Drosophila melanogaster* were detected in the recent global expansion from Africa. Colonization of the macaque monkey, by Portuguese sailors in the 1500s, was detected on the Island of Mauritius, and the Phillippines and/or Indonesia were identified as the possible sources. Similarly, range expansion of *Canis latrans* in North America since around 1900 was detected. Among the examples, only the expansion of *Drosophila buzzatii* in Europe, from South America, was not detectable. Apparently, an extreme bottleneck occurred in the initial Iberian colonization, where only one mtDNA haplotype was found (Rossi et al. 1996), and this haplotype was also the most frequent one in South American flies, as well as being interior in the phylogenetic tree. Moreover, mutation has not created any new, widespread haplotypes in Europe, perhaps because the haplotypes were defined by restriction site variations, which have low mutation rates. Hence geographically widespread young haplotypes (tip clades) are not observed (Templeton 1998).

Most examples of successful identification, whether through the use of phylogenetic trees or spatial analysis without trees (see Templeton 1998; Smouse 1998; Goldstein and Harvey 1999), of events of population expansion and fragmentation are for those events that are relatively recent. Such events are essentially by definition temporary, nonequilibrium processes. We may expect that the signatures in geographic patterning are diminished for events from very long ago. On the other hand, if events are too recent, and there is insufficient time for mutation to create new variants, not all of ancient features will be manifest in geographical genetic patterns, and different models and expectations must be purported. Finally, some aspects of the inference of origins of polymorphism, ancestral sources of variation, or the location of origin of a species itself, are very clearly manifest and testable. These are sometimes contradicted in present patterns of genetic variation. For example, in red pine, *Pinus resinosa*, the geographic region that has by far the greatest levels of chloroplast DNA diversity (Echt et al. 1998; Walter and Epperson 2001) simply cannot be the center of origin (of postglacial populations), because this area was buried by the Wisconsin glacier. The most likely explanation is that this area is recently admixed between two or more lineages, derived ultimately from separate refugial populations. Nonetheless, this violates an assumption often made, that centers of diversity are also centers of origination, and it points out that admixture can also be an important ancient event.

CHAPTER 4

Spatial and Space–Time Statistics

> There are two nonequivalent ways of constructing or specifying nearest neighbor models for representing stochastic dependence of an array of variables on a rectangular plane lattice . . . neither can claim a natural place as the "proper" method of specifying random fields. . . . Ideally random field models should be deduced from plausible space–time processes that describe the system's evolution. The random field then emerges as the spatial equilibrium form or as a cross-section through the evolving structure.
>
> —Robert Haining (1979)

A variety of statistical methods are available for measuring the degree of spatial differentiation, spatial patterns of genetic variation, and the significance of genetic variation among populations. In addition, there are various methods for cross-correlating spatial patterns of genetic variation with other factors, such as variables representing environmental conditions or environmental selection. Measures of spatial patterns are central to genetic survey data, and they have wide-ranging uses as inferential tools. In some instances the primary objective is to quantify spatial autocorrelations with the aim of determining whether the data violate the standard statistical assumption that the data elements of samples are independent identically distributed, and hence whether standard (nonspatial) methods are acceptable. At the other extreme, spatial patterns may be studied to estimate the parameters, for example, the migration rates, in genetic isolation by distance or other space–time processes.

This chapter begins with an overview of the distribution theory for spatial autocorrelation and related statistics, including simple spatial autocorrelation statistics; specialized spatial statistics used to detect directionalities and nonstationarities in spatial patterns; and other methods, such as wombling, which can be used to detect regions where the spatial pattern of genetic variation changes abruptly, potentially because of regional barriers to gene flow. Consideration is extended to indicators of localized (as opposed to global) autocorrelations (LISAs), and measures of overall correlations, such as the Mantel test and its multivariate extension. The next topic regards issues inherent in the use of purely spatial data for making inferences about space–time processes. Issues include the specification of appropriate model type, correlations among spatial statistics, and the relationships of values of spatial statistics to the process parameters in space–time processes. Specification of the latter may be obtained using probability theory and computer simulations. In many instances, the use of space–time data is ideal because it is possible to avoid several difficulties, and this is best illustrated by spatial time series analysis. Space–time data require considerably more effort, but they lead to natural specifications of how migration changes genes frequencies. It may be anticipated that high-throughput genotyping will become more efficient, and augmented with increasing feasibility of obtaining DNA from old materials, including ancient fossils. In the future, the effort required for sufficient space–time genetic data often will be reduced. The next topic is special genetic measures such as F_{ST} and measures of kinship. The final topic regards the information content of data collected for recently developed molecular assays, such as DNA sequences and microsatellite genotypes, and statistics that are explicitly based on these.

SPATIAL AUTOCORRELATION STATISTICS

Statistical measures of patterns in real populations are usually based on the spatial distributions of genetic values in *pairs* of populations, as is most of the theory on genetic isolation by distance (Malécot 1948). It appears that much of the information in a pattern is captured by pairwise measures, and that measures based on triplets, quadruplets, etc. add little. Pairwise spatial autocorrelation statistics such as Moran's I statistic have a number of highly favorable properties. Because the usual analysis of Moran's I statistic is sample-based, it avoids the pitfalls for multilocus, so-called estimates (they are not good estimates, because they are biased with unknown distributions) of theoretical kinship coefficients, which arise from the unknown, a priori expected values of gene frequencies, allowing its distribution to be determined under the null hypothesis. The nuanced critical differences among estimators can be more precisely described in the context of kinship of pairs of *individuals* rather than populations, so this is treated in detail in chapter 8.

In general, unless otherwise specified, it will be assumed that for each allele at each genetic locus, an allele frequency X_i is observed in each population in a system of n studied populations. It can generally be assumed that X_i is the result of sampling from a realization of a *stochastic process*, and that both X_i and realized values have the same (a priori) expected values q for all i. It may or may not be necessary to transform X_i, but the mean observed value X can be subtracted from each, giving $Z_i = X_i - X$. For the remainder of the chapter, unless specified otherwise, Z_i is the mean-adjusted transformed or untransformed gene frequency observed in population i.

To understand the distribution theory of spatial correlation statistics, it is important to first consider precisely what is meant by the spatial distributions on which they are based. Spatial stationarity and what can be termed "spatial regularity" are subtle but critical features that have been widely ignored in the spatial autocorrelation experimental literature in genetics. While they are not important under the null hypothesis of a random distribution, they are part and parcel of nearly all other inferences about processes from spatial measures. Spatial regularity can be exemplified by a lattice in the two-dimensional case shown in figure 4.1. For a lattice, it is straightforward to define a "spatial lag structure," which represents the degree of spatial proximity among populations. Among the many possible lag structures, the example in figure 4.1 defines the following spatial lag orders: "0," which is the given population itself; its eight nearest neighbors at spatial lag order "1"; its sixteen second nearest neighbors labeled "2"; etc. Spatial stationarity is a spatial extension of the concept of (temporal) stationarity in stochastic process theory. For example, in time series stationarity means that the joint distribution of values at different time periods depends only on the length of time separating them. Analogously, spatial stationarity means that the joint distribution of values at populations depends only on their spatial

5 5 5 5 5 5 5 5 5 5 5

5 4 4 4 4 4 4 4 4 4 5

5 4 3 3 3 3 3 3 3 4 5

5 4 3 2 2 2 2 2 3 4 5

5 4 3 2 1 1 1 2 3 4 5

5 4 3 2 1 0 1 2 3 4 5

5 4 3 2 1 1 1 2 3 4 5

5 4 3 2 2 2 2 2 3 4 5

5 4 3 3 3 3 3 3 3 4 5

5 4 4 4 4 4 4 4 4 4 5

5 5 5 5 5 5 5 5 5 5 5

FIGURE 4.1. Populations, each denoted by a number, on a regular grid or lattice. Also shown are the lags for one of many possible lag structures, about the population labeled "0." In this example, the eight populations labeled "1" are defined as first-order neighbors of population "0," those labeled "2" are second-order neighbors, and so on. Some relationships involving the two populations with underlined numbers are discussed in the text.

relationships, independent of their absolute locations. If genetic variables or their transformations may be treated as normally distributed, then "weak stationarity" is sufficient, requiring only that the means, variances, and covariances are temporally and spatially stationary. If the mean adjusted genetic values all have expected value zero, then weak stationarity (which is the definition of stationarity used throughout the book unless specified otherwise) obtains for a space–time population-genetic process if the covariance between Z_i at time t and Z_{i+a} at time $t - k$ depends only on a and k. If sampled at the same time, then the covariance depends only on the spatial lag a (Hooper and Hewings 1981). One way to think of this is to visualize how stochastic effects are "propagated" in the system over time. For example, in a system that is temporal order one and spatial lag order one, i.e., where migration occurs only among first-order neighbors, it will take two generations for a population to affect its second-order neighbors.

Although spatial stationarity or even spatial regularity will not be precisely met in many real systems of populations, it is important to consider the conditions under which they may be approximated. Without some approximation of spatial stationarity it is not possible to make detailed inferences about the parameters of the processes generating a spatial pattern. Spatial stationarity also allows us to gain statistical power, by ensuring spatial replication. One could consider using the spatial lag operators: $L^{(1)}Z_i$ defining a weighted average of the genetic values at first-order neighbors (i.e., the "ones" in figure 4.1); $L^{(2)}Z_i$, a weighted average for the second-order neighbors, etc., and calculating covariances of $L^{(r)}Z_i$ and $L^{(s)}Z_j$. The problem is that most lag operators will not preserve spatial stationarity, assuming it existed among the Z_i themselves; i.e., covariances (and correlations) between two weighted averages will depend not only on r and s (Hooper and Hewings 1981) but also on the weights. For example, consider the relatively low genetic correlation expected by isolation by distance between populations denoted by underlines, $\underline{3}$ and $\underline{4}$ in figure 4.1, which have population 0 as a third-order and fourth-order neighbor, respectively. Other populations labeled 3 and 4 are themselves nearest neighbors, and should have higher correlations. F_{ST} suffers this problem, because it is based on weighted averages of hierarchical groups. There seems to be no reason to use hierarchical groupings unless migration and other processes actually operate in a hierarchical manner. Moreover, if the populations are not regularly spaced it is not usually possible even to define lag structures, let alone expect stationarity among lag-weighted averages. Thus, *spatial stationarity generally will not obtain without spatial regularity*. Difficulties are avoided by using what is termed the finest lag structure (Hooper and Hewings 1981), which in essence means that all pairs of populations are specified separately. Then, it makes more sense to speak in terms of spatial backshift operators, one for each spatial dimension. For example, if $B^{(r)}$ shifts horizontally r units and $B^{(s)}$ shifts horizontally s units, then $B^{(r)}\,B^{(s)} = B^{(r+s)}$. In figure 4.1, population $\underline{3}$ is 2 units to the left of population 0, population 0 is four units left of $\underline{4}$, and $\underline{3}$ is six units left of $\underline{4}$.

MORAN'S *I* STATISTIC

Moran's I statistics, in either their unweighted or weighted form, are in terms of the finest spatial lag structure. With weights w_{ij} defined

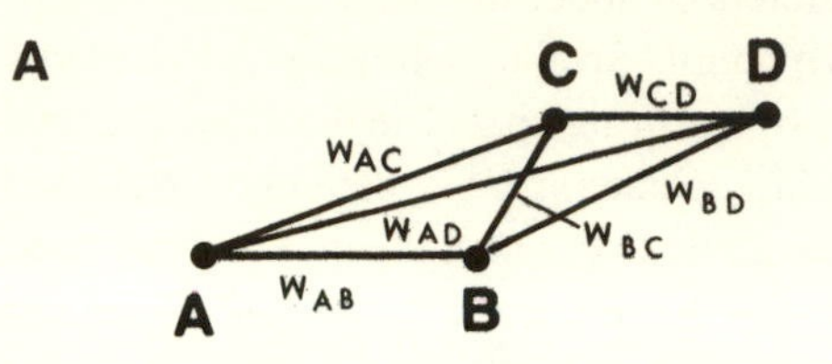

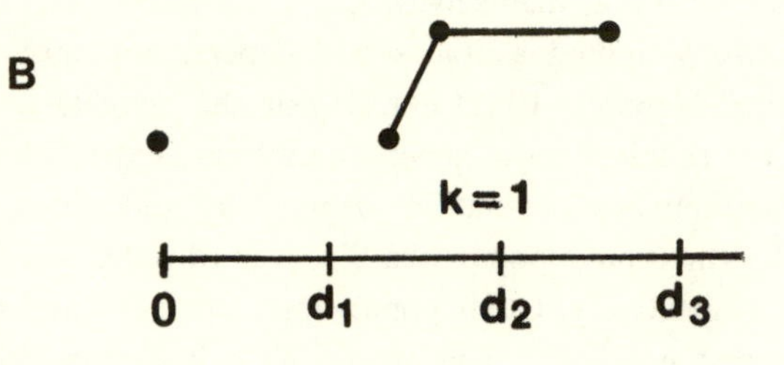

C

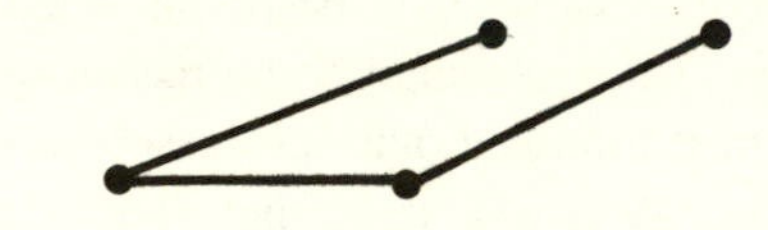

k=2

D

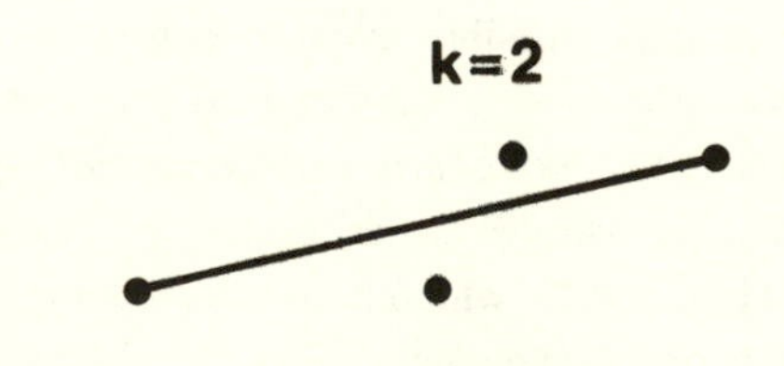

k=3

E

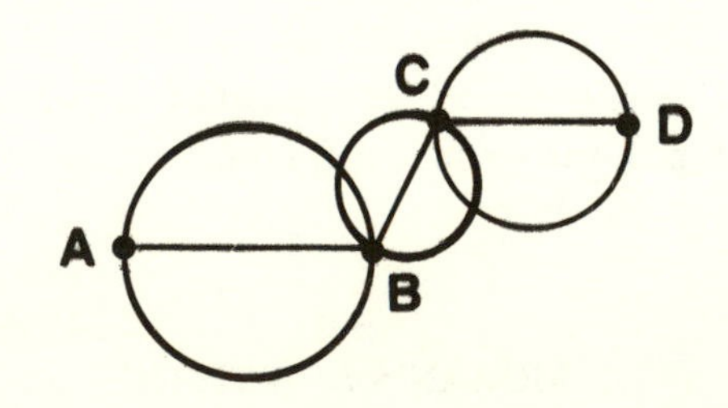

for populations i and j (figure 4.2A), and usually $w_{ij} = w_{ji}$, Moran's I statistic is written

$$I = \frac{n\sum_{i}\sum_{j} w_{ij} Z_i Z_j}{W\sum_{i} Z_i^2} \tag{4.1}$$

where n is the number of populations, and W is the sum of all weights (Cliff and Ord 1981). Note that this definition specifies a single statistic (for each allele). The specification allows great generality in using prior or independent information on the nature of the strengths of statistical (as opposed to process interactions, see discussion below on SAR vs. STAR models) interactions among locations. In practice this is difficult. Examples of the types of weights that might be considered include cases where the w_{ij} are decreasing functions of the distances between populations. Values of w_{ij} for specific pairs might be developed to reflect physical barriers, as is discussed below in detail.

FIGURE 4.2. Representations of weights and distance classes among a generic set of four populations, A, B, C, and D. (A) A schematic simply showing all of the weights w_{ij} among all pairs of populations i and j, where it is assumed that $w_{ij} = w_{ji}$. For example, w_{AD} is the weight between populations A and D. The length of the line is meant to be arbitrary and does not necessarily imply the size of the weights, which can be chosen by the investigator. (B) Distance classes for distances among populations A, B, C, D in (A). The value of d_1 is the upper bound on the first distance class, d_2 that for distance class two, and d_3 for distance class three. Lower bounds are essentially the next lower number. For example, the lower bound for distance class three is slightly greater than the value of d_2. The lines show the joins for distance class one (i.e., weights have value 1.0 for B-C and C-D, zero else). (C) The joins for distance class two (A-C, A-B, and B-D are 1.0, zero else). (D) The only join for distance class three (A-D). (E) The Gabriel connections graph. Note that the circles defined by A-B, B-C, and C-D, do not contain any other population, thus these are Gabriel connected. The circle (not shown) defined by A-C contains B and that defined by A-D contains both B and C. Thus populations A and C are not directly connected, and neither are A and D. The circle defined by B and D contains C, therefore B and D are not connected. Figure by R. Walter.

In cases where the structure is expected to be fairly regular spatially and with approximate spatial stationarity, the approach of forming mutually exclusive classes of pairs of populations appears to be ideal. Then, all pairs of populations are classified, usually in terms of a range of Euclidean distances between pairs of populations, and (distance) classes k are formed. That is, based on knowledge of the distribution of distances, a reasonable number of distance classes can be defined, each usually as a range of distances of separation. More statistical analyses can be pursued if the classes are mutually exclusive, e.g., in cases where the distance classes do not overlap. Regardless, the equation for specific (distance) class k is written

$$I(k) = \frac{n \sum_i \sum_j w_{ij}(k) Z_i Z_j}{W_k \sum_i Z_i^2} \tag{4.2}$$

where $w_{ij}(k)$ are binary (0, 1) variables specifying the inclusion (1) or exclusion (0) of the pair of populations i and j in class k, and W_k is the sum of the weights or twice the number of joins or pairs placed into class k (Sokal and Oden 1978a; Cliff and Ord 1981). Moran's I statistic is thus a product moment type of correlation coefficient between the pairs of populations within a class. However, it may not always freely range from -1.0 to 1.0, and for some extreme spatial distributions it could range outside those bounds.

The distributions and moments of I statistics can be described in terms of three measures of the weighting structure, and if the latter is based on distance then the three measure the distributions of the populations (not their genes per se) in space:

$$S_0 = \sum_{(2)} w_{ij}$$

$$S_1 = \frac{1}{2} \sum_{(2)} (w_{ij} + w_{ji})^2$$

$$S_2 = \sum_{i=1}^{n} \left(\sum_{j=1}^{n} w_{ij} + \sum_{j=1}^{n} w_{ji} \right)^2 \tag{4.3}$$

where $\Sigma_{(2)}$, is a double summation, once over all i populations and once over all j, with the exclusion that j cannot equal i. Note that all pairs of populations are included twice (once with w_{ij} and once with w_{ji}) (Cliff and Ord 1981). In the case of unweighted Moran's I statistics for each distance class k, the w_{ij} are binary (0, 1) indicator variables (δ_{ij}) and we denote the three measures $S_0(k)$, $S_1(k)$, $S_2(k)$. In this case, $S_0(k)$ (which is always the same as W_k) is equal to twice the number of joins or pairs total for distance class k, $S_1(k)$ is four times the number of joins ($2W_k$), and $S_2(k)$ is a second-order function of the spatial distribution of populations.

Two null hypotheses, normality and randomization (Cliff and Ord 1973), of random spatial distributions of gene frequencies may be tested. The differences between the two hypotheses are subtle but have considerable importance. The first is less often useful, and is based on the presumption that all population gene frequencies (adjusted by the a priori expected gene frequency) are independent identically distributed normal variates that are sampled from a (usually stochastic) process. If the underlying space–time process (even if there were no limits to migration) is stochastic, then the a priori expected value is usually unknown. Cliff and Ord (1981) showed, using a theorem from Pitman (1937) and Koopmans (1942), that under the normality assumption, Moran's I is independently distributed from its denominator (equation 4.2). This avoids the problems usually encountered in trying to find the expected values of ratios, and those of powers of ratios, between two dependent random variables, here in the numerator and denominator of I. Under the normality form of the null hypothesis, the expected value of I is $-1/(n-1)$ and the variance is

$$\sigma_N^2 = \frac{n^2 S_1 - nS_2 + 3S_0^2}{S_0^2(n^2-1)} - \frac{1}{(n-1)^2} \tag{4.4}$$

Note that the test for significance of I is not a test for significant *differences* in gene frequencies among populations, but rather of significance of the pairwise aspect of the spatial *pattern* of gene frequencies.

The second form of the null hypothesis makes only the following assumption: precisely, that the expectation of Moran's I statistic is over the n-factorial permutations of the observed data. Thus it is explicitly sample-based, and it simply purports that an observed Moran I statistic, a function of a class of pairs, is not significantly different from that expected if the pairs were randomly selected from the "n choose two," i.e., $n(n - 1)/2$, pairs in total. Cliff and Ord (1973) found that the expected value of I is the same as with the normality assumption, $-1/(n - 1)$, but the variance is different. This sample-based assumption is precisely what allows the distribution to be known, and this is a major advantage over some other so-called kinship measures. Applied to gene frequencies, Moran's I statistic is a measure of kinship, and it is simply one form of averaging pairs of populations, as suggested, for example, by Malécot (1973b), and yet it is one with a known distribution under a null hypothesis of no spatial autocorrelation, no spatial pattern or no genetic isolation by distance. The moments of the distribution can be found under the randomization hypothesis because then the denominator of I (equation 4.3) is fixed, and the expectations of powers of the numerator are straightforward. The variance of I under the randomization hypothesis $\sigma_R{}^2$ is

$$\sigma_R^2 = \frac{n[(n^2 - 3n + 3)S_1 - nS_2 + 3S_0^2] - b_2[(n^2 - n)S_1 - 2nS_2 + 6S_0^2]}{S_0^2(n-1)^{(3)}} - \frac{1}{(n-1)^2} \tag{4.5}$$

where $(n - 1)^{(3)}$ denotes $(n - 1)(n - 2)(n - 3)$, $b_2 = m_4/m_2{}^2$, and m_2 is the mean squared value of the Z_i $(\Sigma Z_i{}^2/n)$ and m_4 is the mean value of the fourth powers of the Z_i $(\Sigma Z_i{}^4/n)$ (results adapted from Cliff and Ord [1981]). Calculations for equation 4.4 or 4.5 are complicated by the fact that S_2 must be determined from the spatial relationships among populations, but this can be implemented by computer, for example, in the SAAP (Exeter Software) program of D. Wartenberg.

A test statistic of the null hypothesis of no autocorrelation under the presumption that the data are normally distributed can be achieved by establishing the asymptotic normality of I. Cliff and Ord (1981) developed a general set of necessary and sufficient conditions, but these are not simple to apply. The simplest sufficient conditions that demonstrate generality for practical purposes are that if the weights are symmetric (i.e., $w_{ij} = w_{ji}$), the sum of weights to any and all populations is finite, and the limit of S_1/n is less than infinity as n goes to infinity, then asymptotic normality is achieved. For symmetrical binary weights, this simply means that no population is joined with an infinite number of other populations. In finite samples, the statistic $(I - \mu_1)/\mu_2^{1/2}$ (where μ_1 and μ_2 are either the known or estimated mean and variance under the null hypothesis) has an asymptotic normal distribution with mean 0 and variance 1, i.e., a standard normal distribution. Under the randomization null hypothesis, the same condition together with a mild condition that is unlikely to be violated for genetic data, are sufficient for $(I - \mu_1)/\mu_2^{1/2}$ to again be asymptotically standard normal (Cliff and Ord 1981).

Moran's I statistic could also be applied to all pairs of sampled genotypes (usually diploid), rather than the gene frequencies observed in each population. First each individual genotype can be converted to an allele frequency (1.0, 0.5, 0.0 if the diploid genotype is homozygous for that gene, heterozygous for that gene, or else, respectively). A similar conversion can be made for haploid data. Then the Z_i are mean-adjusted converted genotypes of individuals, and in general there will be many Z_i for each population. Usually, this would greatly increase the number of pairs or joins (by the square of the mean number of sampled individuals per population), and greatly increase the required computational effort. Presumably statistical power for the randomization null hypothesis would be greatly increased. Certainly in this case normality of the data cannot be assumed. Moreover, if there are multiple sampled individuals within populations, then the null hypothesis test will be affected by differences among populations as well as the spatial pattern among populations.

CORRELOGRAMS

A set of Moran's I statistics, even those for mutually exclusive and exhaustive distance classes (with binary weights), are nonetheless interdependent. Such a set of statistics is termed a correlogram. To test a correlogram for overall significance it is necessary to characterize the dependence of the statistics in the correlogram. A test of significance of a correlogram is a summary test for the existence of nonrandom spatial patterning at all spatial scales. Essentially such a test is without a priori designation of the shape or nature of spatial patterns, unlike the Mantel test. The only presumption is the designation of classes, which if based on ranges of distances introduces essentially no prior form. Oden (1984) showed that generally under either the normal or randomization null hypotheses, the expected value of the product of two Moran's I statistics, I_W and I_Y, each based on a different class of the same (allele frequency) data, is

$$E[I_W I_Y] = \frac{n^2 S_{(WY)1} - nS_{(WY)2} + 3S_{(W)0}\, S_{(Y)0}}{S_{(W)0}\, S_{(Y)0}(n^2 - 1)} \tag{4.6}$$

where $S_{(W)0}$ and $S_{(Y)0}$ have the usual interpretation for the weights w_{ij} for I_W and y_{ij} for I_Y,

$$S_{(WY)1} = \frac{1}{2}\sum_{(2)}(w_{ij} + w_{ji})(y_{ij} + y_{ji}) \tag{4.7}$$

and

$$S_{(WY)2} = \sum_{i=1}^{n}\left(\sum_{j=1}^{n} w_{ij} + \sum_{j=1}^{n} w_{ji}\right)\left(\sum_{j=1}^{n} y_{ij} + \sum_{j=1}^{n} y_{ji}\right) \tag{4.8}$$

Note that this is similar to the value for $E[I^2]$ and it depends only on a function of the weights. Importantly, one needs only to substitute $S_{(W)0} \cdot S_{(Y)0}$ for S_0^2, $S_{(WY)1}$ for S_1, and $S_{(WY)2}$ for S_2 in equation 4.4 to obtain equation 4.6. One can do the same for the randomization hypothesis and make the same substitutions in equation 4.5 to obtain the expected cross-products for the randomization hypothesis (Oden 1984). Thus for both forms of null hypotheses the covariances can be found by subtracting off the square of the mean, $1/(n - 1)^2$. In the case of a set of k statistics the k by k variance–covariance matrix $\mathbf{V}$ and mean vector $\boldsymbol{\mu}$ (containing k elements of $-1/(n - 1)$) can be found. Under the usually met assumption that the I statistics are approximately normally distributed, Oden (1984) showed that the following statistic, Q, should be approximately distributed as a chi-square with degrees of freedom equal to the rank of matrix $\mathbf{V}$:

$$Q = (I - \mu)^{T} H \Lambda^{-1} H^{T} (I - \mu) \tag{4.9}$$

where **Λ** is the matrix of nonzero eigenvalues of **V** and **H** is the matrix of eigenvectors for **Λ** (Oden 1984). In the case of binary weights for a correlogram where the distance classes are mutually exclusive and exhaustive (i.e., all pairs of points are represented in one of the distance classes), Q has $k - 1$ degrees of freedom. Oden (1984) also examined the properties of Q under several alternate hypotheses, and compared its performance to the simply-computed Bonferroni and Šidák (1967) methods for obtaining significance levels for multiple test statistics. He found that the Bonferroni and Šidák methods are preferable when there are few distance classes and weak spatial patterns, but the Q test is more powerful under the opposite conditions. Further, although Oden found that a test based solely on the first (shortest) distance class was inferior when autocorrelation is largest at higher order spatial lags, it is *nearly equivalent* in statistical power and other performance metrics when autocorrelation is greatest at the shortest lag, which is usually the case under isolation by distance. The statistical computing programs SAAP (D. Wartenberg) and PASSAGE (Rosenberg 2002) calculate significance tests for correlograms.

MANTEL TESTS FOR CORRELATIONS OF GENETIC DISTANCE WITH GEOGRAPHIC DISTANCE

Mantel (1967) developed a general model for testing the association of one set of pairwise measures with another: in geographical genetics one would be a measure of genetic distance (or similarity) and the other a measure of spatial distance (or proximity) (Sokal 1979b). If **Y** is an $n \times n$ matrix of measures of genetic distance, and **X** an $n \times n$ matrix of measures of physical distance, among combinations (i, j) of n populations, then the Mantel statistic, here denoted M (often denoted by Z in the literature), is the sum of the products of the elements X_{ij} and Y_{ij}, except for the diagonal elements $i = j$:

$$M = \sum_{i \neq j} X_{ij} Y_{ij} \tag{4.10}$$

In fact, distance measures are usually defined as zero for the same population (i.e., $i = j$), so that normally the diagonal elements would add nothing to M. Smouse et al. (1986) used this fact to remove the dependency of M on the scales of the distance measures, and obtained normalized measures. For the off-diagonal elements, if X is the mean value of X_{ij} and Y the mean value of the Y_{ij}, and SS(X) the sum of squared deviations of the X_{ij} from X, SS(Y) that for the Y_{ij}, then the normalized value

$$r_{XY} = \frac{M - nXY}{[\mathrm{SS}(X) \cdot \mathrm{SS}(Y)]^{1/2}} \qquad (4.11)$$

is a Pearson product-moment correlation coefficient (Smouse et al. 1986). Similarly, $(M - nXY)/\mathrm{SS}(X)$ is a coefficient of regression of the Y_{ij} on the values of X_{ij}.

The distribution of the Mantel statistic M is difficult to obtain. Even though M can be normalized to a regression form, standard regression tests are invalidated by the fact that the Y_{ij} are not independent and neither are the X_{ij} (Mantel 1967, Sokal 1979b). This problem exists for the various other measures of correspondence of two sets of distance values (Gower 1971; Guttman 1968; Sneath and Sokal 1973). One solution is to base a test of significance on a distribution of M values, constructed by Monte Carlo randomization of permuted rows and columns of one distance matrix, keeping the rows and columns of the other matrix fixed (Sokal 1979b). The mean M^* and variance V^* of this distribution are calculated, and the value of $(M - M^*)/V^{*1/2}$ is used as a standard normal deviate test statistic. The test is usually one-tailed: i.e., statistical significance occurs when the observed value exceeds the critical value corresponding to the desired level of significance (usually 0.05), where the critical value matches the upper tail probability. One-tail tests generally are preferred because the alternate hypothesis is that there is a positive association of the two distance variables, i.e., in the present context a positive association of genetic distance with geographic distance (Smouse et al. 1986).

The Mantel coefficient is generally used with specified forms of genetic and geographical distance. However, the latter can also be constructed as distance classes, like Moran's I statistics. Although it can be used to test for spatial patterns, its mathematical probability distribution is generally unknown, unlike spatial autocorrelation statistics.

The normalized Mantel coefficient can be extended to regression of a genetic distance matrix $\mathbf{Y}$, simultaneously on several h other distance matrices, $\mathbf{X}_1$, $\mathbf{X}_2$, . . . , $\mathbf{X}_h$ (Smouse et al. 1986). In most experimental settings the a priori designations of multiple predictive distances are likely to be themselves correlated. It may be of considerable interest to determine which of the $\mathbf{X}_i$ are the best predictors and which ones add little predictive power once others are accounted for. This is analogous to standard problems of multiple, stepwise, and conditional regression, but it is considerably more complex because of nonindependence of the elements within each matrix $\mathbf{X}_i$. Smouse et al. (1986) provide a detailed description of a complex set of paramount issues, including whether the $\mathbf{X}_i$ should be treated as fixed effects or not, computational procedures, and various methods (Dow and Cheverud 1985; Hubert 1985; Sokal and Thomson 1987) for managing the interdependency of the $\mathbf{X}_i$. Importantly, Oden (1992) found that the test statistic of Dow and Cheverud fails when there is spatial autocorrelation, the very feature of interest, although this test could be used for other contexts.

MEASURES OF GEOGRAPHIC DISTANCES AND SPECIAL WEIGHTING SCHEMES

Both Moran's I statistic and the Mantel statistic allow quite general designations of weights and measures of geographic distance. Skilled investigators can take advantage of such flexibility to increase statistical power. Conversely, a poor choice of weighting schemes could give misleading results. Issues of weighting differ from but are intimately connected with spatial regularity and spatial stationarity, which are important for all but the null hypothesis test. Specific forms of weighting schemes are needed for systems where there is either highly irregular spacing of populations or other causes of irregularities in migration rates where rates of migration do not closely follow the distances among populations. The weights chosen in spatial autocorrelation analysis (figure 4.2) are not directly comparable to rates of migration among populations. This problem is analogous to issues regarding the relationship of purely spatial statistical models versus the underlying space–time processes, which can be most precisely illustrated using the comparison of the autoregressive structure in spatial versus space–time autoregressive models discussed later in this chapter.

In systems where the populations are fairly uniformly or "regularly" distributed, simple measures of distance may be used. In many cases, binary placements of pairs into distance classes, in the standard I correlogram, has favorable properties. For the Mantel test one can either specify the geographic distances, or use binary weights to calculate M separately for each distance class. In such cases M and I have very similar properties. Other measures of geographic distance can be used. The obvious choice of distance is the Euclidean distance among populations, and either this can be put directly in the geographic distance matrix X for M, or a measure of its inverse can be used as w_{ij} for calculating a weighted I statistic. Other measures of distance include Manhattan distances, which are measured on a grid. "Distance" (and rate of migration) might be influenced by the length of the border between two regional populations, as well as distances from their centers (e.g., see Cliff and Ord 1981, pages 17–18).

In general, if dispersal or gene flow probabilities follow a certain function on distance, it may be tempting to choose weights with the same form, but this generally will be incorrect. For example, suppose dispersal follows an exponentially decreasing function on distance. In systems of populations that exist in only one spatial dimension, the theoretical models of isolation by distance, whether in terms of gene correlations or probabilities of identity by descent, do indeed show an exponential decrease of genetic spatial correlation with distance (although the parameters of the two exponential functions may differ). However, if dispersal followed a Gaussian or other nonexponential form the correlations would still show a contradictory exponential decrease. In systems in two-spatial dimensions, including most terrestrial plants and animals, the decrease of spatial genetic correlations with distance is not exponential, but the decrease in probability of dispersing may be exponentially decreasing with distance. In general, the form of the function of dispersal on distance has no *direct* correspondence with the shape of the correlation function on distance.

Some systems of populations are spaced so irregularly that the influence of irregularities may not be ignored. An extreme example may be some animal species that are so effective at finding populations they always go to the nearest-neighbor populations (regardless of where they are), and never migrate directly to more distant populations. For another example, freshwater aquatic and riparian species may move along streams and drainage networks of watersheds (e.g., review by Templeton 1998). In such cases, migration rates will depend more directly on the distribution of populations per se, than on the distances among populations. In addition to irregularity of the actual distribution of populations, the distribution of sample sites may be irregular. The superposition of both complicates the situation further.

Perhaps most importantly there may be absolute physical barriers to gene flow. For example, perhaps migration can occur only across saddles in a mountain range, not across the peaks. Such barriers can be addressed by considerations of "connectivity." The simplest idea of connectivity is binary, where if two populations are connected then the number one is assigned, if not then zero. In some systems there may be a great deal of prior information on the connections that should be established, and in the extreme case we may know precisely all of the pairs of populations that ought to be connected or not. However, in many other cases a more objective way of determining connectivity is needed. A number of connectivity criteria have been developed, and these have widely differing properties.

The simplest connectivity criterion is to connect each population only to its nearest (spatial lag one) neighbor(s). When populations are naturally distributed in continuous space, each population should have only one true nearest neighbor. Such a criterion is usually too severe, and it generally would not define spatial lags with order greater than one. Normally greater connectivity is desired. Another idea could be to force connectivity to each population's h nearest neighbors, etc., but this would usually lead to very strange connectivity. Matula and Sokal (1980) examined the geometric and other properties of criteria, and found that Gabriel connectivity (Gabriel and Sokal 1969) has the precise form desired. Two populations are Gabriel connected if and only if the circle defined by treating the two as at opposite ends of its diameter does not contain any other population (figure 4.2E). An equivalent and computationally useful criterion is that the distance separating them is less than the sum of their distances from any other given population. Gabriel connectivity insures that there is a "tree" of connections among all populations, i.e., that one can "move" among all populations by moving along the connections, and importantly this tree is uniquely defined for any finite set of population locations (Matula and Sokal 1980). Without this it might not be possible to define higher order spatial lags. Gabriel connectedness also has the desirable condition that there can be no overlap among the geographic regions defined by connections (the "planarity" condition).

One way of incorporating connectivity into Moran's *I* statistic is to consider the minimum integral number of connections or "edges" there are between each pair of populations and to use these as spatial lags or "distance classes." For example, there are three edges between populations *A* and *D* (*AB*, *BC*, and *CD*) in figure 4.2E. It is important to note that even though Gabriel connectivity avoids geometrical inconsistencies in defining the connections among pairs, it is in general very unlikely that classed lags (numbers of edges) would preserve spatial stationarity. An especially interesting Gabriel graph was presented in Matula and Sokal (1980) (figure 4.3). Note that although the original system of populations exists in two-dimensional space, the Gabriel-connected populations in this example form what is essentially a system with one-spatial dimension. Another option for incorporating connectivity in Moran's *I* statistic is to measure the geographic distance along the connections, i.e., the distance between a pair of populations is the sum of the distance lengths of the connections that provide the shortest path between them. This is a very strong example of the effect of spatial irregularity. Correct use of the method implies that the assumption that gene flow happens only among connected populations must be sure, otherwise distances can be grossly miscalculated. For example, in figure 4.3 the Euclidean distance between populations 1 and 2 is much smaller than the distance traversed along the Gabriel-connected sequence {1, 4, 3, 7, 6, 5, 2}.

DIRECTIONAL AUTOCORRELATION AND SPECIAL WEIGHTING SCHEMES

The flexibility of joining and weighting schemes allows specification for autocorrelation in different spatial dimensions (e.g., Cliff and Ord

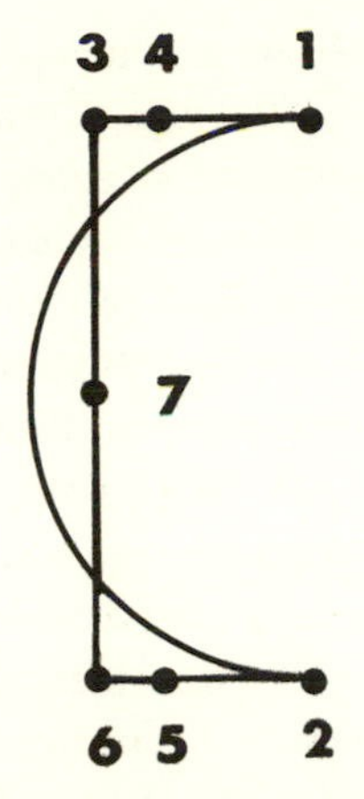

FIGURE 4.3. A Gabriel graph among seven populations (numbered 1–7) in two-dimensional space. Populations with lines drawn between them are Gabriel connected. Also shown is the half of the circle defined by populations 1 and 2. Interestingly, if population 7 is omitted, then populations 1 and 2 become Gabriel connected. Note that the path of connections going between populations 1 and 2 is much longer than the direct distance between them. Note also that the graph of Gabriel connections among the seven populations forms a system with essentially one spatial dimension, even though the original system of populations exists in two-dimensional space. The graph is patterned after one presented in Matula and Sokal (1980). Figure drawn by R. Walter.

1973; Jumars et al. 1977). For data on populations in two-dimensional space, the standard correlogram of I statistics as a function of distance classes represents a collapse to a one-dimensional representation. There can be various reasons to have expected or observed directional trends in the data. For example, there may be a narrow band of populations, extending say northeast to southwest, that have relatively high gene frequencies, caused by asymmetries in the rates of migration or other reasons (Epperson 1993a,b). Although there may be substantial autocorrelation at distances less than the width of the band, at the somewhat longer distance classes much of the autocorrelation could be missed. Oden and Sokal (1986) developed a procedure generally suited to detect directional trends in data sets in two dimensions. The procedure does not require a priori designation of either the directions or centers of such trends. It imposes without loss of generality an X-Y coordinate grid mapped onto the system of populations, and uses both the geographic distance as well as the slope of the line (in the X-Y coordinates) connecting each pair of populations. In the case of unweighted Moran's I statistics, instead of using only a range of distances to form classes of pairs, a second classification based on ranges of slopes is added. Slopes range from 0 to 360 degrees. For each standard distance class there are subclasses based on ranges of slopes, i.e., for each class of slopes there are subclasses of ranges of distances, and for each slope/distance class a Moran's I statistic is calculated. Once the classifications are designated, further flexibility is allowed by choosing other (nonbinary) weights, completely analogous to the case of I statistics based solely on distance.

It is informative to display the results of Oden and Sokal's directional analysis in the form of a graph with sectored concentric rings or annuli, where the distances of the inner and outer outlines of the ring from the center represent the range of distances for the class, and the lines sectoring the rings represent the range of slopes (Oden and Sokal 1986). An example was shown in chapter 2 (figure 2.6). Because pairs of populations are defined by both distance and slope, the number of distance class ranges usually must be reduced to have sufficient numbers of pairs in all distance/direction classes. Some distance-scale resolution is lost in return for dimensional resolution. How the two-dimensional correlograms may be used to detect various forms of directionality in systems with two spatial dimensions is fairly complex, and may depend on the geometrical shapes of areas with concentrations of unusual gene frequencies (Oden and Sokal 1986). Nonetheless, it is clear that if a particular dataset is suspected to show substantial directionality of any type, directional correlograms should be analyzed as well as the distance-based "standard" correlogram.

Directionality can be created in patterns from a situation that might be described as semispatial regularity, where the strength of process interactions, migration rates, are in many respects regular, but have some degree of irregularity. For example, in a regular lattice of uniformly spaced populations (figure 4.1) migration rates along one dimension (e.g., north–south) might be consistent among all populations, but different from those in the other, perpendicular, direction (east–west). In other cases one dimension may be aligned with the axis of the vector for prevailing winds, the slope of a mountainside, or along a stream or watershed. Directionality within a dimension still maintains symmetry in the spatial correlations aligned in that dimension (Epperson 1993a,b). Differences between two dimensions, or "dimensional directionality," are determined by differences in migration rates (e.g., contrasting north–south versus east–west). In principle, if the direction of directional differences is known or can be reasonably posited, then the two axes are determined (as is the orientation of an appropriate *X-Y* coordinate system). Then all pairs of populations can be classified by the joint numbers (or ranges of numbers) of spatial lags separating them in both dimensions. For example, a class may be defined for all pairs of populations separated by lag distance x to $x + a$ (measured in the X dimension) and by y to $y + b$ (measured in the Y dimension).

SPATIAL STATISTICS FOR LOCALIZED AND SYSTEMIC SPATIAL NONSTATIONARITY

Local environmental selection and localized events can create local concentrations or autocorrelations of gene frequencies that are distinguishable from "global" autocorrelations. The latter are based either on weighted sums of the products of all pairs of populations (weighted I statistics or the Mantel statistic) or on defined sets of pairs of populations (autocorrelation correlograms). Localized autocorrelations also imply a form of spatial nonstationarity, where there are nonhomogeneous processes acting over the landscape. One appropriate measure of local spatial autocorrelation is given by

$$I_i = \frac{nZ_i \sum_j w_{ij} Z_j}{\sum_j Z_j^2} \tag{4.12}$$

(Anselin 1995). This formula is very similar to Moran's I statistic (equation 4.1), except that only pairs that include population i are counted in the numerator, and W is missing from the denominator. A local spatial autocorrelation coefficient is calculated for a given population i, which serves as the focal point for structure. As before, we can use general weights, or simply use binary values to set up correlograms, but in most cases we would want to use fairly short distances, because of the emphasis on focal points and localized autocorrelation.

The expected value of I_i under the randomization null hypothesis is $-w_i/(n - 1)$, where w_i is the sum of the weights to population i (Sokal et al. 1998a). Note that if I_i were divided by w_i its expected value would be equal to that of Moran's I. This statistic has two important properties: (1) it indicates autocorrelation focused about each population i; and (2) the sum of the I_i is proportional to I with the same set of w_{ij}. These are properties Anselin (1995) has proposed be required for any measure considered to be a local indicator of spatial association, or LISA (Sokal et al. 1998a).

In computer simulations of nearest-neighbor migration, which causes global and spatially stationary autocorrelation, but with added "hotspots" of various processes expected to cause localized autocorrelations, Sokal et al. (1998a) illustrated the important result that *tests based on LISAs were not reliable in the presence of substantial global autocorrelation*. However, LISAs still may be useful as an exploratory tool for subsequent investigations, especially when there are genes for several loci that are responding to the same local condition. Tests for local autocorrelation are reliable when there is essentially no global autocorrelation, but only if the permutational variance is used. The variance under the randomization hypothesis is given by Sokal et al. (1998b) but it is not very useful, because standard deviate test statistics based on it are not asymptotically normal. Instead, Sokal et al. (1998b) suggest creating a sample-based distribution from random permutations of the data. Thus, in a nearly panmictic system of populations, the operation of localized factors could be reliably tested for. Specifically, two different types of permutation tests were studied; one was the usual permutation of all pairs, analogous to randomly permuting rows and columns as described above for the Mantel test; the other was random permutation of only the $n - 1$ rows and columns apart from row i and column i. Sokal et al. (1998b) found little difference between total randomization and conditional (on fixing row i and column i) randomization, although there are theoretical reasons to expect the conditional permutation to be slightly more reliable. Other methods of detecting localized concentrations and defining localized groups were reviewed by Legendre and Fortin (1989).

Localized spatial nonstationarities may affect not only the local spatial autocorrelations, but may also cause more specific localized changes in spatial distributions. Primary among these may be various local barriers to gene flow. If gene flow is locally limited, greater genetic differentiation should exist in the immediate vicinity. Other factors could cause unusually abrupt spatial genetic differentiation in and around a local area, including selection favoring a gene in that local area. Such localized features are a form of spatial nonstationarity in the space–time process, and they may cause detectable nonstationarities in spatial data.

Barbujani et al. (1989) refined a technique called "wombling" (Womble 1951) that uses measures of differential gradients in spatial genetic patterns. It is useful to consider the distribution of genes in populations as a "surface" on the plane. The locations of the populations sampled will determine the *X-Y* coordinates in the plane, and a third dimension represents their gene frequencies. The surface may have peaks of especially high gene frequency and valleys of particularly low. In areas immediately around a barrier to gene flow, the slope of the surface should be steep. The method of Barbujani et al. (1989) employs a rather involved set of criteria for finding the average slope at center points between populations. It then calculates the steepness of the slope and its direction. Localized regions containing many steep slope values pointing in a parallel fashion can identify areas where differentiation is abrupt. Moreover, these may be interpreted by investigators, who may also a posteriori examine the surface for geographic features such as impassable mountain ranges. Various criteria may be developed for separating out signals of actual process barriers from stochastic and statistical variations (Barbujani et al. 1989).

Barbujani et al. (1989) studied wombling in simulation studies of standard genetic isolation by distance in systems with two spatial dimensions. Although the simulation was of individual genotypes within a population, the salient features should also apply to gene frequencies in systems of discrete populations. Typically, a few nominally significant slopes were observed scattered over the surface. These must be spurious because the simulated stochastic process was spatially stationary, which indicates that the nominal significance level cannot be taken at face value. When localized influences were added to the simulation, they were generally detectable. Both selective clines and reinforced special patches of gene concentrations resulted in higher concentrations of significant slopes, which also were aligned parallel. Nonetheless, it appears that confident detection of barriers to gene flow and similar-acting features will usually require multivariate genetic data (i.e., multiallelic and/or multiple locus), and, in particular, the slopes for different loci should be aligned (Barbujani et al. 1989). In a remarkably strong display of the sensitivity of wombling techniques, Barbujani and Sokal (1990) found that of 33 localized regions identified as having abrupt changes in gene frequencies in human populations in Europe, nearly all corresponded to geographic boundaries of language and often with substantial physical barriers as well.

At the other end of the spectrum of spatial nonstationarities are effects that exist on a large scale, systemic or nearly encompassing the entire system of populations. If such factors or effects vary smoothly across the space of a system, they can be called trends and modeled with simple polynomials (Cliff and Ord 1981). Trends pose important problems in geographic analysis, and in many practical applications in geology and geography (Whitten 1975; Haggett et al. 1977) the goal is to remove the effects of trends from predictive models based on spatial autocorrelations. In population genetics we are often interested in the trend effects of a selective gradient, but there are other population genetic factors that operate as trends, such as recurrent immigration from an outside source entering from a given direction. Empirical goals may range from demonstrating natural selection to removing the trend in order to study local migration patterns and rates. Trend surface analysis appears to offer some promise as a means to partition the spatial autocorrelation into two different spatial scales. However, if there are *process* interactions between two types of space–time processes, the separation may not be complete or exact. In particular, it is not clear that the removal of trends in spatial data actually restores spatial stationarity in the isolation by distance process.

Trend surface analyses are usually developed as polynomial regression models of systems with two-spatial dimensions:

$$q_{xy} = \sum_{i=0}^{s}\sum_{j=0}^{v}\beta_{ij}x^i y^j + \varepsilon \tag{4.13}$$

where q_{xy} is the value of the variable (in population genetics, the gene frequency) at a population whose location is defined by the x, y Cartesian coordinates in the plane. Trend surface analysis typically uses linear (both s and v are either zero or one, but both are not zero) or quadratic (typically $s = v = 2$), but cubic models are also used. Naturally, linear and quadratic models would describe only surfaces that have smooth changes; i.e., there cannot be many peaks or valleys in the predicted surface. In contrast to standard multiple regression, the stochastic error terms ε, identically distributed normal random variables, are not generally independent in spatial data and neither are the residuals $\tilde{\varepsilon}$, which are measured in the usual way. Generally, if there is positive autocorrelation in the original data, the standard regression variance estimator is biased *downward*, thereby inflating the observed R-square measure of fit of the data to the polynomial (Cliff and Ord 1981). Spatial autocorrelation creates difficulties for which precise solutions are not easily obtained. Somewhat ad hoc suggestions include adding more regressor terms, or trying to add spatial autoregressive terms to account for the autocorrelated structure (Cliff and Ord 1981). Tests for autocorrelation among the residuals are important, and Cliff and Ord (1981) provided arguments that only the Moran's I statistic under the normality assumption was likely to provide appropriate tests. Moran's I statistic is calculated in the same way as before (equation 4.1):

$$I = \frac{n\sum_i\sum_j w_{ij}\tilde{\varepsilon}_i\tilde{\varepsilon}_j}{W\sum_i \tilde{\varepsilon}_i^2} \tag{4.14}$$

where i and j are retained here to denote populations i and j. The moments of I under the null hypothesis are generally different and more complex, but a general formula was developed by Cliff and Ord (1981). It is important to emphasize that even if there is no spatial autocorrelation among the errors ε the residuals $\tilde{\varepsilon}$ are autocorrelated under the null hypothesis (Cliff and Ord 1981). The standard normal test statistic is formed in the same way, and it is asymptotically normal (Cliff and Ord 1981). If (excess) autocorrelation is observed among the residuals, it may imply either that more regressors are needed or there is an autoregressive structure (Cliff and Ord 1981). The latter would be expected in any population genetic process in which there is limited migration among populations (Epperson 1993a).

Bocquet-Appel and Sokal (1989) examined some of the properties of linear and quadratic trend surface analysis. Residuals produced by models of pure isolation by distance (using the same type of individual-based simulations mentioned earlier) exhibit high autocorrelations, similar to those produced among the original gene frequency values. In models with added recurrent immigrations from outside the population into one side of the system, a cline is superimposed with the isolation by distance, and the long-distance negative correlations typical of clines of gene frequencies did not occur among the residuals. Indeed, the residuals for the two types of models were nearly identical. Thus Bocquet-Appel and Sokal (1989) conclude that trend surface analysis successfully removed trends.

MULTIPLE GENETIC CHARACTERS

Genetic surveys typically generate data for multiple alleles and loci, inducing several differing additional statistical issues. Spatial distributions of frequencies of alleles of a genetic locus are correlated both in the processes (in fact, they are stochastically interdependent) and in samples (statistically correlated). For a locus with two alleles, the pattern of frequencies of one allele contains the same information as that for the other allele. Patterns for three or more alleles are correlated, but in complicated ways. To the author's knowledge, even under the null hypothesis there is only one existing way to precisely calculate the stochastic covariances among spatial correlations for alleles of the same locus for diploid data. The method is based on all pairs of sampled diploid individual genotypes (Epperson, n.d. (a)). Because it is based on each genotype having a unique spatial location, it is better applied to the spatial distribution of diploid genotypes within a population, and hence it is developed in chapter 8.

In principle, it could be applied to data among populations, but empirical studies usually have samples of multiple individuals from each population. Hence, it apparently confounds genetic *variation* among populations with the *pattern* of gene frequency differences among populations. Thus there remain statistical problems for averaging over alleles. One solution is to use only one allele per locus, another is to have many alleles so that the covariances are negligible.

Spatial correlations for different unlinked loci are generally stochastically independent, and approximately so even if they are tightly linked. Spatial autocorrelations should also generally be approximately independent for unlinked and linked loci, as long as the system of sampled populations is fairly large, regular in spatial distribution, and near equilibrium. In the following discussions, it is assumed that different genetic variables are for different, independent loci unless otherwise indicated. Various standard methods may be used for analyzing sets of independent spatial statistics (for example, Moran's I for only the first distance class) with no particular problems. For example, if the statistics are also assumed to be identically distributed (thus they are independent identically distributed [iid]), then the observed variance of the statistics may be used. Assorted standard tests may be correctly applied, for example, t tests for differences between values for two sets of loci. The situation for tests based on entire spatial patterns, for example, testing for differences among spatial correlograms, is much more involved. The reason is that in such tests the null hypothesis usually allows that there are spatial patterns and spatial autocorrelations, and hence there are correlations among the values for different distance classes for the same locus. The problems arise from the lack of specification of those correlations, and they depend on the spatial pattern itself, which in turn depends on the process. Even when loci are subject to exactly the same stationary process, they will not have the same spatial distribution of allele frequencies, but they are expected to have the "same" spatial correlations. For a simple example, consider two diallelic loci in two populations with a certain amount of drift, mutation, and migration. There is usually no basis for expecting particular alleles, say allele "one" for both loci, to both have higher frequency in one population and lower in the other population. However, the degree of differentiation among populations is expected to be the same for the two loci, at least within stochastic and statistical error. In short, we do not want to test that the spatial patterns are the same, rather that the spatial autocorrelations are the same (Sokal 1979b, 1986). Unfortunately, analytical transformation of patterns to correlograms are difficult. This area requires more research.

A particularly important issue is to detect differences among correlograms that may indicate that some loci, contrasted with others, have been subjected to natural selection. Sokal and colleagues have developed a number of nonparametric methods for measuring the difference between two correlograms that have the same or very similar distance classes. For example, the "Manhattan" distance (Sokal 1986) is the average absolute difference of Moran's I statistics for comparable distance classes. Once sets of distance measures have been calculated for all pairs of correlograms, further statistical analyses may be conducted, for example, UPGMA hierarchical clustering or k-means clustering (Sokal 1986). In such analysis, care must be taken that noninformative distance classes are not included; for genetic isolation by distance it is especially important to omit the distance classes for very large distances (e.g., Sokal and Wartenberg 1983). Space–time Monte Carlo computer simulations of the distributions of single-locus genotypes can address the variation that occurs among spatial autocorrelations for loci subject to the exact same process (usually determined simply by the dispersal rate, in an isolation by distance process), and for loci subject to nonidentical processes but in the same system of populations. For example, among multiple loci, i.e., sets of replicate simulations (Epperson 1990a), some with selection, some without, simulations that had the same parameter values clustered together, using UPGMA.

Sokal and Wartenberg (1983) developed a Mantel test for contrasts among replicated sets of simulations of genetic isolation by distance. They showed that differences in dispersal parameters cause statistically significant differences among correlograms, whereas simulations with the same parameters are similar. The two "distance" matrices that Sokal and Wartenberg (1983) used for their Mantel test were (1) the matrix of the pairwise areas between correlograms, and (2) a matrix that had elements that specified the contrast. In general, the Mantel test can be used by clever designations of contrasts that reflect the null hypothesis of interest. The investigator must have considerable knowledge about the kinds of contrasts that will reflect the paramount spatial structural differences, yet remain invariant with respect to the other differences (discussed above) that occur among the actual spatial *distributions* for genes that are subject to the same process. Generally, such Mantel tests cannot rely on normal distribution theory; rather, permutational variances must be computed. Care must be taken that the appropriate restrictions are placed on which rows and columns are to be permuted.

A good example of the use of appropriate restricted permutations to reflect the null hypothesis and alternate hypothesis in forming a proper statistical test is presented in Sokal and Thomson (1987). In this example, two different join matrices (using binary weights) were constructed, to reflect two different hypotheses, H_1 and H_2, of underlying patchy ecological habitats (and implicitly the effects of these on the genotypes). To analyze which of the two hypotheses is "closer to the truth" two tests may be conducted. For one test, H_1 is considered the null hypothesis, and here the pairwise distances under H_1 form one matrix and the connectivity of H_2 forms the other. Only the permutations of populations within the groups defined by H_1 should be allowed, to form a proper test distribution (Sokal and Thomson 1987). For the second test the roles of H_1 and H_2 are reversed.

Correlations among I statistics for different distance classes, alleles, and loci, and how these reflect structural features of spatial patterns remain important issues, and further research is needed. Moreover, the precise linkage between the features of the underlying space–time processes to invariance of relationships among Moran's I statistics for different distance classes needs further development. A related concern is how to measure "cross-correlations" among multiple loci (see, for example, section 5.4 of Cliff and Ord [1981]), while properly taking into account spatial autocorrelations and genetic isolation by distance.

SPATIAL MODELS AND STATISTICAL ANALYSES OF SPATIAL PATTERNS

The spatial versus space–time extensions of *time series* models (Epperson 1993a, 2000a) best illustrate the distinction between purely spatial and space–time stochastic and statistical models. Space–time statistical models can be directly tied to space–time processes, and thus avoid many of the problems involved in specifying autocorrelation or autoregressive structure (Epperson 1993a,b). However, the use of space–time statistical models has the logistically nontrivial requirement of space–time data. In the 1970s statistical geographers developed a number of theoretical models, the first of which were *purely spatial* models, alternatively called spatial processes and random fields, including direct extensions of time series models. The first of these, following the Box and Jenkins (1976) approach, treated a single *time-like* spatial dimension; an example might be the flow of a river. Next came extensions to two time-like spatial dimensions; an example might be the directional spread of a contagion over a surface (Haining 1977, 1978, 1979; Bennett 1979).

It is a nontrivial problem to extend autoregressive time series models to spatial axes in systems where there are interactions in *both* directions within dimensions (Bennett 1979), because there is no longer a *sequence* of autoregressive effects of one location on another, and hence specification of the autoregressive structure is confounded. Indeed, there are at least three approaches to so-called spatial autoregressive (SAR) models. These include conditional autoregressive, simultaneous autoregressive, and covariance-structured autoregressive (Bennett 1979; Cliff and Ord 1981; Upton and Fingleton 1985). Like time series, SAR models can be made for stochastic processes and for statistical analysis. But, for the discussions that follow, let $\mathbf{Z}$ be a $1 \times n$ vector of *observed* genetic values (e.g., gene frequencies), either transformed or not but mean adjusted, one from each of n populations at some point in time. For the simultaneous autoregressive spatial statistical model, $\mathbf{Z} = p\mathbf{WZ} + \mathbf{e}$, where $\mathbf{W}$ is an $n \times n$ matrix of relative *statistical* (not stochastic) interactions between populations, p is an overall weight, and $\mathbf{e}$ is a vector of error terms. Note the "simultaneous" nature of this equation, because $\mathbf{Z}$, which is an instantaneous snapshot of the spatial genetic distribution at the time of sampling, occurs on both sides of the equation. This is quite different from the effects of migration, which happen over time, especially from one generation to the next (Epperson 1993a). To see this, the above equation should be compared to Equation 4.15 below. It is not at all clear how spatial statistical interactions may be determined from the migrational interactions, even if the latter were known. This problem is analogous to the general difficulty of assigning weights, for example, to Gabriel-connected graphs, even in spatially regular systems. The same problem occurs with the other two forms of purely spatial autoregressive models. In essence, these statistical models assume that the spatial statistical interactions (including spatial autocorrelations) are instantaneous.

It is doubtful that biological systems have process interactions that are instantaneous rather than spatial–temporal, because the variables of interest in biological systems (in our case gene frequencies) interact and change over time (Epperson 1993a). To understand this further we may contrast the biological system with a two-dimensional "random field," where, for example, there may be interactions among electrical charges. The charges may interact effectively instantaneously and produce a steady state. The much longer timescales of observation made on the steady state could be described with spatial models (e.g., Larrimore 1977). The problem, in population biological systems, of relating space–time autoregressive structures to spatial ones is similar to the problem of specifying the relationships between spatial correlations for different distance classes, which has blocked the development of exact statistical tests for comparing I correlograms to one another or to one produced by the same space–time stochastic process. The primary spatial interaction in population genetics is via migration, and the statistical effects of migrational interactions are not instantaneous; they expand in spatial range over time. For example, in a system with migration only among nearest neighbors it takes two generations for a population to interact with its second-nearest neighbors, three generations for its third-nearest neighbors, etc. Moreover, even if we could conceive a way to fold space–time interactions into spatial ones, it would still seem that spatial models are hopeless as statistical methods. For example, we would not be able to distinguish direct migration between populations b units apart (e.g., migration directly from population at x to $x + b$) from migrational effects among nearest neighbors lagged b time periods apart). These considerations illustrate the need for understanding the underlying space–time process, which extends spatial processes to include a temporal dimension. Because many geographically distributed variables also have interactions that operate over time, statistical geographers were early to recognize the need for space–time processes, and it is equally important in population genetics. Over several decades, leading statistical geographers, many themselves working with purely spatial processes, made repeated calls for studies of the underlying space–time processes for geographical variables (Whittle 1954; Bartlett 1971; Haining 1977, 1978, 1979; Bennett 1979; Bennett et al. 1985).

The limitations of purely spatial statistical models are also apparent in the effort to separate the effects of migration from those environmental, location-dependent, selection. It might seem that such selection might be incorporated by adding location-dependent linear regression terms, with a matrix $\mathbf{X}$ (an $n \times k$ matrix of values representing values of k different environmental factors at each of the n locations) and a vector $\mathbf{b}$ (a vector of regression coefficients representing the degrees of influence on the gene frequencies), giving $\mathbf{Z} = p\mathbf{WZ} + \mathbf{Xb} + \mathbf{e}$ (Epperson 1993a,b). Although statistical significance of nonzero estimates of the elements of $\mathbf{b}$ would suggest that selection is acting (Epperson 1993b), not all of the effects of selection would be represented by estimates of $\mathbf{b}$. Selection at one population would affect not only that population, but also, over time, the spatially proximal populations, because of migration. To the author's knowledge, environmental selection has not been analyzed in this way. However, Upton and Fingleton (1985) used this model to study spatial patterns of blood groups in counties of Ireland, and successfully identified significant factors of both location-dependent influences (independent evidence of degree of recent immigration from England) and interactions between spatially proximal counties (from local migration between Irish counties).

In contrast, space–time models that include both absolute location-dependent factors and relative location interactions, can, in principle, completely separate the space–time autoregressive effects of migration from the effects of location-dependent selection (Epperson 1993a,b). It appears that one of the most important considerations of spatial pattern analyses, to partition the effects of differential selection from those of gene flow, cannot be completely accomplished by purely spatial analysis of spatial data, without a detailed understanding of the underlying spatial–temporal process, analyzed theoretically either with mathematical analytics or with Monte Carlo computer simulations.

SPACE–TIME ANALYSIS

By adding the temporal dimension to data analysis, it is possible to recover the full details of the underlying space–time processes of population genetics. The most promising and to date best developed space–time statistical models appropriate for genetics of discrete populations are spatial time series models. The models include so-called space time autoregressive (STAR) without or with (STARMA) moving average terms, additional regressor variables (STARR or STARMAR), and possibly integrated (STARIMAR, etc.). Like time series they can be used both to model theoretical stochastic processes and as statistical models for analyzing data. Statistical models are the subject here, and the theoretical justification for how spatial time series models subsume and greatly extend nearly all migration models for systems of discrete populations is covered in chapter 5. As statistical models the added MA terms may model the statistical effects of "stochastic migration" (Epperson 1994), and the addition of regressors and integration also add still more flexibility. STAR and STARMA models have the proper structure needed to directly relate statistical interactions to stochastic process interactions, in contrast to the purely spatial SAR models, which generally allow little insight into how small-scale interactions develop larger scale spatial patterns.

The STAR statistical model applied to migration processes in populations genetics can be written

$$Z_{x,n} = \left(1 - \sum_{b \neq 0} l_b - k\right) Z_{x,n-1} + \sum_{b \neq 0} l_b Z_{x+b,n-1} + a_{x,n} \qquad (4.15)$$

where $Z_{x,n}$ is the (transformed or not) mean-adjusted gene frequency observed at a population located at x (x is a scalar for the one spatial dimension case and a vector of coordinates for multidimensional space) and at time (usually generation) n (Epperson 1993b). The identity of the autoregressive structure to that of stochastic processes of migration may be seen in the fact that if one substitutes the stochastic variables for gene frequencies $z_{x,n}$ for observed values $Z_{x,n}$ in equation 4.15, then the stochastic equation 5.21 is yielded. Similar substitutions of observed values for stochastic variables are straightforward for other STARIMAR models. Each l_b (for $b \neq 0$) is a migration rate from all appropriately defined populations at $x + b$, to population x, b is a scalar or vector of values for the spatial lags, and k is a systematic force (e.g., mutation or selection) that ensures temporal stationarity. The variable $a_{x,n}$ may be considered as an error term that may have both statistical and stochastic (genetic drift from generation $n - 1$ to n) components. Thus, the observations at any time n are simply weighted linear combinations of preceding gene frequencies and random effects.

Equation 4.15 is not restricted to simple tests of the null hypothesis of no spatial autocorrelation, in part because it is expressed in term of the finest lag structure. Other spatial lag weightings could be employed (Hooper and Hewings 1981, but use of the finest lag structure maintains spatial stationarity (if it existed among the $Z_{x,n}$), which is important for the estimation of migration rates. Other work rewrites the interactions in terms of some weight matrices W_b, for each order of spatial lags b, or alternatively a weighted "neighbor" joining matrix. In many cases the W_b are symmetric, but they need not be, and they may reflect time lags (for models with temporal order greater than one) as well as spatial lags. They are analogous to those used in spatial autocorrelation analysis (e.g., Epperson 1993b). This alternative directly relates all of the interactions, with various temporal and spatial lags, to a matrix of weights that expresses the *relative* strengths of interactions over space within the same time lag. There is an ambiguity in the choice of overall weights of the same temporal order versus the relative ones, and this is an important distinction between stochastic and statistical models. The general STARMA statistical approach has recently been developed into a statistical computing program by Lee and Epperson (n.d.). A goal of any space–time analysis is to reduce the number of parameters, relying on spatial stationarity. In practice it is usually appropriate to consider a quite limited set of lags, where the interactions occur only between locations separated by limited distances.

STARMA procedures can be used to obtain many statistical objectives previously unavailable in population genetics. Statistical objectives can be classified into four steps (Pfeifer and Deutsch 1980a; Bennett 1981):

1. *Process identification*—identification of the interactions via spatial proximity, e.g., the maximum distance over which there are direct interactions (via migration) among the variables (gene frequencies)
2. *Estimation of parameters* (particularly the migration rates)
3. *Model fitting*—tests of fit of the statistical model (and significance of parameter estimates) (from steps 1 and 2) to data using remaining degrees of freedom
4. *Forecasting*—projecting the process into the future using fitted models

The first three steps may be iterated as a space–time extension of the Box–Jenkins (Box and Jenkins 1976) method for time series.

Several statistical objectives fall under the rubric "process identification." As applied to population genetic systems the most important aspects are to determine (1) the maximum spatial distance (or lags) over which direct migration occurs; and (2) the number of time lags. In many cases (e.g., if there are no seed banks, in plants, or other causes of overlapping generations), the maximum temporal lag will be one. Often the maximum spatial lags will be few or even one (strict nearest neighbor migration). Importantly, this step of the procedure identifies the numbers of parameters (the numbers of l_b) needed in the model (equation 4.15). Thus, identification also specifies or formulates the autoregressive structure of the process, which, as was discussed above, is problematic using purely spatial models and data. In fact, the identification step requires the space–time correlations, which require space–time data.

The combination of space–time correlations and the so-called partial space–time correlations form a set that is unique to a completely specified STARMA stochastic model. Partial space–time correlations are correlations between variables (separated in space or time or both), conditioned on all spatially and temporally intermediate variables. They may be ordered in several different ways, but ordering based firstly on spatial lags and secondly on temporal lags (Martin and Oeppen 1975), rather than vice versa (Pfeifer and Deutsch 1980a–c), is more appropriate for low temporal order processes. Similarly, the set of estimated correlations have properties that fit the process within statistical error.

There are various ways of averaging among pairs of observed gene frequency values in estimating the space–time correlations. For example, in the spatial time series literature the estimated value of the variable at a location x may be a weighted average of the value for neighboring locations, and then the covariance (since it is a function of the products) may be a complex function of the products of the weighted values (e.g., Pfeifer and Deutsch 1980d). A certain degree of efficiency may be gained. However, to simplify the exposition of STARMA methods, we will be concerned here only with unweighted measures (analogous to unweighted Moran's I statistics), as well as in terms of the "finest lag structure" (Hooper and Hewings 1981). Then all pairs of observed gene frequencies are classified according only to the spatial lags b between populations and the number of generations t separating them in time. This is analogous to a space–time correlogram extension of Moran's I statistic. Use of the finest lag structure leads to consistent estimators of space–time covariances and correlations. Thus, let $w_{ij}(b, t)$ be indicator variables such that they equal one if two populations are separated by b spatial lags and t temporal lags, and zero otherwise. Note that this is a temporal extension of the $w_{ij}(k)$ used for spatial I correlograms. Then, a proper estimate of the space–time covariance $\gamma_{b,t}$ is given by

$$\hat{\gamma}_{b,t} = \frac{1}{c} \sum_{x} \sum_{n} Z_{x,n} Z_{n+b,n-t} \tag{4.16}$$

where the summations are over all appropriate pairs, which total c in number (Hooper and Hewings 1981). Thus far the development is completely analogous to Moran's I statistic. As an example, consider r populations on a (single-dimension) line, sampled for T generations. There are rT total observed values. Then, for example, the number of pairs c for equation 4.16 for pairs of populations separated by two spatial lags and by three generations is $(r - 2)(T - 3)$.

There are also several possible choices for estimates of the variances of the $Z_{x,n}$ (Martin and Oeppen 1975). For example, one could estimate the variance of gene frequencies at any given population but across time, or one could fix a time and measure the variance among populations. This can be problematic when using other lag structures, but with the finest lag structure approach developed here, if there is spatial and temporal stationarity, then the best estimate of the variance is the same for all $Z_{x,n}$ and thus is denoted $\hat{\gamma}_{0,0}$; hence it should be based on all the data, $\Sigma_x\Sigma_n Z^2_{x,n}$ (recall that the $Z_{x,n}$ are already adjusted for the grand mean value). Thus, the estimator of space–time correlations is

$$\hat{\rho}_{b,t} = \frac{\hat{\gamma}_{b,t}}{(\hat{\gamma}_{0,0}\hat{\gamma}_{0,0})^{1/2}} \tag{4.17}$$

Using the above definitions, equation 4.17 is clearly a space–time analog of Moran's *I* statistic (equation 4.2), with *c* playing the role of W_k. Other estimates of variances and covariances could have somewhat different values. Formulas for various estimators of the partial space–time correlations are much more complex (Martin and Oeppen 1975; Pfeifer and Deutsch 1980a,c; Hooper and Hewings 1981), and are not displayed here.

Estimates of the spatial and space–time correlations can be compared to theoretical values that have well-defined properties under different types of processes, and currently in the author's lab, we are developing statistical criteria for such comparisons (Lee and Epperson, n.d.). Most importantly, pure STAR processes (with no shared stochastic migration effects), have theoretical partial space–time correlations that cut to zero beyond the maximum spatial distances or the maximum temporal lags of the processes. The spatial lags beyond which the partial space–time correlations become and remain zero identify the maximum distance at which direct migration occurs. Thus, for example, in the common situation of a Markovian (nonoverlapping generations, no age structure or seed banks) STAR system, if migration is only among nearest neighbors, then the partial space–time correlation function becomes zero for all spatial lags greater than one and for all temporal lags greater than one, at least within statistical error. The correlation function for a STAR process tails off in a distinctive manner (according to the parameters) in space and time. For example, in one-dimensional migration models, whether strict-nearest neighbor or not, the correlation function decreases smoothly with a rate that depends on the rates of migration (Epperson 2000a).

For the next step, estimation of migration and other parameters, there are several available approaches. One of the simplest to employ is to substitute the estimated spatial and space–time correlations into the space–time extensions of the Yule–Walker equations for time series (Epperson 1993a). For example, for a temporal order one (Markovian) STAR model:

$$\hat{p}_{b,n} = \sum_{m} \hat{l}_m \hat{p}_{b+m,n-1} \tag{4.18}$$

(except for $n = 0$, $b = 0$). The summation is over all spatial lags m for populations that contribute migrants plus the rate of no migration $m = 0$. In this case the following equation holds for the estimated variance:

$$\hat{\sigma}_z^2 = \hat{\sigma}_a^2 \left(1 - \sum_{b} \hat{l}_b \hat{p}_{b,1}\right)^{-1} \tag{4.19}$$

Equations 4.18 and 4.19 are linear and can be solved explicitly in terms of the estimates of the parameters l_b (Epperson 1993a), or they can be numerically evaluated using standard matrix methods. Equations of the type in equation 4.18 can be can be used to estimate migration rates and, in principle, the strengths of systematic forces such as natural selection. In addition, because σ_z^2 can be estimated from data, equation 4.19 can be used to estimate σ_a^2, and, for example, if there is no sampling error (i.e., complete censussing) this should equal $1/8N$, where N is the population size or variance effective population size (Crow and Denniston 1988). Although the Yule–Walker estimates are correct, they are not very efficient; i.e., they may have relatively large standard errors.

Other estimators are based directly on equation 4.15 and analogous equations for other STARMA models, and include such standard approaches as least squares (Pfeifer and Deutsch 1980a) and maximum likelihood (Larrimore 1977). The likelihood equations are quite complex functions of the l_b and considerable recursive computations are required. Conditional maximum likelihood estimators (MLE) are useful and good approximations when the number of observations is large. The equations are much simpler for the likelihoods conditioned on substituting the average values at each location for the values (not measured) preceding the time period of observations. A second approximation uses conditional least-squares, which also requires less computation. For the case of STAR models, the distributions of conditional least squares estimates are asymptotically equivalent to standard linear regression estimates, which can also be used as an approximation for samples of moderate size. Although the computational effort required for the various estimators apparently was a major obstacle to the use of STARMA models in the 1970s, this poses little problem today.

Once estimates of the parameters are obtained, several important tests of the statistical model identified can be conducted. One of the most important is to test the fit of the data to the correlation structure of the identified model. This can be tested by calculating the spatial and space–time autocorrelations and partial autocorrelations for the residuals. These statistics have well-known distributions (e.g., Cliff and Ord 1981). If the model is precisely correct, the residuals should have a nonautocorrelated "white noise" distribution, unlike those from purely spatial analysis, as, for example, was shown in trend surface analysis. Nonzero partial correlations or nonzero correlations at any given temporal and spatial lags would suggest that additional parameters are needed (i.e., representing additional interaction and/or shared stochastic inputs, respectively, at those temporal and spatial lags), following the logic outlined for the identification phase, which was based on the original observed data. Another important test is for statistical significance of the parameter estimates. Parameters with nonsignificant estimates may be dropped in the next iteration of the modeling procedure. After testing goodness of fit of the data to the initial model, if necessary the procedure can be iterated, starting with corresponding modifications of the identified model, until the model passes all tests.

After the first three steps have gone through the needed iterations, the model can be used to forecast the future values of the variables in the system. This simply involves using the best-fitted identified model form, with its MLE parameter estimates, applied to the observed values (or estimates of the actual values) for the most recent generation(s) (only one, the present generation, if Markovian) (Bennett 1979, 1981). STARMA analysis may also incorporate additional features, including missing data points (Larrimore 1977), seasonal and otherwise cyclic causes of variation (Pfeifer and Deutsch 1980e), and transformation of data. Additional regressors or covariates (e.g., Upton and Fingleton 1985) can be included for environmental selection.

MEASURES OF KINSHIP

For geographically distributed population genetic survey data, a number of measures of "kinship" among populations have been developed. Efforts have been made to relate these so-called kinship measures to the covariances, correlations, or average probabilities of identity by descent, in theoretical processes of genetic isolation by distance models. Although they are generally not the same as the theoretical measures, they have been widely used in among-population studies, and thus they deserve brief discussion here. A variety of pairwise population kinship measures have been developed by Morton (Morton et al. 1968, 1971), mostly based on the observed gene frequencies q_i in populations at a fixed time. One common measure of genetic relatedness R_{ij} of two populations i and j is given by $(q_i - q)(q_j - q)/q(1 - q)$, where q is the observed (possibly weighted) average gene frequency in the system of populations (Cockerham 1969; Morton 1975). Values of R_{ij} may be averaged over various groups of *pairs* of populations, for example, all pairs of populations within a certain area. More directly tied to genetic isolation by distance are cases where the groups of pairs are defined by the geographic distance separating pairs. Then, for statistical analysis of real systems, ranges of distances must be chosen. This procedure is fully analogous to choosing distance classes for spatial autocorrelation analysis. Indeed, Barbujani (1987) has shown that for a diallelic locus

$$R = \frac{n-1}{n}\frac{I}{F_{\mathrm{ST}}} \tag{4.20}$$

where I is Moran's I statistic, and F_{ST} is the usual estimator of Wright's (1943) theoretical value, given by equation 4.22. If the number of populations n is large, then $R \approx I/F_{\mathrm{ST}}$. Problems of estimating F_{ST} are described below. Moreover, R is a ratio of two random variables, and its distribution is unknown, as is its relationship to theoretical values of kinship. However, Moran's I statistic has a known distribution under the randomization null hypothesis (equation 4.5), and in this case Moran's I statistic can be converted to kinship measures.

There are also multiple-locus and multilocus measures of R (e.g., Rousset 2000), although both suffer the same difficulties as do single-locus measures of R. Multiple-locus measures involve weighting first over loci (and alleles) and second over pairs of populations, whereas Moran's I statistics for multiple-locus data are weighted over pairs of populations and then over (alleles and) loci. They use the same information. Truly multilocus measures may use additional information if it is present—essentially such information is linkage disequilibrium, which will generally be near zero for neutral loci in outcrossing populations (Epperson 1995b).

Many experimental studies have fitted the following model in terms of observed averaged pairwise measures of kinship $R(d)$ as a function of geographic distances of separation:

$$R(d) = (1 - L)ae^{-bd} + L \tag{4.21}$$

where a, b, and L are constants, and these studies have often found a good fit (e.g., Jorde et al. 1982; Morton 1982). However, this may be in part due to the fact that there are three parameters to fit and a limited number of distance classes. This also assumes there is an exponential function, which is untrue for multidimensional genetic isolation by distance models. Moreover, even the (linear) manner in which the correction (some would say "fudge") factor L enters into equation 4.21, as well as the genetic meaning of L, have been questioned (Harpending 1973; Malécot 1973b; Nei 1973b). Also popular are measures generally termed genetic distances, most notably those developed by Nei (1972, 1973a,b), which may be considered inverse measures of similarity (Malécot 1975). However, most of these are tied to specific theoretical processes that do not include migration.

MEASURES OF OVERALL DEGREE OF GENETIC DIFFERENTIATION OF A SYSTEM

The focus of this chapter is on the spatial and spatial–temporal *pattern* of genetic differentiation. However, measures of overall differentiation, as well as tests for the absence of any difference among populations, are also of interest in geographic studies. Regarding the former, F statistics are particularly popular. In this section we treat F statistics only as measures of differentiation in either Wright's island models or truly hierarchical models, where in the first instance the level of gene flow is the same among all populations in the system, or in the second, among all populations within the same hierarchical groups. There is little to recommend the use of F statistics designed for pairwise averages. In real systems that have a single hierarchical level, corresponding conceptually to the original theoretical model by Wright, the usual estimator of F_{ST} is:

$$F_{ST} = \frac{\sum_{i=1}^{n}(q_i - q)^2 / (n-1)}{q(1-q)} \tag{4.22}$$

where q_i is the gene frequency in population i, and q is the average gene frequency. Note that the numerator of equation 4.22 is the usual estimate of the variance σ^2. As was noted by Wright (1951), this F_{ST} is not the same as the inbreeding coefficient F_{ST} in theoretical, stochastic processes of genetic isolation by distance (Wright 1943). One could question whether the value given by equation 4.22 is even an appropriate estimator, since its expectation is not the theoretical F_{ST}. Even the more sophisticated estimator θ of Weir and Cockerham (1984) is biased, but less so. Part of the problem is that the a priori distribution of q under Wright's island model is not known unless the migration rates and population sizes are known; moreover, the actual q may be very different from its a priori expected value. In addition, q may be a random variable and it is generally difficult to find the expected value of a ratio of two nonindependent random variables.

Because the distribution of the estimator F_{ST} (equation 4.22) is not known, it is usually not used directly in tests for significant genetic differences among the populations in a system, and there is evidence that such tests might be seriously over-liberal, rejecting the null hypothesis too often when it is true (Epperson and Li 1996). In other words, random samples from populations that all have the same expected frequencies may be rejected too frequently. Instead, tests such as Workman and Niswander's (1970) are usually based on chi-square test statistics of the homogeneity of sample gene frequencies among populations. Workman and Niswander (1970) showed how important it is to use the weighted values if the sample sizes are not equal among populations. If the estimator of the variance in gene frequencies is weighted by the sample sizes, and q is the average gene frequency weighted by the sample sizes, then substituting weighted values for the unweighted ones in equation 4.22 gives an estimator of F_{ST} that corresponds to a chi-square contingency test statistic for the dependence of gene frequency on population. This method makes the implicit assumption that each sampled gene is independent, and this is clearly not true if the fixation indices within populations are nonzero. For example, in the case of three alleles, if fixation indices are zero for all alleles and for all populations, then the data may be considered trinomial, and the contingency table is of dimensions 3 by r. In the two-allele case, the usual contingency chi-square test statistic, under the above assumptions, is equivalent to $2nF_{ST}$, where $2n$ is the total number of genes sampled for all populations and F_{ST} is taken to be calculated with values weighted by sample sizes. This chi-square test statistic has $r - 1$ degrees of freedom. Similar procedures lead to contingency test statistics for k alleles, which then have $(k - 1)(r - 1)$ degrees of freedom. In addition, if loci may be treated as independent, then multiple locus data can be summarized by adding the chi squares across loci, and also adding their degrees of freedom (Workman and Niswander 1970).

The preferred estimator of Wright's F_{ST} was developed by Weir and Cockerham (1984). The statistical derivation of this estimator is based on correlation coefficients and this allows unambiguous extensions to multiple alleles and loci. Although the distribution of the estimator is unknown, methods are given for resample estimates of the variance. If there are multiple loci, it is preferable to resample over loci rather than populations. Weir and Cockerham (1984) conducted simulations that showed that at least under some circumstances their estimator is less biased than the usual one (equation 4.22). Their estimator can be extended to multiple levels of hierarchy.

Under the specific conditions of Wright's island model there is a relationship between F_{ST} and the product Nm of the size of the populations N and the migration rate m:

$$F_{ST} \approx \frac{1}{1+4Nm} \tag{4.23}$$

(Wright 1951). Technically, this assumes that there are infinite numbers of populations, and the system is not spatially explicit (because there are uniform migration rates m), but systems with finite numbers of populations have similar form. It is popular to substitute estimators of F_{ST} into equation 4.23 and solve for $4Nm$, even though remarkably little is known about the properties of such estimators of the combined rates of migration and drift.

Rogers and Eriksson (1988) derived estimators for the effects of several forms of genetic drift in migration models, based on theoretical relationships developed in Rogers (1988), which considers genetic drift before, during, and after migration. Only drift after migration was considered by Wright (1951). Drifts prior to and during migration are considered stochastic migration effects. The interested reader should consult these two papers for details because the estimation formulas are too complex to write here. Note, however, that stochastic migration effects, especially when kin-structured, alter the relationship between N, m, and F_{ST} suggested by equation 4.23. Chesser and Baker (1996) analyzed the effects of differences in paternal and maternal effective population sizes.

OTHER ESTIMATORS

Other statistics are specifically designed to use additional information that may be contained in DNA sequence data or special types of data. Analyses of allele frequency data are variously based on the infinite alleles mutation model, the k-alleles model, or no mutation model at all, and these can be applied to DNA sequence data and other haplotype data. The other methods are typically based on the infinite sites mutation model. The few existing investigations (Ewens 1974) indicate that there will not be more information for DNA data in a system of populations unless the total number of individuals in the system is very large. It is often unlikely that there are many, sufficiently high-frequency (i.e., likely to be sampled) alleles that have multiply mutated relative to others still segregating in the system.

One method, developed by Slatkin and Maddison (1989) uses sequence data to estimate a gene genealogy for nonrecombining sequences (such as mitochondrial and chloroplast DNA), then, assuming that the tree is correct, finds the minimum number of migration events that must have occurred. All genes must coalesce at some time, and they must be in the same population to do so. Such methods do not use geographic information in reconstructing the phylogeny. This is technically incorrect, but it may be adequate if the coalecences of the genes occur on a longer time scale than the migration processes. In other words, the mutation process must be much slower than the migration process, and the probability of a coalescence among variants much smaller than the probability of migration (Nordborg 1997).

Although the coalescence theory of Kingman (1982a,b) configured the coalescences for groups of genes, most of the coalescence theory is in terms of pairs of genes, or can be collapsed to a pairwise model (Hudson 1990). Usually a haploid population model is used, with no selection and, importantly, usually no recombination. Under the infinite sites mutation model the number of mutational differences S between two genes is approximately distributed as a Poisson variable with parameter $\mu E[T]$. This result is based on the assumptions that all pairs of genes are random samples of some process and that the lengths of times between new mutation events have exponential distributions. We will consider here that the data are DNA sequences with nucleotide point mutations, although other haploid data types could be modeled similarly. Then μ is the mutation rate for the entire sequence length and $E[T]$ is the expected coalesence time (Hudson 1990), in units of generations. Malécot (1973b) originally created geographic models that include migration effects in the calculation of pairwise coalescence probabilities. Yet most of the statistics developed for coalescence of geographically distributed data have ignored geography in calculating the pairwise coalescences, or have assumed very simple models, such as Nordborg's (1997) two-population model. In simple models of n randomly sampled genes from a single randomly mating diploid population of $2N$ genes, the expected value of S is $4N\mu\ \Sigma_i 1/i$, where the summation over i is from 1 to $n - 1$, measured in units of $2N$ generations (Hudson 1990). Hence, assuming that there cannot be two simultaneous coalescences (e.g., Fu 1997), we may use the value of S to find the estimated coalescence times among all *pairs* of genes, and this results in an estimated gene genealogy among all n sampled genes. For example, with Slatkin and Maddison's (1989) approach this phylogeny together with the locations of the sampled genes is used to estimate the minimum number of migration events. In addition, estimators for hierarchical systems of populations based on nucleotide diversity have been developed by Holsinger and Mason-Gamer (1996), and estimators for the special stepwise mutations believed to occur in microsatellite loci have also been developed. The development of structured coalescent models is currently a very active research area. Particularly noteworthy is Nagylaki's (2002) analysis of the time and place of the most recent common ancestor (see chapter 3) in the one-dimensional stepping stone migration model, using diffusion methods, and a similar work by Wakely (2001).

Bertorelle and Barbujani (1995) developed a summary measure of spatial patterns especially designed for molecular genetic data. They adopted the "Analysis of Molecular Variance, AMOVA" approach of Excoffier et al. (1992), which calculates covariances of genetic states for haplotype data. Typically this is DNA sequence data with multiple polymorphic sites, and such sites may be base pair differences, insertions, or restriction length polymorphisms (RFLPs). Usually the inferences are designed for *nonrecombining* haplotypes. The statistic of Bertorelle and Barbujani (1995) is in essence a multilocus/site measure. The analog of a single-locus Moran's I for haploid data could be configured by setting one allele to a genotypic value Z, of 1.0 and $Z = 0$ for else. Thus haplotypes could be treated as alleles, although this ignores possible additional information there may be if there is linkage disequilibrium among the polymorphic sites. In contrast, Bertorelle and Barbujani (1995) follow Excoffier et al. (1992), and construct a *set* of binary values for *each* polymorphic site. Although in general there may be many alternate states for a polymorphic site, for example, the four different base pairs (hence in this case there would be four binary indicators for each base pair site), under the infinite sites mutation assumption there can be only one mutation at each site and hence a maximum of two states, which can be coded arbitrarily zero or one, as a single binary variable. Nonetheless, let q_{ik} be zero or one reflecting the state of site k in sample haplotype i, let S be the number of sites, and n the sample size. Let the q_k be the mean values of q_{ik} in the sample, and let $Z_{ik} = q_{ik} - q_k$. The statistic developed by Bertorelle and Barbujani (1995) may be written

$$I = \frac{n\sum_i \sum_j w_{ij} \sum_{k=1}^{S} Z_{ik} Z_{jk}}{W \sum_i \sum_{k=1}^{S} Z_{ik}^2} \qquad (4.24)$$

where the usual defintions of w_{ij} and W apply. Bertorelle and Barbujani (1995) call their measures AIDAs (autocorrelation indexes for DNA analysis), because of the apparent similarity with Moran's autocorrelation statistics (equation 4.1). The statistic also bears some similarities to that of Smouse and Peakall (1999). However, it does not have the same distribution as Moran's I, even under the null hypothesis of randomization. Bertorelle and Barbujani (1995) point out that AIDA has the same expected value, $-1/(n - 1)$, but that its variance is different and is, in fact, unknown. They suggest that a test distribution should be formed by Monte Carlo randomizations of the data, and demonstrated, using articificially designed patterns, that AIDAs can detect spatial gradients. It is not obvious how the theoretical values in models of isolation by distance, whether in terms of probabilities of identity by descent or genetic correlations, are related to AIDAs. To the author's knowledge, the distribution of AIDAs under isolation by distance has not been characterized, although this could be done with computer simulations.

CHAPTER 5

Theory of Genetics as Stochastic Spatial–Temporal Processes

> A difficulty arises about the probability that a locus (or a gamete) bears a given allele a: this probability generally cannot be supposed constant, even if it is defined as an a priori probability with respect to "foundation generation" F_0 and "geographical region" R. The probability is modified in each step (from generation F_n to generation F_{n+1}) by outside immigration, mutation, and selection.
>
> —Gustave Malécot (1973b)

Studies of geographic distributions of genetic variation and past geotemporal events make reference, either explicitly or implicitly, to an underlying spatial–temporal process. For example, methods for surveying population structure, to study the fitness or selective importance of particular genes, rely on the outcome of spatial–temporal processes. Linkage disequilibrium between markers and disease genes may be caused by spatial structure, and then special methods such as the transmission disequilibrium test (Spielman et al. 1993) must be used. Studies using selectively neutral markers to measure general levels of gene flow and studies tracing past founding events, migrations, and other episodic events operate in a spatial–temporal context. Naturally, space–time processes are complex, and there are wide-ranging methods of statistical analysis (chapter 4). Most studies seek to parse off particular features of greatest interest, for example,

1. The identification of regions where changes in gene frequencies in existing spatial patterns are abrupt may reflect barriers to gene flow (e.g., Barbujani and Sokal 1991a,b), whether the underlying system is in equilibrium or not.

2. The preponderance of a particular haplotype both in a recently founded population and in only one of several potential source populations may usefully identify the latter as the source of the former, in a nonequilibrium system.
3. The use of existing spatial patterns to infer the locations of ancestral populations from long periods past, such as the "out-of-Africa" theory of the origins of modern humans.

Yet it remains unclear what conditions are required for the valid separation of particular bits of spatial or temporal patterns from other possible influences of the space–time process in which they are embedded. Migrations into a population may involve several source populations, which are also changing genetically over time.

In this chapter, the main object is the general formulation of the theoretical basis for spatial–temporal population genetics, especially processes with limited rates and distances of migration among discrete populations. While it would be ideal to have a single, master equation that incorporates all important space–time processes of isolation by distance and auxiliary processes, no such equation has yet been developed. This chapter begins with two general types of "master models" that are complimentary and particularly important and flexible. One (equation 5.2) is based on genealogies (identity by descent and coalescence), and the other (equations 5.21 and 5.37) is based on the spatial time series framework for correlations of gene frequencies. The models are unifying in the sense that the basic assumptions are minimal, and thus general features may be examined, although they still do not subsume *all* particular systems. From first principles, results for the various cases are deduced according to specified added assumptions. These results are followed by examination of space–time measures of correlations and genealogies, and then some related issues.

The most basic assumption made in this chapter is the existence of a system of discrete, identifiable populations that may exchange migrants. Systems in which the distributions of individuals are highly clumped spatially yet exhibit no clear delineation of populations or subpopulations have rarely been modeled. Moreover, what might be considered highly changeable metapopulations, again with indiscernible or minimally discernable populations, have rarely been modeled in a spatially explicit way. Hence such systems are not discussed in this chapter. The systems covered in this chapter also focus on stochastic processes. These are paramount for neutral markers. However, loci under selection also may be modeled usefully as stochastic processes, and the role of chance may remain significant despite strong selective pressures. The models require some degree of "spatial regularity" of systems, i.e., regularity in how populations interact over space (chapter 4). Any system must have some regularity in order to make sense out of it. The simplest example is where the rate of migration among populations depends primarily on the distance or lags separating them. Examined models are expressed in terms of probabilities of identity by descent, gene frequency covariances, and coalescence. Models for quantitative traits per se are not examined but existing theoretical studies indicate that spatial correlations for additive genetic components of quantitative traits exhibit very similar properties and have a similar form of decrease with distance between populations (Rogers and Harpending 1983; Lande 1991; Nagylaki 1994).

SPATIAL PROBABILITIES OF IDENTITY BY DESCENT

Precise stochastic models are available in terms of the probabilities of identity by descent among genes from the same or different populations, in a spatially regular system. The simplest view of such models is the lattice model, where it is posited that populations exist on the nodes of a regular lattice (figure 5.1). This is more of a mental crutch or visualization than a strict requirement. What is actually required is that there are well-ordered sets of populations that have migration rates according to sets of spatial lags (Hooper and Hewings 1981). An alternative view, appropriate for many systems, is where there are identifiable "first- (spatial) order neighbor" populations, "second-order neighbors," and so on. (Spatial lag structures may also be defined in terms of spatial hierarchies of populations [e.g., Sawyer and Felsenstein 1983], but then spatial stationarity is rarely maintained [chapter 4]). In contrast, for extreme cases of virtually no spatial regularity, for example, where all pairs of populations have *different* migration rates, essentially no generalized results may be obtained, although one may reiterate migration matrices (if known) to project the

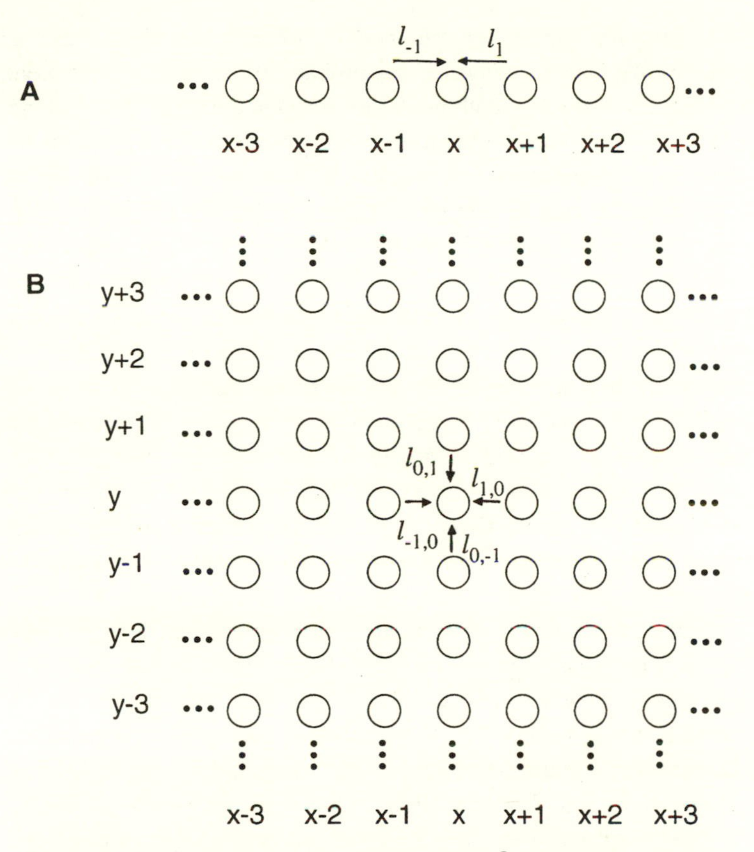

FIGURE 5.1. Lattice representations of populations (○), located on (A) a line and (B) a two-dimensional lattice. Also shown are the migration parameters for the strict stepping stone migration pattern: (A) l_{-1} and l_1; and (B) $l_{0,1}$, $l_{1,0}$, $l_{-1,0}$, and $l_{0,-1}$. For (A) x represents the location of the population, and for (B) x and y represent the two coordinates locating the populations. Arrows point in the direction of migrations for the labeled migration parameters. Figure by R. Walter.

system forward or backward (Malécot 1950, 1973b; Bodmer and Cavalli-Sforza 1968; Imaizumi et al. 1970; Rogers and Harpending 1986).

The key concept of genetic isolation by distance is the probability of identity by descent, $\varphi_n(x, w)$, between two genes, one chosen at random from one population x and the other chosen at random from another population w at generation n (e.g., Malécot 1948, 1967). The x and w can be interpreted as locators, scalars in the case of a system with one spatial dimension, vectors of coordinates in multidimensional systems. We may also consider the probability of identity by descent, $\varphi_n(x, x)$, for two genes both randomly selected from the same population x. The additional assumption is usually made that $\varphi_n(x, x)$ is the same for all populations and thus can be denoted $\varphi_n(0)$. Note that we also implicitly assume that generations are discrete and nonoverlapping. The simplest interpretation of probabilities of identity by descent is where there is no mutation, and then they are the probabilities that two genes are descended from identical genes in the ancestral population, n generations in the past. If there is mutation or other factors that change identity, then identity by descent means both that the two genes are descended from identical genes in the ancestral population and that neither have experienced mutations. If every mutation gives rise to an allele that never before existed, the key assumption of the infinite alleles mutation model (Crow 1954), then the probabilities imply identity in allelic state, because if two genes are identical by descent they are also the same allele.

In the lattice model of a system of populations, it is generally desirable to fix the migration rates as being constant in probability over generations. Thus let l_{xz} be the rate of dispersal from population z to population x, or technically the probability that a gene in population x at generation n is derived from a gene in population z in the previous generation $n - 1$ (Malécot 1975). For present purposes it is further assumed that the probabilities of migrations for each gene are independent of the others, so that, for example, the model immediately following would not apply to kin-structured migration (e.g., Fix 1978).

To develop general recurrence relationships for the probabilities of identity by descent, $\varphi_n(x, w)$, between pairs of genes, under the assumptions made thus far, the ways in which two genes are descended over a generation are considered. If two genes, one from population x the other from w, at generation n, are derived from two *different* populations, z and u ($z \neq u$), in the previous generation, then their probability of identity by descent is $\varphi_{n-1}(z, u)$. The associated probability of such event is the product of l_{xz} and l_{wu}. Summing over all possible different source populations, we have

$$\sum_{z \neq u} \sum_{u} l_{xz} l_{wu} \varphi_{n-1}(z, u)$$

Further, we must consider the probability that two genes at generation n, perhaps in different (say x and w) or the same populations, are derived from the same population z at generation $n - 1$ (the probability for this is $l_{xz}\, l_{wz}$) and they are identical by descent. This is a bit more complicated, and it may depend on the mating system within populations. The problem is completely analogous to the probability of identity by descent of two randomly chosen genes in a single-population model. Initially, we consider an "ideal" population that is monoecious random mating with fixed size N. There is a $1/N$ chance that both genes are from the same individual, and then the probability of identity by descent is $[1 + \varphi_{n-1}(0)]/2$. There is a probability of $1 - 1/N$ that they are from different individuals and then the probability of identity by descent is $\varphi_{n-1}(0)$. For each single-source population we have $\varphi_{n-I}(0) + [1 - \varphi_{n-1}(0)]/2N$, and summing over all source populations:

$$\sum_z l_{xz} l_{wz} \left[\varphi_{n-1}(z,\ z) + \frac{1 - \varphi_{n-1}(z,\ z)}{2N} \right]$$

It is possible to convert to this case other mating systems such as dioecious random mating, variation in numbers of progeny per individual, as well as various regular forms of inbreeding within populations (Malécot 1973a, 1975). In such mating systems the probability of identity by descent of the two genes within a diploid individual at generation n, f_n, differs from the probability of identity by descent of the two genes from different individuals within the same population $\varphi_{n-1}(0)$. These conversions frequently use the fact that the inbreeding coefficient can often be set equal to $\alpha + (1 - \alpha)\varphi_{n-1}(0)$, where α is the within population correlation between uniting gametes (Wright 1965, Malécot 1975). Nonetheless, the recurrence is

$$\varphi_n(x,\ w) = \sum_{z \neq u} \sum_u l_{xz} l_{wu} \varphi_{n-1}(z,\ u) + \sum_z l_{xz} l_{wz} \left[\varphi_{n-1}(z,\ z) + \frac{1 - \varphi_{n-1}(z,\ z)}{2N} \right]$$

This can be rewritten in a more compact form:

$$\varphi_n(x, w) = \sum_z \sum_u l_{xz} l_{wu} \varphi_{n-1}(z, u) + \sum_z l_{xz} l_{wz} \frac{1 - \varphi_{n-1}(0)}{2N} \quad (5.1)$$

In equation 5.1, we may substitute an effective population size $N_e = N/(1 + \alpha)$ for N. Note that in this case N_e is effective relative to identity by descent, corresponding to the "inbreeding effective number" (Crow and Denniston 1988). It is also possible to modify equation 5.1 to make it appropriate for uniparentally inherited haploid systems, such as mitochondria or chloroplast genes, and to adjust it for regular changes in migration rates, such as sinusoidal or cyclic, by making scaled adjustments to the products of the migration rates (Malécot 1973a). As noted earlier, kin-structured migration is one example where equation 5.1 cannot be appropriately modified, because the joint probabilities of two genes dispersing is not the product of the two individual ones.

Equation 5.1 must be modified if there is mutation, and with the usual assumption that the two genes mutate independently it is easy to modify:

$$\varphi_n(x, w) = (1-k)^2 \left[\sum_z \sum_u l_{xz} l_{wu} \varphi_{n-1}(z, u) + \sum_z l_{xz} l_{wz} \frac{1 - \varphi_{n-1}(0)}{2N} \right] \quad (5.2)$$

where k is the mutation rate for the infinite alleles mutation model. Equation 5.2 holds if k instead represents other "outside systematic forces," a function of reversible mutation rates or immigration from outside the system, or some forms of selection, or combinations of these (Malécot 1948).

There is little opportunity for analytical results unless there is spatial regularity. Moreover, for concreteness we will assume that the migration rates depend only on the relative locations. So-called boundary populations, populations along the margins of a system, can be dealt with by "reflecting" migration against the boundaries, or by connecting the "ends" of a system (Malécot 1948; Maruyama 1971; Felsenstein 1975), giving a circle in the one-dimension case and a "torus" in the two-dimension case. Analytical tools such as diffusion,

Fourier transform, and inversion into infinite "shocks" representations (all of these are discussed in this chapter) are favored by the use of relative locations, which may be viewed in terms of spatial lags (for lattice models) or more generally as lags of spatial order.

Building on the rather minimal assumptions made to this point, there are two different ways to proceed further toward analytical results. Most of the modeling efforts have made the additional assumption that it is only the *absolute value* of the spatial lags that matters. A violation of this assumption is illustrated by the following example: migration rates between two populations separated by two spatial units in one spatial dimension, say from x to $x + 2$ (from left to right), may differ from that from x to $x - 2$ (right to left) (Figure 5.1). This has been termed directionality within a dimension (e.g., Epperson 1993a,b). This assumption favors the analytical utility of diffusion methods and methods based on Fourier transforms. In contrast, the spatial time series method of inversion into infinite shocks (Epperson 1993b) does not require this assumption. Nonetheless, with the assumption that only the absolute values of spatial lags affects migration rates, a recursion can be found for the probability of identity by descent of two genes, from different populations that are separated by spatial lags (contained in the vector or scalar y) $\varphi_n(y)$ for generation n:

$$\varphi_n(y) = (1-k)^2\left[\sum_z \sum_u l_z l_x \varphi_{n-1}(y+z-x) + \sum_z l_z l_{y+z}\frac{1-\varphi_{n-1}(0)}{2N}\right] \tag{5.3}$$

where l_z is the migration rate from population $x + z$ to population x. The so-called stepping stone migration function (Kimura and Weiss 1964; Maruyama 1970) is a special case; for example, in the one-dimension case $l_1 = l_{-1} = m$ and $l_0 = 1 - 2m$, and dispersal probabilities are zero else. Note that $\varphi_n(y)$ is also the sum of the probabilities of the various possible coalescence events between the designated pairs of genes. A pairwise coalescent probability $\pi_p(y)$, also called kinship chain, is the probability that, counting backward from n to $n - 1$, $n - 2$, . . . , the two genes first have a common ancestor p generations earlier (i.e., at generation $n - p$). Thus,

$$\varphi_n(y) = \sum_{p=1}^{p=n} (1-k)^{2p} \pi_p(y) \tag{5.4}$$

(Malécot 1973a, 1975).

Under general conditions, if k is larger than zero the probabilities of identity by descent reach stationary distributions, as the number of generations n becomes large. Equation 5.3 becomes

$$\varphi(y) = (1-k)^2 \left[\sum_z \sum_x l_z l_x \varphi(y+z-x) + \sum_z l_z l_{y+z} \frac{1-\varphi(0)}{2N} \right] \tag{5.5}$$

The assumption that only the absolute values of the spatial lags of separation determine the migration probabilities corresponds to the *homogeneous* migration case sensu Malécot (1973a). Then the Fourier transform of the migration probability rule is $L(\alpha) = \Sigma_y \alpha^y l_y$, where $\alpha = e^{-i\theta}$ in the one-dimension case. The precise specification of α is compounded in multispatial dimension processes. The Fourier transform of the left hand side of equation 5.5 is defined by

$$K(\alpha) = F[\varphi(y)] = \sum_y \alpha^y \varphi(y) \tag{5.6}$$

Thus equation 5.5 leads to

$$K(\alpha) = \frac{1-\varphi(0)}{2N} \frac{(1-k)^2 L(\alpha) L(1/\alpha)}{1-(1-k)^2 L(\alpha) L(1/\alpha)} \tag{5.7}$$

Malécot (1973a). As long as there are bounds to the distances of migration, then $L(\alpha)$ and $L(1/\alpha)$ are rational fractions.

In the case of one spatial dimension the inversion of the Fourier transform is straightforward, because α is a single complex number in the roots of the equation:

$$\frac{1}{(1-k)^2} = L(\alpha) L\left(\frac{1}{\alpha}\right) \tag{5.8}$$

If we let σ^2 be the variance in the distance of migration, and if k/σ^2 is small, then the inversion leads to

$$\varphi(0) = \frac{1 + [1 - (2k)^{1/2}/\sigma]^r}{4N\sigma(2k)^{1/2}\{1 - [1 - (2k)^{1/2}/\sigma]^r\} + 1 + [1 - (2k)^{1/2}/\sigma]^r} \tag{5.9}$$

and

$$\frac{\varphi(y)}{\varphi(0)} = \frac{[1 - (2k)^{1/2}/\sigma]^y + [1 - (2k)^{1/2}/\sigma]^{r-y}}{1 + [1 - (2k)^{1/2}/\sigma]^r} \tag{5.10}$$

(Malécot 1973a, 1975). Equations 5.9 and 5.10 are for case of a circle of r populations. The case of an infinitely long linear system can be obtained by letting r go to infinity, giving the exponential decrease with distance:

$$\varphi(0) = \frac{1}{1 + 4N\sigma\sqrt{2k}} \tag{5.11}$$

and

$$\frac{\varphi(y)}{\varphi(0)} = e^{-y\sqrt{2k/\sigma^2}} \tag{5.12}$$

(Malécot 1948, 1973a, 1975). The same result can be found by other means, including diffusion methods. An exponential decrease is also observed for the spatial correlations of gene frequencies (figure 5.2).

Unfortunately, the two-dimension case is much more difficult. The Fourier transform is in two dimensions (i.e. "α" is α_1, α_2, etc.). The formula for the inverse of the transform involves a double integral (in the complex number plane), and no simple closed form has yet been found (Malécot 1950, 1971, 1973a). It is possible to express results in terms of Bessel functions, and to obtain approximations using elliptic integrals (Malécot 1950, 1971, 1973a; Kimura and Weiss 1964) for certain ranges of parameter values. In particular, if k is small and distance is large (and now assuming that the migration rates are the same for all directions in both spatial dimensions), then

$$\varphi(y) \sim \frac{a}{\sqrt{y}} e^{-y\sqrt{2k}/\sigma} \tag{5.13}$$

(Malécot 1948). It is important to note that this is *not* an exponential decrease with distance (y) because of the factor $y^{-1/2}$. Equation 5.13 does not give the absolute value of the probabilities of identity by descent, which requires specification of the constant a, but it does show the form of its decrease with distance for long distances of separation.

There are some available means for characterizing $\varphi(0)$ in the two-dimension case:

$$\varphi(0) \sim \frac{1}{1 + 4N\pi\sigma^2/K(\alpha)} \tag{5.14}$$

where $\sin \alpha = (1 + k/2\sigma^2)^{-1/2}$, and the Bessel function $K(\alpha)$ changes very slowly with k. When $k/2\sigma^2$ is small (say less than 0.003) the approximation in equation 5.13 is close (Malécot 1973a), but only for large y.

Similar methods can be used for systems with three spatial dimensions, in which case we have for long distances of separation

$$\varphi(y) \sim \frac{a}{y} e^{-y\sqrt{2k}/\sigma} \tag{5.15}$$

(Kimura and Weiss 1964; Malécot 1973a,b). There is a remarkable similarity of equations 5.12, 5.13 and 5.15 in terms of y^{-c}: $c = 0$ in the one-dimension case, $c = ½$ in the two-dimension case, and $c = 1$ in the three-dimensional case. This clearly shows the important principle that the rate of decay with distance increases as the number of spatial dimensions increases.

SPATIAL AND SPACE–TIME GENETIC CORRELATIONS

The probabilities of identity by descent are related to the covariances in gene frequencies among populations, which are more easily estimated using real data. Precisely, the corresponding recurrence for covariances is

$$\sigma_n(x, w) = (1-k)^2 \left\{ \left[1 - \frac{\delta(w-x)}{2N} \right] \sum_z \sum_u l_{xz} l_{wu} \sigma_{n-1}(z, u) \right\} + \delta(w-x) \frac{C - C^2}{2N} \tag{5.16}$$

where C is the equilibrium gene frequency and δ is Kronecker's delta (it equals zero except when $w = x$, i.e., $\delta(0) = 1$, zero otherwise) and at equilibrium

$$\sigma(x, w) = (1-k)^2 \left\{ \left[1 - \frac{\delta(w-x)}{2N} \right] \sum_z \sum_u l_{xz} l_{wu} \sigma(z, u) \right\} + \delta(w-x) \frac{C - C^2}{2N} \tag{5.17}$$

(Malécot 1971). In terms of spatial lags for the homogeneous migration case,

$$\sigma(y) = (1-k)^2 \left\{ \left[1 - \frac{\delta(y)}{2N} \right] \sum_z \sum_x l_z l_x \sigma(y + z - x) \right\} + \delta(y) \frac{C - C^2}{2N} \tag{5.18}$$

(Malécot 1971). Correlations can be obtained by dividing by the standard deviations, and the results for formula for the correlation functions on distance are analogous to those outlined above for the probabilities of identity by descent.

Although Malécot (1948) first developed these results, Kimura and Weiss (1964) later rederived some of the same results using a slightly different model. They assumed that migration occurs only among nearest neighbors (two neighbors in a one-dimensional system, four in a two-dimensional system), the so-called stepping stone migration model (Kimura 1953). They further ignored the possibility that two genes may come from the same population, and hence the incremental variance that may be added. A generalization of Kimura and Weiss (1964) starts with the recursions of a posteriori or "conditional" expected values (conditioned on the values of q for generation $n - 1$):

$$q_{x,n} = \left(1 - \sum_{b \neq 0} l_b - k \right) q_{x,n-1} + kC + \sum_{b \neq 0} l_b q_{x+b,n-1} + \xi_{x,n} \tag{5.19}$$

As before, each l_b (for $b \neq 0$) is a migration rate from all appropriately defined populations at $x + b$, to population x. The $\xi_{x,n}$ are the "sampling errors" associated with sampling of genes during genetic drift each generation, and thus they have zero means and variances σ^2:

$$\sigma^2 = \frac{(1 - q''_{x,n})q''_{x,n}}{2N} \tag{5.20}$$

Here $q''_{x,n}$ = the expected value of $q_{x,n}$ (conditioned on $q_{x,n-1}$ for all x), which equals the right-hand side of equation 5.19 without the term $\xi_{x,n}$, and at equilibrium the a priori or unconditioned expected value of $q''_{x,n}$ is C, i.e., the "equilibrium" value of the mean gene frequency in the system (Malécot 1971). At equilibrium, the expression in equation 5.20 has a priori expected value $C(1 - C)/2N$, and this is not necessarily the same as the variance among populations at any time period (Malécot 1973b), although the two are often confused in the literature. Analysis of equation 5.19 has also demonstrated the exponential decrease in spatial correlations with distance for systems with one spatial dimension (figure 5.2).

It is yet unclear whether it is necessary to transform for stabilizing the conditional variances, which depend on the gene frequencies (equation 5.20). This dependency can be greatly reduced by use of the arcsine square root transformation (Fisher and Ford 1947; Bodmer 1960). Malécot (1972) maintained that it is not needed for equations such as 5.18 and 5.19, and the solutions that follow, because the a priori expected value is constant among populations, and everything cancels when considering the correlations rather than the covariances. However, technically it is necessary in order to use inversion theorems from an alternative approach known as spatial time series theory, in space–time autoregressive (STAR) models, because the theorems rely on the assumption that the conditional variances are homogeneous. We shall see that such theorems are quite useful in calculating spatial and space–time correlations for general models, and that they can be further extended to systems with shared stochastic inputs (STARMA). Bodmer and Cavalli-Sforza (1968) used the transformation to good advantage in establishing the important result that moderate variations in the sizes of populations (in the de-

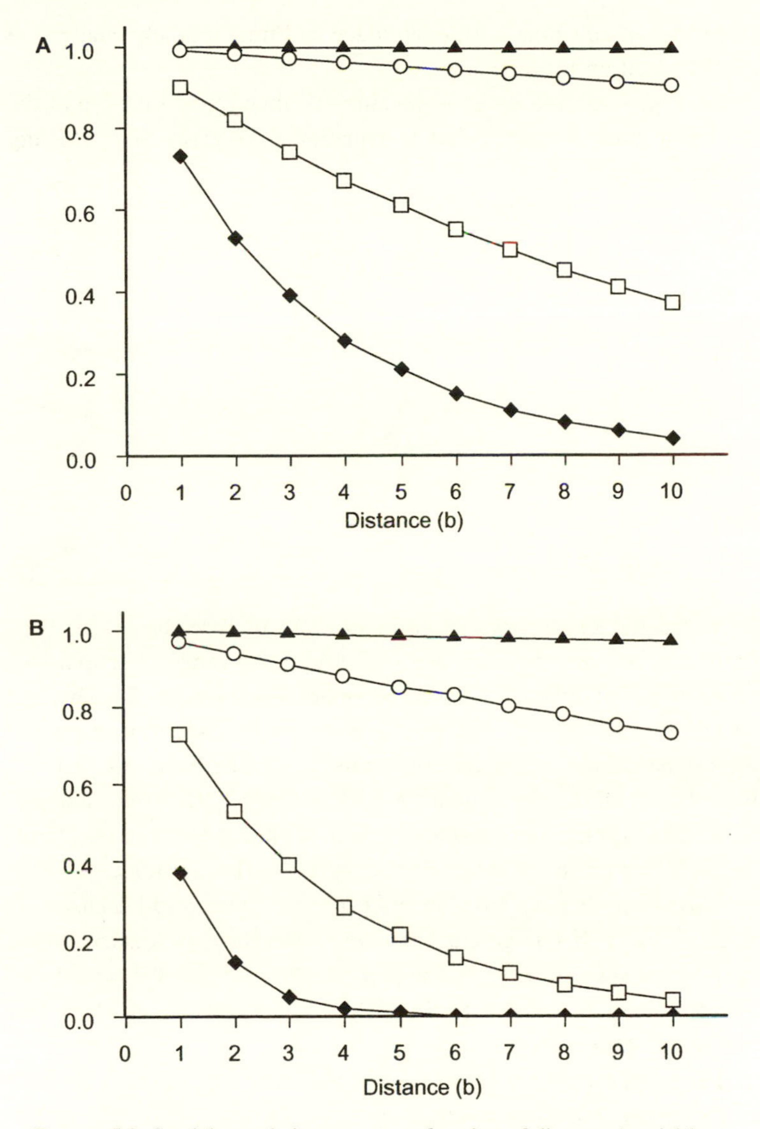

FIGURE 5.2. Spatial correlations $p_{b,0}$ as a function of distance (spatial lags b) in systems of populations in one spatial dimensions with the isotropic strict stepping stone pattern of migration: (A) $l_{-1} = l_1 = 0.1$; (B) $l_{-1} = l_1 = 0.01$. Various values of k are 10^{-7} (▲), 10^{-5} (○),10^{-3} (□), and 10^{-2} (◆). Adapted from Epperson (1993a).

nominator of equation 5.20) over space and time scarcely change the results based on mean values.

Let $z_{x,n}$ = arcsine $\sqrt{q_{x,n}}$, − arcsine $\sqrt{C}$; then transformation of the stochastic process represented in equation 5.19 gives the following stochastic equation:

$$z_{x,n} = \left(1 - \sum_{b \neq 0} l_b - k\right) z_{x,n-1} + \sum_{b \neq 0} l_b z_{x+b,n-1} + a_{x,n} \tag{5.21}$$

(Epperson 1993b). Letting l_0 =

$$1 - \sum_{b \neq 0} l_b - k$$

then

$$z_{x,n} = \Sigma \, \Sigma \cdots \Sigma l_{(b_1, b_2, \ldots, b_k)} z_{(x_1 + b_1, x_2 + b_2, \ldots, x_k + b_k), n-1} + a_{x,n} \tag{5.22}$$

Note that the summations in equation 5.22 include the vector 0 = (0, 0, . . . , 0), and they are taken over a certain range of spatial lags for each spatial dimension k, i.e., b_k ranges from $-l_{1k}$ to l_{2k}. This is a STAR process. Technically, the assumption is made that migration rates depend only on relative rather than absolute locations, but this formulation allows the migration rates to vary both within and between dimensions. The variable $a_{x,n}$ is a stochastic variable representing a "white noise" process, with expected value zero and variance $\sim 1/8N$. Note that $z_{x,n}$ has a priori expected value zero because it is mean-adjusted. If k is greater than zero, then these processes always have a stationary distribution (Epperson 1993b). Note that for sake of brevity we allow x and b to be vectors for systems with more than one spatial dimension.

The space–time covariances ($\sigma_{b,t}$) and correlations ($p_{b,t}$), as well as the spatial covariances ($\sigma_{b,0}$) and correlations ($p_{b,0}$), can be defined solely in terms of the spatial (b) and temporal (t) lags separating pairs of population in space and time (Hooper and Hewings 1981): $\sigma_{b,t} = E(z_{x,n} z_{x-b,n-t})$ for $t = 0, n$ and all spatial lag vectors b; $p_{b,t} = \sigma_{b,t}/\sigma^2_z$, where $\sigma^2_z = E(z_{x,t}^2)$ for any x and t. Note that the purely spatial and purely temporal correlations are symmetrical, $p_{0,t} = p_{0,-t}$ and $p_{b,0} = p_{-b,0}$, and some of the space–time correlations are symmetrical, $p_{b,t} = p_{-b,-t}$, but for anisotropic processes $p_{b,t}$ may not equal $p_{-b,t}$ (for $t \neq 0$, $b \neq 0$) (Epperson 1993b).

To characterize the spatial correlations for *general* processes is not simple. For example, equation 9.13 is difficult to numerically evaluate if k is small, because it involves summing very high powers of numbers very near 1.0. For gene frequencies in samples from discrete populations (but not for samples of individual genotypes within populations) this problem can be solved not only for spatial but also for space–time correlations, using methods of *inverting* stationary STAR processes into what are called infinite parameter space–time moving average (STMA) processes. Process equation 5.22 is inverted to

$$z_{x,n} = \Psi(B_x, B_n)a_{x,n} \tag{5.23}$$

(Epperson 1993b). Each $z_{x,n}$ is the sum of weighted (by $\psi_{b,t}$) random inputs $a_{x,n-t}$, from the past and present generations. B_x and B_n are backshift operators in space and time, respectively, that locate inputs (from populations) according to spatial lags and time elapsed. The individual moving average coefficients $\psi_{b,t}$ each represent the component influence of the population with spatial lags b away from x and temporal lag t away from n, on the value of $z_{x,n}$. $\Psi(B_x, B_n)$ is the generating function for the $\psi_{b,t}$ (Bennett 1979), and various algorithms may be used to calculate them for a specific process. By taking the expected value of products of equation 5.23, the variance and correlations may be found:

$$\sigma_z^2 = \sigma_a^2 \sum_m \sum_{k=0}^{\infty} \psi_{m,k}^2 \tag{5.24}$$

$$p_{b,n} = \frac{\sum_m \sum_{k=0}^{\infty} \psi_{m,k}\psi_{m+b,k+n}}{\sum_m \sum_{k=0}^{\infty} \psi_{m,k}^2} \tag{5.25}$$

Technically, the summations go to infinity, but one of the advantages of this approach is that even in cases where k is small, unlike equation 9.13, closely approximate values of $\sigma^2{}_z$ and $p_{b,t}$ can be computed with moderate summation limits in most cases.

It can also be shown, using equation 5.22, that

$$p_{b,t} = \sum_m l_m p_{b+m,t-1} \tag{5.26}$$

(except for $t = 0$, $b = 0$). In general, the summation is taken over all spatial lags m that exchange migrants, and including $m = 0$. In addition,

$$\sigma_z^2 = \sigma_a^2 \left(1 - \sum_b l_b p_{b,1}\right)^{-1} \tag{5.27}$$

where $\sigma^2{}_a = 1/8N$ (Epperson 1993b). Such equations express constraints on the space–time correlations.

Any system with one spatial dimension may be written

$$z_{x,n} = \sum_{b=-l_1}^{l_2} l_b z_{x+b,n-1} + a_{x,n} \tag{5.28}$$

where l_{-b} is the migration rate from population $x - b$ to x (figure 5.1). In the case of the strict stepping stone migration system, migration occurs only from the two adjacent neighbors, possibly with different rates, l_{-1} from the left or negative direction, and l_1 from the right or positive direction. It can be shown that at equilibrium,

$$\psi_{b,t} = l_0 \psi_{b,t-1} + l_1 \psi_{b+1,t-1} + l_{-1} \psi_{b-1,t-1} \tag{5.29}$$

for $|b| < t$, and zero for else (Epperson 1993b). By definition, $l_0 = 1 - l_1 - l_{-1} - k$. Using equation 5.29 iteratively, noting that $\psi_{0,0} = 1$ and all other $\psi_{b,0} = 0$, provides an algorithm for calculating the $\psi_{b,t}$. An example of the $\psi_{b,t}$ coefficients is shown in figure 5.3 for an anisotropic case with $l_{-1} = 0.0405$, $l_1 = 0.0810$, $k = 0.01$ (thus $l_0 = 0.8685$). For this model there is greater migration from right to left than there is from left to right on the line shown in figure 5.1A. Note that the $\psi_{b,t}$ coefficients, for $t > 1$, are larger (have more influence on $z_{x,n}$) on the right side of x than do those on the left side. The same is true for the space–time correlations of $z_{x,n}$: those for populations to the right ($b < 0$) are greater than for those to the left ($b > 0$) (figure 5.4). In fact, some of the space–time correlations with some of the past populations on the right side actually increase as the time lag increases, although they decrease as time lag increases further. Computer programs written in FORTRAN that calculate the

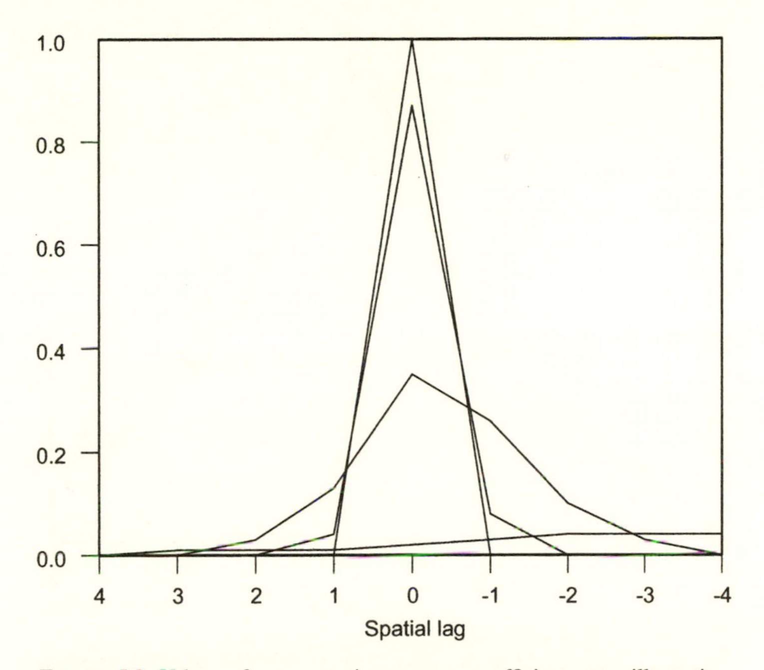

FIGURE 5.3. Values of some moving average coefficients $\psi_{b,t}$, illustrating the relative degrees of influences of stochastic drift inputs from the past, for an anisotropic strict stepping stone migration model with one spatial dimension. Migration rates from right to left ($l_1 = 0.0810$) are twice as great as those from left to right ($l_{-1} = 0.0405$). The X axis represents spatial lags b, and the four curves represent temporal lags t of 0, 1, 10, and 100 generations, respectively, from top to bottom. Note the spatial asymmetry of $\psi_{b,t}$ for $t > 1$ and the near-zero values for $t = 100$.

$\psi_{b,t}$ and $p_{b,t}$ for the specific cases of strict stepping stone migration for one- and two-spatial dimension processes are available from the author.

Finite parameter STAR models for two-spatial dimensions with temporal order one have process equations of the form

$$z_{x,y,n} = \sum_{a=-k_1}^{k_2} \sum_{b=-l_1}^{l_2} l_{b,a} z_{x+b,y+a,n-1} + a_{x,y,n} \tag{5.30}$$

Here $z_{x,y,n}$ is the transformed (and mean adjusted) gene frequency in a population with coordinates x in the first spatial dimension and y in the second dimension (horizontal and vertical, respectively, in figure

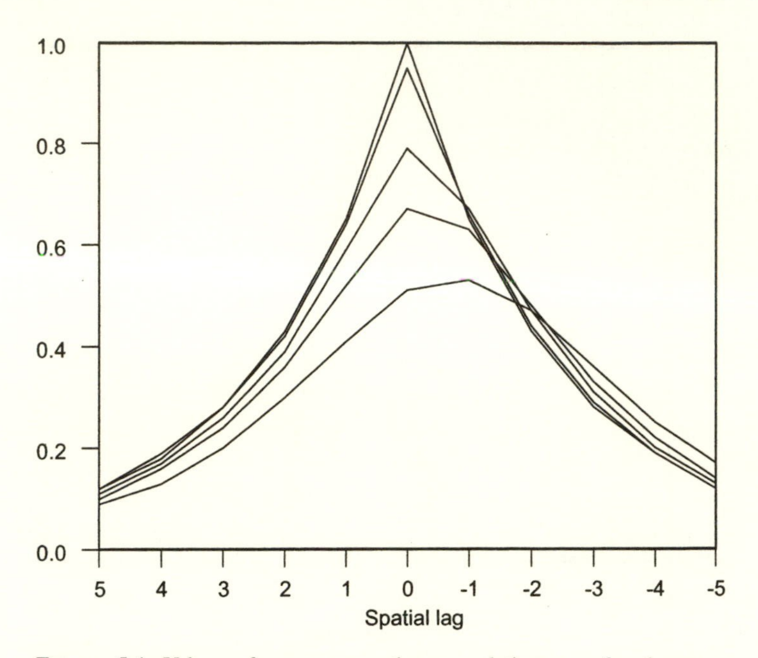

FIGURE 5.4. Values of some space–time correlations $p_{b,t}$ for the same model as in figure 5.3. The *X* axis represents spatial lags *b*, and each of the four curves represent temporal lags *t* of 0, 1, 10, and 20 generations, respectively, from top to bottom. Note the spatial asymmetry of $p_{b,t}$ for $t \geq 1$.

5.1). Parameters l_1 and l_2 are as in the one-dimensional case (i.e., the minimum and maximum lags in the first spatial dimension for populations exchanging migrants with $z_{x,y,n}$), and k_1 and k_2 are the analogous limits for the second spatial dimension.

For the strict stepping stone model migration occurs only from the four nearest neighbors; thus:

$$z_{x,y,n} = l_{0,0} z_{x,y,n-1} + l_{-1,0} z_{x-1,y,n-1} + l_{1,0} z_{x+1,y,n-1} + l_{0,-1} z_{x,y-1,n-1} + l_{0,1} z_{x,y+1,n-1} + a_{x,y,n} \tag{5.31}$$

The MA coefficients can be easily calculated by iterating the formula

$$\psi_{b,a,t} = l_{0,0} \psi_{b,a,t-1} + l_{-1,0} \psi_{b-1,a,t-1} + l_{1,0} \psi_{b+1,a,t-1} + l_{0,-1} \psi_{b,a-1,t-1} + l_{0,1} \psi_{b,a+1,t-1} \tag{5.32}$$

and noting that $\psi_{0,0,0} = 1.0$, and $\psi_{b,a,0} = 0$ for all other *a* and *b*. As an example, consider the MA coefficients for past generations (with lags greater than one), for a model with anisotropic migration rates, for example, $l_{-1,0} = 0.02$, $l_{1,0} = 0.01$, $l_{0,-1} = 0.01$, $l_{0,1} = 0.06$ and systematic pressure $k = 0.01$. There are greater migration rates from left to right (negative to positive in x) and from top to bottom (positive to negative in y) than right to left and bottom to top (figure 5.1B), respectively. Values of $\psi_{b,a,t}$ are greater for $b > 0$, and for $a < 0$. As for the one-dimensional case, the gradient of $\psi_{b,a,t}$ becomes flatter as t becomes large.

For anisotropic migration processes, the spatial correlations are always the same for opposite directions within a dimension, but they may differ between two spatial dimensions. Anisotropy in only one dimension causes asymmetry in the space–time correlations even in the same dimension (Epperson 1993b). Examples of some spatial and space–time correlations are shown in table 5.1. Notice that the space–time correlations are greater for past populations that are to the left or "above" present populations, in fitting with the MA coefficients. In cases of special interest, where migration occurs only in

TABLE 5.1. Equilibrium spatial correlations $p_{b,a,0}$ and space–time correlations for temporal lag five, $p_{b,a,5}$, for a two-dimensional strict stepping stone migration model with $k = 0.01$, $l_{-1,0} = 0.02$, $l_{1,0} = 0.01$, $l_{0,-1} = 0.01$, and $l_{0,1} = 0.06$

	$p_{4,a,0}$	$p_{3,a,0}$	$p_{2,a,0}$	$p_{1,a,0}$	$p_{0,a,0}$	$p_{-1,a,0}$	$p_{-2,a,0}$	$p_{-3,a,0}$	$p_{-4,a,0}$
$p_{b,-3,0}$	.00	.01	.02	.05	.08	.05	.02	.01	.00
$p_{b,-2,0}$	.01	.02	.04	.10	.18	.10	.04	.02	.01
$p_{b,-1,0}$	.01	.02	.07	.18	.41	.18	.07	.02	.01
$p_{b,0,0}$	.01	.03	.08	.26	1.00	.26	.08	.03	.01
$p_{b,1,0}$	.01	.02	.07	.18	.41	.18	.07	.02	.01
$p_{b,2,0}$	.01	.02	.04	.10	.18	.10	.04	.02	.01
$p_{b,3,0}$	.00	.01	.02	.05	.08	.05	.02	.01	.00
	$p_{4,a,5}$	$p_{3,a,5}$	$p_{2,a,5}$	$p_{1,a,5}$	$p_{0,a,5}$	$p_{-1,a,5}$	$p_{-2,a,5}$	$p_{-3,a,5}$	$p_{-4,a,5}$
$p_{b,-3,5}$	.00	.01	.03	.06	.10	.06	.03	.01	.00
$p_{b,-2,5}$	.01	.02	.05	.12	.22	.11	.04	.02	.01
$p_{b,-1,5}$	.01	.03	.07	.20	.46	.19	.07	.02	.01
$p_{b,0,5}$	.01	.03	.09	.26	.70	.23	.08	.03	.01
$p_{b,1,5}$	.01	.02	.07	.17	.32	.15	.06	.02	.01
$p_{b,2,5}$	.01	.02	.04	.09	.14	.08	.04	.01	.01
$p_{b,3,5}$	.00	.01	.02	.04	.06	.04	.02	.01	.00

Source: Adapted from Epperson (2000a).

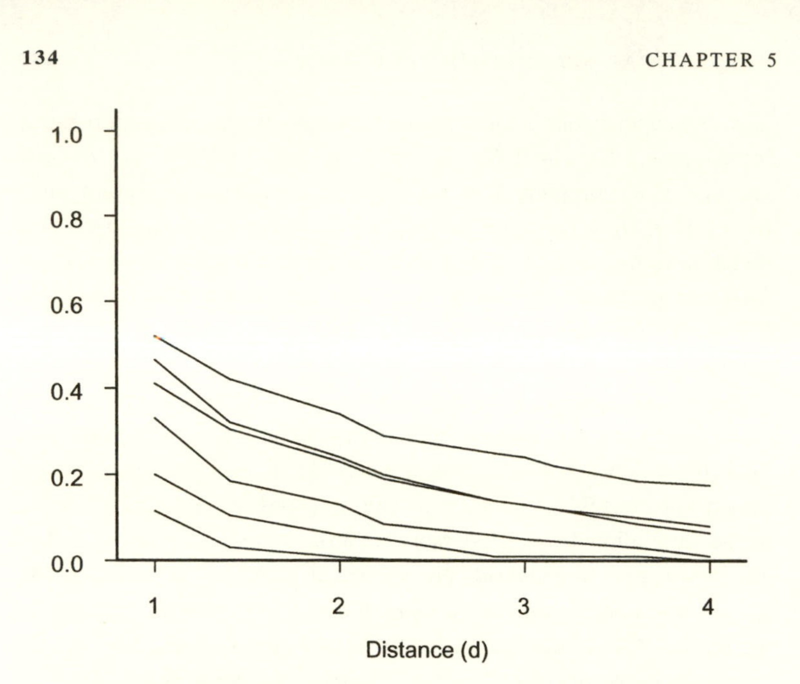

FIGURE 5.5. Spatial correlations $\rho_{b,a,0}$, as a function of distance $d = (a^2 + b^2)^{1/2}$, where a and b are the spatial lags, in systems with two spatial dimensions. All models have isotropic strict stepping stone migration, $l = l_{-1,0} = l_{1,0} = l_{0,-1} = l_{0,1}$. The six cases are (from top to bottom) $k = 0.001$, $l = 0.1$; $k = 0.001$, $l = 0.0202$; $k = 0.01$, $l = 0.1$; $k = 0.01$, $l = 0.0202$; $k = 0.1$, $l = 0.1$; and $k = 0.1$, $l = 0.0202$. Adapted from Epperson (1993b).

one direction within each dimension, the MA coefficients in one direction (for each dimension) are zero, yet the correlations for short distances remain large. Also of special interest are cases where migration occurs exclusively in one dimension, which conditions the collapse of a two-dimensional system to a one-dimensional system. The spatial and space–time correlations are zero for spatial lags in the dimension without migration, and those for the dimension with migration follow that of the one-dimensional models (Epperson 1993b).

Figure 5.5 shows how spatial correlations appear in isotropic two-dimensional systems. Greater migration rates result in greater spatial (and space–time) correlations and lower spatial variance. Greater systematic pressure strongly reduces spatial correlations as well as the spatial variance. The decreases of spatial and space–time correlations with distance do *not* appear to be particularly close to exponential, contrary to a common misconception, although they are usually monotonic or nearly so. The generally lower correlations compared to the one-dimension case can be understood by visualizing the paths of how local stochastic "disturbances" percolate faster through a two-dimensional lattice. Another view is that matrices of correlations are positive definite, which places greater constraints on the values of the correlations for the two-dimensional case (e.g., Taneja and Aroian 1980; Pfeifer and Deutsch 1980b).

Partial correlations of a space–time process can reveal definitive characteristics. Although there are alternative ways of defining partial correlation coefficients, one most convenient is to define them, in terms of the finest lag structure, as correlations between $z_{x,n}$ and $z_{x+b,n-k}$ *conditioned* on all of the spatially *and* temporally "intermediate" variables (sensu Hooper and Hewings 1981). For example, the intermediate variables for the one-dimensional case are the set of $z_{y,n-s}$ such that $y = x, x + 1, \ldots, x + b$, and $s = 1, \ldots, k$, except for $y = x + b$ and $s = k$ (where b and k are positive). Such partial correlations are zero for pairs of populations separated by more than the greatest spatial lag between populations exchanging migrants, or separated in time by more than one generation, for first-order temporal models. The immediate importance of this is that the distance beyond which partial correlations are zero in real systems indicates the spatial limits on migration (Epperson 1993b, 2000a).

Much more general models of dispersal than have hitherto been characterized could be analyzed by inverting more general STAR models. For example, thus far in this chapter we have considered only processes with temporal order one, i.e., Markovian processes. In other words, the assay period is considered discrete and depending only on the values of gene frequencies in the previous time period or generation. However, STAR methods can, in principle, be extended to any finite number of time lags or temporal orders, although it makes the construction of computational formula such as equation 5.32 more difficult. This could be important for species, such as humans, where there are overlapping generations. It may also be particularly important in those plant species that have seed banks that outlive their germinated cohorts. Such processes may cause time delays in the migrational processes that bind populations together. STAR models can also be extended to cyclic migrational patterns, where, for example, rates of migration follow seasons or other cyclic changes.

The parameter k, the recall coefficient, has the same interpretations as for the previously discussed models of the type Malécot created. The parameter could represent a *global* force of selection, mutation, immigration into the system from outside, or combinations of these. However, importantly, other, *regional*, controlling factors can also be incorporated in the spatial time series framework. To the degree that these factors (including some forms of environmental selection) can be linearized they can be included into the recursion equations. Take the example of a single environmental factor that varies according to specific populations as selectively pulling with strength r_x the population gene frequency toward some (transformed as for $z_{x,n}$) value $z_{x,\infty}$; then the recursion equations could be written

$$z_{x,n} = \left(1 - \sum_{b \neq 0} l_b - k - r_x\right) z_{x,n-1} + \sum_{b \neq 0} l_b z_{x+b,n-1} + a_{x,n} + r_x(z_{x,n-1} - z_{x,\infty}) \tag{5.33}$$

In principle, such processes could be inverted and the spatial and space–time correlations calculated for any configuration of r_x.

Equation 5.19 suggests that the gene frequencies prior to drift are (conditionally) deterministically dependent on the migration parameters. One way to generate this is to assume that the gene frequencies in groups of migrants coming into a population are exactly those in the source populations from whence they came (Epperson 1993b). There may be other ways, especially after transformation or consideration of the a priori expectations. However, it is clear that those equations (patterned after Kimura and Weiss 1964) as well as Malécot's models do assume that the individual migrants travel distances independently.

STOCHASTIC MIGRATION

Nonindependence of migrants, which cannot be modeled using equations such as equations 5.2 and 5.16, and stochastic migration can be modeled, in a spatially explicit way, in modified STAR models known as STARMA or space–time autoregressive moving average spatial time series models (Epperson 1994). STARMA processes incorporate sets of "shared" and "unshared" linear stochastic inputs (MA) that may arise from stochastic sampling in the formation of groups of migrants. Any particular source population may give rise to emigrants that are inbred or consanguineous with respect to the source population. In the simplest case emigrants may represent a small group *randomly sampled* from the source population, and thus there is additional genetic drift within this group. In other cases, the formation of a group of emigrants may not be random; it may be controlled by more systematic processes, such as emigration of a family or other kin-structured group. The gene frequencies in the group are correlated, with intraclass (intra-emigrant-group) correlation coefficient α, with respect to the source population. The genetic drift in such groups is an additional stochastic input to the system.

The precise manners with which stochastic migration events enter into the models depend on the details of how groups of emigrants are formed and how they are distributed into different migrant groups going to different populations (Epperson 1994). For concreteness, let us first consider the example where populations are distributed in only one spatial dimension, and location is indexed by an integer x along a line. As before, let $q_{x,n-1}$ denote the gene frequency in each population x at generation $n - 1$ in the adults, after genetic drift but before migration. Then, from each adult population, groups of effectively presampled or predestined emigrants may be sampled in a manner depending on the system, and the gene frequencies in the emigrant groups are $I_{x,n-1}$. We may also want to consider the gene frequencies in the nonemigrants, $NI_{x,n-1}$, particularly if the emigrants do not reproduce locally before emigrating (Epperson 1994). Other important details include the rules that govern how each migrant group going to a specific population is chosen from an emigrant group.

To further specify a clear example, let us assume that the life cycle has the following features (figure 5.6). Each migrant group will move to different populations (at locations $x - b$) each located at different spatial lags b from x, where they will form a proportion l_b of the juveniles. These, the proportion k from the outside systematic pressure, and the remainder from the resident contributions, form the

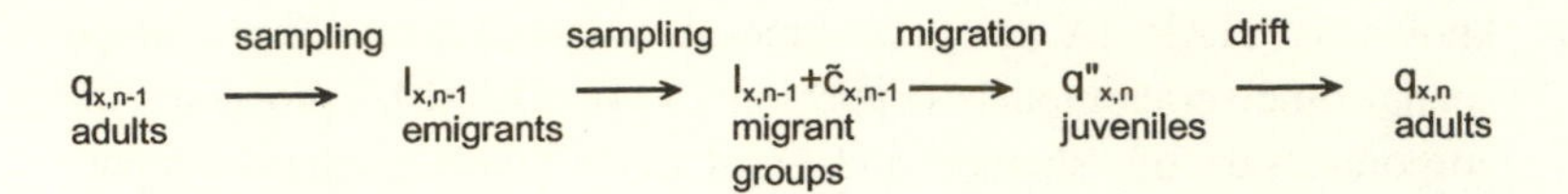

FIGURE 5.6. Model of discrete generation life-cycle. Let $q_{x+b,n-1}$ denote the gene frequency in each population x at generation $n - 1$ in the adults, after genetic drift but before migration. Then, from each adult population, groups of effectively presampled or predestined emigrants may be sampled in various manners, depending on the system, and the gene frequencies in the emigrant groups are $I_{x+b,n-1}$ (which may be equal to $q_{x+b,n-1}$, $q_{x+b,n-1}$ + unshared stochastic inputs, or $q_{x+b,n-1}$ + shared stochastic inputs). Separate migrant groups (from $x + b$ going to other populations, e.g., to x with rate l_b) form from the emigrant group (and for each group there may be additional unshared stochastic input). Following dispersal, the juveniles in the next generation have gene frequencies $j_{x,n}$ or $q''_{x,n}$. Finally, adults are sampled from juveniles, with added genetic drift and have gene frequencies $q_{x,n}$.

juveniles at x in the next generation (n), with gene frequencies $j_{x,n}$ (figure 5.6). In the strict stepping stone migration model there are only two nonzero migration rates l_1 and l_{-1}. Finally, genetic drift acts within each population to form the gene frequencies in the next adult generation, $q_{x,n}$. The general equations for such processes are

$$E(q_{x,n}) = \left(1 - \sum_{b \neq 0} l_b - k\right) q_{x,n-1} + kC + \sum_{b \neq 0} l_b q_{x+b,n-1} \tag{5.34}$$

where E is the expected value conditioned on the previous generation of the process (i.e., given $q_{x,n-1}$, for all x) (Weiss and Kimura 1965; Bodmer and Cavalli-Sforza 1968; Epperson 1993b). Equation 5.34 is as before, but the conditional variances and covariances are different. We let

$$l_0 = 1 - \sum_{b \neq 0} l_b - k \tag{5.35}$$

and the following stochastic equation obtains:

$$q_{x,n} = kC + \sum_{b} l_b q_{x+b,n-1} + \sum_{b} l_b \tilde{c}_{x+b,n-1} + \xi_{x,n} \tag{5.36}$$

(Epperson 1994). The first two terms (the deterministic components) together with the last term, constitute the usual model with deterministic gene frequencies in migrant groups (MDGF) of Kimura and Weiss (1964) and Bodmer and Cavalli-Sforza (1968). As before, $\xi_{x,n}$ is the stochastic input caused by drift within populations and it has mean zero and binomial variance. In stochastic migration models there are additional stochastic inputs, $\tilde{c}_{x+b,n-1}$, which have variances that depend on some function $f(l_b)$, and they may be shared among many populations. After subtracting off the a priori expected value C, and transforming the stochastic variables, $z_{x,n}$ = arcsine $(q_{x,n}^{1/2})$, we have the transformed process:

$$z_{x,n} = \sum_b l_b z_{x+b,n-1} + \sum_b l_b c_{x+b,n-1} + a_{x,n} \qquad (5.37)$$

These processes are STARMA processes of temporal order one. For the MDGF model the conditional variance of $z_{x,n} = 1/8N + O(1/N^2)$. The situation for processes with multiple stochastic migration inputs is more complex. The relative magnitudes of the multiple stochastic inputs $a_{x,n}$ and $c_{x,n}$ are approximately the same as are $\xi_{x,n+1}$ and $\tilde{c}_{x,n}$ in the untransformed process (Epperson 1994).

A remarkable feature about such models is that processes with all three components—genetic drift, shared stochastic migration effects, and unshared stochastic migration effects—can be treated as a combination of separate "partial" processes that can be simply "added" to models with the MDGF assumption (Epperson 1994), greatly simplifying analyses and calculations of correlations. The key feature in determining how spatial and space–time correlations are affected is how different stochastic sampling effects are shared among different populations. Stochastic migration effects that are independent for different migrant groups simply add to a modified drift input $a'_{x,t}$, still with zero means but different variances $(\sigma^2_{a'})$, and they do not affect the spatial and space–time correlations (Epperson 1994, 2000a). In equilibrium models the expected spatial correlations do not depend on N or on the fixation index within each population (Epperson 1993a,b, 1994), whereas F_{ST} depends on the variance and hence N (Slatkin and Barton 1989). Many different forms of stochastic migration can be modeled by combining shared and unshared inputs. In contrast, shared stochastic migration effects can cause dramatic changes in the shape of the functions of spatial and space–time correlations on distances of separation. This indicates the inadequacy of previous model predictions for correlations in natural systems where the MDGF assumption is substantially violated. Presumably, Malécot's models of identity by descent are also inadequate when the migrants are highly shared among receiving populations.

The above example can be extended to general patterns of migration, as long as each population exchanges migrants with a finite number of other populations. It can also be extended to systems with two spatial dimensions or three (or more) spatial dimensions, by simply recognizing the scalar locator and lag indices as (possibly vectorial) locators in multidimension space.

Examples of shared stochastic migration effects, transformed as $c_{x,n}$, include cases where emigrants from a source population represent a presampled group from the adults, and where this stochastic effect is shared in migrant groups going to different recipient populations. For example, perhaps only a small proportion of a population of a plant species may have mature seeds during a time when interpopulation dispersal is most likely, perhaps following seasonal factors. Shared stochastic migration effects appear to usually increase spatial correlations (for short distances), although they also appear to generally change the shape of the correlation function on distance, p_b. For example, p_b may not be smoothly decreasing with distance b even in the strict stepping stone migration one-dimensional model (figure 5.7). As is shown in the next section, even more dramatic effects occur when there are fissions of populations (Fix 1975, 1978, 1993) and "negatively shared" stochastic migration effects.

Stationarity for STARMA models depends only on the form of the autoregressive (AR) terms (and hence the corresponding STAR model), not the moving average (MA) terms. In the spatial time series literature $\Phi(B_x, B_t)$ is the backshift operator function for shifting spatially (generically written B_x but implying operators B_i for all dimensions) and temporally (B_t) while properly weighting by the strengths of interactions (in population genetics these are the migration rates). These should not be confused with the probabilities of identity by descent, generally denoted by φ. Stationarity requires that the inverse of $\Phi(B_x, B_t)$, $\Phi^{-1}(B_x, B_t)$, converges when $|B_t| \leq 1$ and all spatial $|B_i| \leq 1$ (Taneja and Aroian 1980). A sufficient (but not necessary) set of conditions obtains when all of the l_b have the same sign, a finite number of them are nonzero, and $|\Sigma l_b| < 1$ (Epperson 1993b). The very nature of migration models insures stationarity, and addition of shared stochastic migration effects does not change this (Epperson 1994). Invertibility into infinite STMA representations requires similar conditions for convergence of $\Theta^{-1}(B_x, B_t)$, which in the present models is $\Phi^{-1}(B_x, B_t)$ $(\Sigma\, l_b\, B_x^{-b}\, B_t)$, which has the same convergence properties as does $\Phi^{-1}(B_x, B_t)$. Thus, convergence and inversion are also a direct result of the population biological meaning of the migration parameters (Epperson 1993b, 1994).

Inversion of equation 5.37,

$$z_{x,n} = \Phi^{-1}(B_x, B_t)\left(\sum_{b\neq 0} l_b B_x^{-b} B_t\right)c_{x,n} + \Phi^{-1}(B_x, B_t)a_{x,n} \quad (5.38)$$

(Epperson 1994), shows that $z_{x,n}$ is the sum of *two* infinite moving average processes, which can be denoted $\psi^{(c)}_{b,n}$ and $\psi^{(a)}_{b,n}$. For ease of exposition, we consider below that all unshared stochastic migration inputs have been folded into $a'_{x,n}$ and we drop the prime; thus $a_{x,n}$. It is also useful to realize that $\psi^{(c)}_{b,0} = 0$ for all b (including 0), $\psi^{(a)}_{0,0} = 1$, and $\psi^{(a)}_{b,0} = 0$ for $b \neq (0, 0, \ldots, 0)$, and that for $n > 0$, $\psi^{(c)}_{b,n} = \psi^{(a)}_{b,n} - l_0\psi^{(a)}_{b,n-1}$ (Epperson 1994). For example, $\psi^{(c)}_{b,1} = l_b$ for $b \neq 0$, and $\psi^{(c)}_{0,1} = 0$. Moreover,

$$\psi^{(c)}_{n,k} = \sum_{b\neq 0} l_b \psi^{(a)}_{n-b,k-1} \quad (5.39)$$

Thus it is possible to compute all of the $\psi^{(c)}_{b,n}$ directly from the $\psi^{(a)}_{b,n}$, which in turn can be found from STAR analogues. Using the fact that the expected values of all cross-products between $c_{x,t}$ and $a_{x,t}$ are zero, it can also be shown that

$$\sigma_z^2 = \sigma_c^2 \Sigma\Sigma\ \psi^{(c)2}_{m,k} + \sigma_a^2 \Sigma\Sigma\ \psi^{(a)2}_{m,k} \quad (5.40)$$

Where $\psi^{(c)2}_{m,k}$ is the square of $\psi^{(c)}_{m,k}$, and $\psi^{(a)2}_{m,k}$ is the square of $\psi^{(a)}_{m,k}$. Thus shared stochastic migration effects, like unshared stochastic migration effects, always increase the variance of $z_{x,n}$ (Epperson 1994).

To use the analogue of equation 5.25 to calculate spatial and space–time correlations, it is convenient to partition equation 5.38 into two separate processes, one without $a_{x,t}$ and one without $c_{x,t}$, and find the infinite moving average coefficients for each (Epperson 1994). The space–time covariances and variances of these separate processes are defined as $\sigma_z^{(c)2} = \sigma_c^2 \Sigma\Sigma\psi_{n,k}^{(c)2}$, $\gamma_{l,m}^{(c)} = \sigma_c^2 \Sigma\Sigma\psi_{n,k}^{(c)} \psi_{n+l,k+m}^{(c)}$, and $\sigma_z^{(a)2} = \sigma_a^2 \Sigma\Sigma\psi_{n,k}^{(a)2}$, $\gamma_{l,m}^{(a)} = \sigma_a^2 \Sigma\Sigma\psi_{n,k}^{(a)}\psi_{n+l,k+m}^{(a)}$. Defining similarly the separate correlations by simply dividing the covariances by the variances, it can be shown that the correlations for the total process are weighted averages of those for the separate processes:

$$p_{l,m} = \frac{\sigma_z^{(c)2} p_{l,m}^{(c)} + \sigma_z^{(a)2} p_{l,m}^{(a)}}{\sigma_z^{(c)2} + \sigma_z^{(a)2}} \tag{5.41}$$

(Epperson 1994, 2000a). It is also possible to obtain $\sigma_z^{(c)2}$ from $\sigma_z^{(a)2}$, and $\gamma_{l,m}^{(c)}$ or $p_{l,m}^{(c)}$, by substituting $\psi_{n,k}^{(a)} - l_0\psi_{n,k-1}^{(a)}$ for $\psi_{n,k}^{(c)}$, giving $\sigma_z^{(c)2} = \sigma_c^2 \Sigma\Sigma(\psi_{n,k}^{(a)} - l_0\ \psi_{n,k-1}^{(a)})^2$. Note that where N is very large and $\sigma_c^2 >> \Sigma l_b/8N$, $a_{x,t}$ may be omitted. Conversely, where $a_{x,t}$ dominates, the correlations of the process are like those of a STAR process.

Examples of spatial correlations for models with and without shared stochastic inputs are shown in figure 5.7 for strictly nearest-neighbor migration cases in one-dimension models, and where migration is isotropic (i.e., both nearest neighbors contribute equal numbers of migrants). For this case,

$$z_{x,n} = \sum_{b=-1}^{b=1} l_b z_{x+b,n-1} + l_{-1}c_{x-1,n-1} + l_1 c_{x+1,n-1} + a_{x,n} \tag{5.42}$$

In general, σ_c^2 and σ_a^2 may depend respectively on the amount of shared and unshared stochastic migration effects, and σ_a^2 also depends on the drift variance. The spatial correlations ($p_{b,0}^{(c)}$) for a specific case for the separate process generated by $\{c_{x,n}\}$ only, for several sets of parameter values, are shown in figure 5.7. Spatial correlations for STAR processes ($p_{b,0}^{(a)}$ in the present context) with the same parameter values are included for comparison. In every case examined, $p_{b,0}^{(c)} > p_{b,0}^{(a)}$, at least for the small values of b included in the comparisons (Epperson 1994). Thus it appears that the addition of shared stochastic migration generally tends to *increase* spatial correlations in systems with one spatial dimension. Note that in such cases, the spatial correlations are not necessarily a strictly monotonically decreasing function of distance, and apparently this occurs when migration

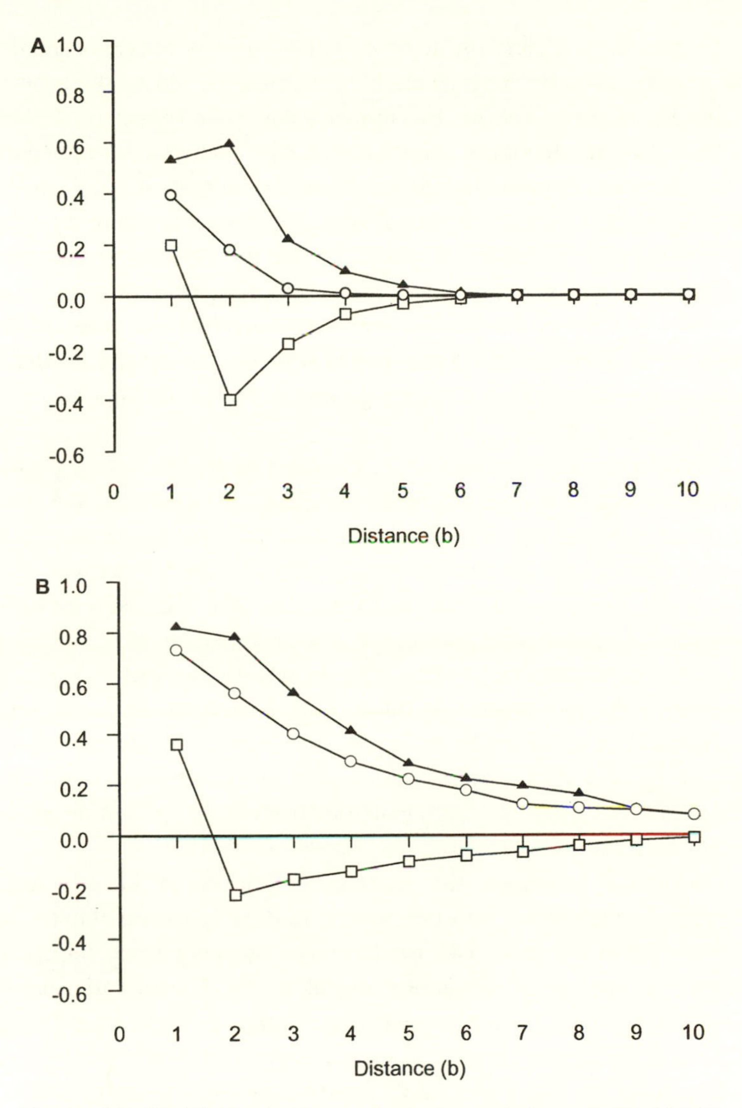

FIGURE 5.7. Spatial correlations $p_{b,0}$ between populations separated by b spatial lags for several isotropic stepping stone migration models for systems with one spatial dimension. For all models the outside systematic force k is 0.01. For (A) the migration rates are $l_{-1} = l_1 = 0.01$, and for (B) they are $l_{-1} = l_1 = 0.1$, from nearest neighbors. Results for three different types of processes are shown: one model (○) has no shared stochastic migration effects (STAR); another has only positively shared stochastic migration effects (▲); and the third (□) has only negatively shared stochastic migration effects. Adapted from Epperson (2000a).

rates are small. Clearly the autocorrelation curve is not exponentially decreasing as some geographers have claimed should be the general rule. The same is true for two-dimensional cases (Epperson 1993b; 1994). Another important feature is that $\sigma_z^{(c)2}$ increases as migration rates increase, opposite the effects of migration rates on $\sigma_z^{(a)2}$ in the STAR models. Such results are general for processes where the relative value rates of the shared effects among populations are the same as those for the migration rates. The results for partial processes can be used to form mixed process simply by adjusting the values of σ_c^2 and σ_a^2. In particular, the degree of kin structure (or equally the intraclass correlation) within emigrant groups increases the importance of σ_c^2 relative to σ_a^2.

Similar results are observed for the strict stepping stone migration model with two spatial dimensions. Figure 5.8 shows for the two-dimensional case the correlations for the STAR (or unshared effects only) models and for the completely shared partial processes. The latter are much greater than the former, and the increases are even greater for models with two spatial dimensions than those observed for the one-dimension models. This is due to the fact that $c_{x,n}$ is shared by the four nearest neighbors (which themselves are diagonal neighbors) of x but not by x. In part such effects may be due to the geometry of a perfect square lattice support. Nonetheless, stochastic migration can cause striking deviations from the paradigm of monotonic decrease for short distances (Epperson 1994).

In mixed processes, the relative importance of kin-structured stochastic migration effects can be seen in the relative weightings (by variances) in equation 5.41 for the strict stepping stone migration model in one spatial dimension (equation 5.42). The variances of gene frequencies can be computed as ratios, $\sigma_z^{(c)2}/\sigma_c^2$ or $\sigma_z^{(a)2}/\sigma_a^2$. Consider the ratio of these two variances $(\sigma_z^{(c)2}/\sigma_c^2)/(\sigma_z^{(a)2}/\sigma_a^2) = V_r$. The ratio of the weights $\sigma_z^{(c)2}/\sigma_z^{(a)2}$ can be obtained by multiplying V_r by σ_c^2/σ_a^2,

$$\sigma_c^2 = \frac{1 + (2l_1N - 1)\alpha}{16l_1N} \tag{5.43}$$

where α is the correlation among members of an emigrant group (Rogers 1987). When N is large enough, $\sigma_c^2 = \alpha/8$. For sake of dis-

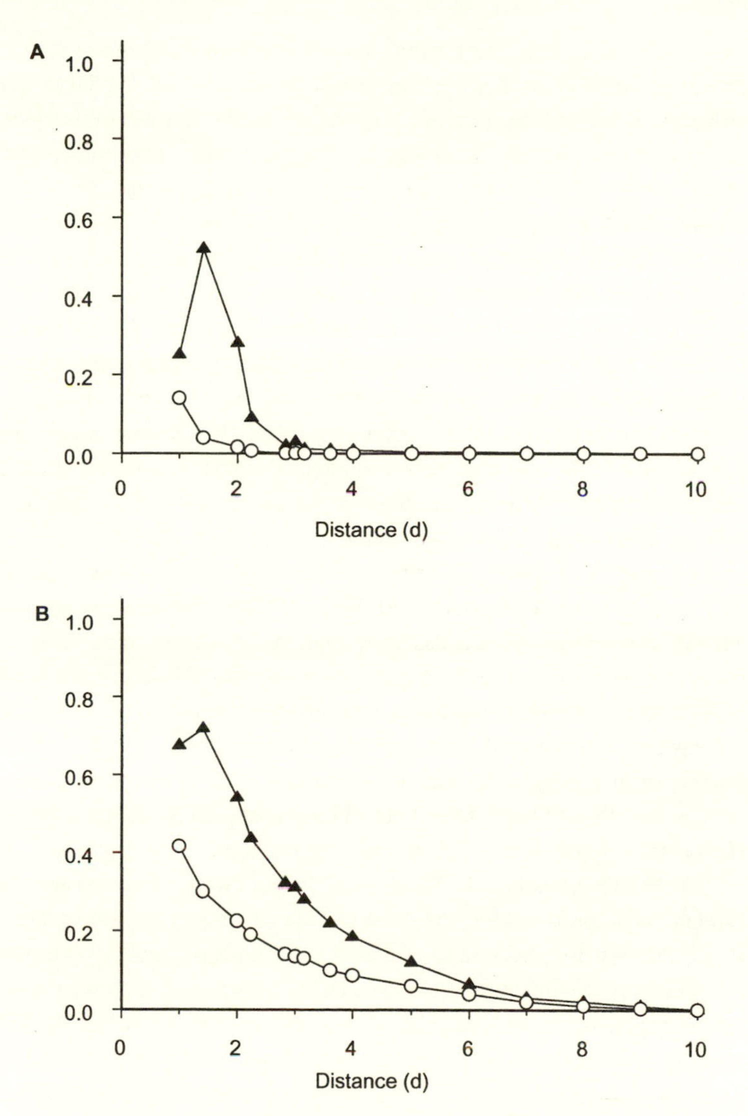

FIGURE 5.8. Spatial correlations $p_{b,a,0}$ as a function of distance $d = (a^2 + b^2)^{1/2}$, where a and b are the spatial lags in systems with two spatial dimensions. Each figure shows two different models: one (○) with no shared stochastic inputs, the usual model of migration with deterministic gene frequencies; and the other (▲) with only shared stochastic migration effects (negligible genetic drift within populations). The models have $k = 0.01$ and have completely isotropic migration rates of 0.002 (A) and 0.1 (B). Adapted from Epperson (1994).

cussion let us assume the kinship among survivors of genetic drift is moderate (thus α in these groups must be very small for large N), and drift is effectively random, i.e., $\sigma_a^2 = 1/8N$. If unshared effects are added to represent sampling of migrants from emigrants, then $\sigma_c^2 >> \sigma_{a'}^2$, because $\sigma_{a'}^2 = (1 + 2l_1)/8N$. The value of V_r is roughly $N\alpha$ if α is not too small and is $1/2l_1$ if $\alpha = 0$. Thus modest kin structure in emigrants causes shared stochastic migration to have greater effect than unshared inputs when $N\alpha$ is greater than the ratio V_r. In this case, the correlations will be more like those for the shared stochastic migration partial process. The value of α can approach 1.0 in cases where the emigrant group has small size. In such cases even when population sizes N are relatively small the shared stochastic migration effects will dominate (Epperson 1994). Generally, shared effects contribute little when $\alpha = 0$; however, even then they can make a substantial contribution when l_1 is on the order of 0.1.

The models of Malécot are based on probabilities or covariances, and unlike those of Kimura and Weiss (1964) do not require the MDGF assumption, which assumes that the stochastic input takes a particular form. The effects of stochastic migration in equations such as equation 5.43 was explicitly examined in work by Rogers and colleagues. They found that the order of stochastic inputs before, during, or after migration was not important as long as the migration events are independent (Rogers and Harpending 1986; Rogers 1988). Hence the degree of correlations among migrants is paramount.

Like STAR models, STARMA models can be expanded to accommodate systems in which the time delays of effects of past generations, through migration and resident contributions, are greater than one generation (Epperson 1993a,b). Other forces and factors of evolutionary genetic processes can also be modeled in generalized STARMA models. If selection acts approximately linearly to change the mean gene frequencies uniformly in the populations in a system, modified STARMA models known as STARIMA (Pfeifer and Deutsch 1980e) can be developed in which selection acts as a trend in the mean. The gene frequencies can be differenced once over succeeding generations, and the differenced values are then used as the stochastic variables (z) in the STARMA framework (equation 5.37), possibly with additional linear effects from population-specific selection (environmental selection) in the STARMAR formulation (equation 5.33 with added shared stochastic inputs). In addition, variables can be differenced multiply. It is also possible to model processes where the parameters change in a cyclic fashion. In general, simple analytic results can be derived if the rates of interactions are fixed or changing regularly (e.g., cyclical). Differencing also favors the use of the STARMA framework for statistical analyses of real systems, which methods are well-developed (chapter 4), because differenced variables should be less prone to violate stationarity assumptions.

STARMA models, as extensions of autoregressive moving average (ARMA) time series models, with their spatial regularity, admit spectral or "frequency-domain" representations (e.g., Oprian et al. 1980; Aroian 1985). The spectrum treats the spatial regularity in terms of waves analogous to the treatment of temporal regularity of ARMA (and STARMA), and it provides an alternative approach to analytic results. STARMA models also admit a generalization of the Wold decomposition theorem (e.g., Hooper and Hewings 1981). In contrast, other results on multivariate ARMA (MARMA) in the time series literature have not especially addressed spatially regular systems. They are analogous to the "migration matrix" approaches discussed below.

NEGATIVELY SHARED STOCHASTIC MIGRATION EFFECTS

Negative correlations between migrant groups from the same source, which could result from various types of population fission events, can be modeled in the case of the strict stepping stone genetic model with one spatial dimension. The process can be viewed as a partial process like positively shared stochastic migration, but to avoid confusion we use $d_{x,n}$ instead of $c_{x,n}$:

$$z_{x,n} = \sum_{b=-1}^{b=1} l_b z_{x+b,n-1} - l_{-1} d_{x-1,n-1} + l_1 d_{x+1,n-1} + a_{x,n} \qquad (5.44)$$

The simplest type of fission process is isotropic ($l_{-1} = l_1$), and here the negative effects are shared equally. As before, the process is actually a combination of two independent infinite MA processes, with coefficients $\psi^{(a)}_{b,t}$ and $\psi^{(d)}_{b,t}$, respectively. The partial process $\psi^{(d)}_{b,t}$ produces very different types of MA coefficients. Although, in this isotropic case, the shock at a location has no indirect or direct effect on future populations at the same location (i.e., the $\psi_{0,t}^{(d)}$ are zero for all t), the spatial and space–time correlations exhibit striking contrasts to other processes. Typical spatial correlograms are shown in figure 5.7. The correlations are positive at spatial lag one, but highly negative at lag two, and they slowly rise back up toward zero in an apparent asymptotic approach.

MIGRATION MATRICES

Spatial correlations can be calculated for complex migration patterns (e.g., Bodmer and Cavalli-Sforza 1968) using the migration matrix approach. This approach is particularly appropriate for modeling a given real system where the actual migration rates between all pairs of populations are known or at least estimated independently, especially if the system contains a small number of populations. Patterns of migration might be quite complex, perhaps with little or no spatial regularity. Because specific migration rates are used for all pairs of populations, problems of special conditions for boundary populations are moot. The major limitation of the migration matrix approach is that because it is specific to a large set of migration parameters it is difficult to determine much generality about migration/drift systems.

Let L be an $r \times r$ square matrix with (nondiagonal) elements representing rates (zero and nonzero) of migration among all pairs of r populations in the system. Malécot (1950, 1973b) showed that at equilibrium with systematic pressure k, the corresponding matrix of probabilities of identity by descent between pairs of populations Φ is

$$\Phi = \sum_{p=1}^{n} (1-k)^{2p} L^p D L_t^p \tag{5.45}$$

where L_t is the transpose of L, and D is a diagonal matrix of elements $(1 - \varphi(x, x))/2N_x$ (Malécot 1973b). Because this formulation allows different sizes of populations, $\varphi(x, x)$ may differ among populations. If the population sizes are approximately constant for all populations ($N_x = N$) and hence so are the $\varphi(x, x)$ ($= \varphi(0)$), then

$$\Phi = \frac{1-\varphi(0)}{2N} \sum_{p=1}^{n} (1-k)^{2p} L^p L_t^p \tag{5.46}$$

(Malécot 1973b). Because the symmetrical matrix LL_t (the product of L and L_t) has the same matrix of eigenvectors as does Φ, the former can be used for finding the latter (Malécot 1973b). Imaizumi et al. (1970) showed that this approach leads to the well-known result for Wright's island model (1943) (where all migration rates are equal): $\varphi(x, y) \cong 1/(4Nm + 1)$, for $x \neq y$, and m is the migration rate.

F STATISTICS AND WRIGHT'S ISLAND MODEL

The F-statistic F_{ST} (Wright 1951, 1965), although widely used as a measure of averaged population genetic differentiation, is at best a partial measure of the information in spatial patterns of variation. If there is no spatial pattern per se to differentiation, then F_{ST} is sufficient in the sense that only a total average measure of population differentiation is needed. Rogers (1988) has clearly illustrated the limitations of F_{ST}, in showing that when there is limited migration distance or a pattern of migration among populations, resulting in a pattern of spatial genetic correlations, then F_{ST} is simply a weighted sum of the within-population correlations (divided by $q(1 - q)$, where q is the overall gene frequency in the system). Thus F_{ST} does not contain information on the correlations among populations, the information that is captured in spatial correlations.

There is an additional important problem of using F_{ST}, even for Wright's (1943, 1951) island model. Wright's island model is simply described—it presumes that the system consists of an array of discrete populations, each constant and equal in effective population size N and each receiving an equal proportion m of migrants each generation. The genetic makeup of the migrants are representative of, "drawn at random from" (Wright 1943), the entire system and this is tantamount to assuming that the rates of migration among all populations are the same. Wright's island model does not allow migration rates to depend on spatial proximity and indeed it is not spatially explicit. However, the system could be partitioned at multiple hierarchical levels, although it is probably rare that migration acts in a hierarchical fashion in real systems. F_{ST} has been assigned various definitions, but the original one by Wright (1943, 1951) was in terms of partitioned inbreeding coefficients in the island model. Thus, F_{IS} is the level of inbreeding within populations (i.e., the inbreeding coefficient of individuals relative to the population in which they live); F_{IT} is that relative to the total island system (i.e., the inbreeding coefficient of individuals relative to the total system); and F_{ST} is $(F_{IT} - F_{IS}) /(1 - F_{IS})$. Malécot (1948) expressed these coefficients in terms of probabilities of identity by descent.

Because inbreeding coefficients are generally not observable, Wright (1965) appealed to Wahlund's formula for heterozygosity in admixed populations, $H = 2q(1 - q) - 2\sigma_{ST}^2$, where q is the mean gene frequency in the system and σ_{ST}^2 is the variance in gene frequencies among populations, in a two-allele system. Wright (1965) reasoned that if "accidents of sampling" are ignored, then F_{ST} should be the variance divided by $q(1 - q)$, although he pointed out that estimates of F_{ST} were different from the theoretical values of F_{ST}. However, for the theoretical values q is the a priori or unconditional expected mean gene frequency of the stochastic process, not the actual mean. This would seem to place considerable constraints on appropriate types of genes and systems, valid largely only when the system has the value of q predicted by an equilibrium and at the same value when the system first existed. If the total gene frequency for the system has drifted the actual q will differ from the a priori expected value. There seems to be no compelling reason for dividing by observed values of $q(1 - q)$. Moreover, doing so makes F_{ST} as an estimator quite unstable, because it is sensitive to the denominator $q(1 - q)$ which in turn is sensitive to changes in q. It may be that the variance σ_{ST}^2 itself is more robust because, although the genetic drift within each population, as for binomial sampling, is dependent on $q(1 - q)$, the variance among populations also depends on the migration rates. The problems of substituting the observed mean for the expected mean are further illustrated by the quote by Malécot at the beginning of this chapter. Similar issues (discussed in chapters 8 and 9) exist for the F statistics used for studies of genetic structure within large continuous populations.

In Wright's island model with an infinite number of equal-sized populations and migration rate m at equilibrium

$$F_{\mathrm{ST}} \approx \frac{1}{1+4Nm} \tag{5.47}$$

(Wright 1951). Systems with finite numbers of populations have similar form. For loci with multiple alleles Slatkin and Barton (1989) define F_{ST} in terms of Malécot's probabilities of identity by descent through the formula

$$F_{\mathrm{ST}} = \frac{f_0 - f}{1-f} \tag{5.48}$$

where f_0 is for two genes randomly chosen from within a population and f is for two genes randomly chosen from the entire system. They used the above-discussed results of Malécot (1948) for the strict stepping stone migration model and an approximate equality of the probabilities for two genes from pairs of populations, according to specific spatial scales, to f_0. In an attempt to move beyond the island model, other models of spatial patterns in genetic isolation by distance have been developed in terms of pairwise or hierarchical measures of F_{ST}. There are results on the form of the function of F_{ST} for averages over all pairs of populations at particular distances of separation, as before building upon spatially regular system assumptions (Slatkin and Barton 1989), but again these are based on hierarchical averaging that is approximate only for special spatial scales. Such averaging seems unnecessary.

Other studies have examined hierarchical F_{ST} parameterizations in theoretical models that incorporate deviations from the migrants with deterministic gene frequency assumption, i.e., stochastic migration effects (Rogers and Harpending 1986; Rogers 1988). When the stochastic migration effects are not directly shared among subpopulations the correlations in the adults are unchanged (Epperson 1994; see also Rogers and Harpending 1986), but the variance in gene frequencies is increased, which affects F_{ST} (Rogers 1988). The magnitude of stochastic migration effects is enhanced when there is a kinship relationship or positive correlation between the members of a migrant group (i.e., kin-structured stochastic migration) (Fix 1978, 1993; Rogers 1987, 1988).

SPACE–TIME PROBABILITIES OF IDENTITY BY DESCENT

Probabilities of identity by descent between genes separated *in time* as well as in space (Epperson 1999a) provide the additional characterization needed to understand how genetic lineages trace through time and space. "Space–time probabilities of identity by descent" are like the space–time correlations of gene frequencies, in that they provide a more complete description, but they differ in being genealogical. Thus, they may allow additional insights to be gained into genealogies in structured populations, especially where past migrational patterns are to be studied. Space–time probabilities also provide the probability theory needed to interpret samples of DNA taken from different time periods, which is becoming increasingly important.

The natural extension of Malécot's definition of spatial probabilities of identity by descent between genes separated in space to the space–time domain is the probability, denoted $\varphi_{n,b}(x, w)$, that two genes are identical by descent (IBD) where one gene, Γ, is selected at random from a population at generation $n - b$ at location x, and the other, Γ', is selected at random from a population at generation n and located at w (figure 5.9). Note that for the purely spatial probabilities of IBD ($b = 0$) $\varphi_{n,0}(x, w) = \varphi_{n,0}(w, x)$. As before, let l_{zw} be the (time-forward) rate of migration from w to z that occurs over one generation. Ignoring mutation for the moment, there are three ways that Γ and Γ' may be IBD: Γ' is directly descended from Γ, in which case it is necessarily identical by descent; Γ' is not directly descended from Γ, but is descended from another individual gene in population x, and the latter gene is IBD with Γ; and Γ' is not directly descended from population x (i.e., it is descended from some other population z), but is nonetheless IBD with Γ. The forms of these probabilities are rather more complex than those for the purely spatial ones in equation 5.1. However, there are useful relationships of the space–time probabilities to the spatial ones. For sake of brevity, the assumption is made that the populations are randomly mating and

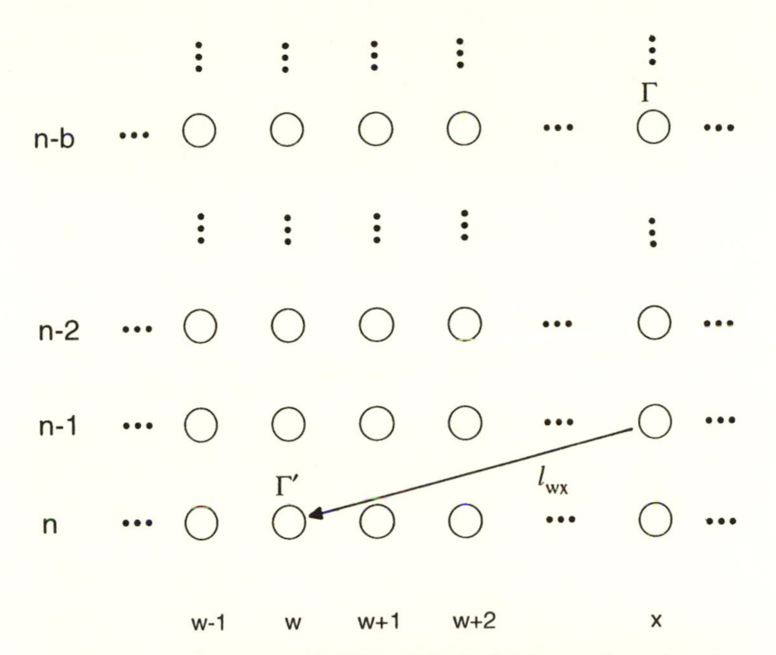

FIGURE 5.9. A space–time probability of identity by descent, denoted $\varphi_{n,b}(x, w)$, is the probability that two genes are identical by descent (IBD), where one gene, Γ, is selected at random from a population at generation $n - b$ ($b > 0$) at location x (where x is a vector of coordinates locating the population), and the other, Γ′, is selected at random from a population at generation n and located at w. The rate of migration that occurs over one generation from x to w is l_{wx}. The probability that the same two genes coalesced s generations prior to $n - b$ is denoted $\pi_{n,b,s}(x, w)$. Figure by R. Walter.

monoecious, although generality can be increased by substituting effective populations sizes in what follows. Let us first consider the equation for a time lag of one time period or generation,

$$\varphi_{n,1}(x, w) = \frac{1}{2N} l_{wx} + \left(1 - \frac{1}{2N}\right) l_{wx} \varphi_{n-1,0}(x, x) + \sum_{z \neq x} l_{wz} \varphi_{n-1,0}(x, z) \tag{5.49}$$

The one-generation lag probabilities $\varphi_{n,1}(x, w)$ can be calculated as linear functions of the spatial probabilities of identity by descent (Epperson 1999a). Similarly, if there is mutation or other systematic force of strength k, then at equilibrium (omitting the subscript for generation n),

$$\varphi_1(x, w) = (1-k)\left[\frac{1}{2N}l_{wx} + \left(1-\frac{1}{2N}\right)l_{wx}\varphi_0(0) + \sum_{z\neq x} l_{wz}\varphi_0(x, z)\right] \tag{5.50}$$

Note that the stationarity conditions are the same as those for spatial probabilities of identity by descent.

It is possible to generate general equations for higher temporal order time lags by considering the same three possible ways and incorporating all possible paths of migration in the intervening generation(s). For example, for two genes separated in time by two generations (temporal lag two), at equilibrium

$$\varphi_2(x, w) = (1-k)^2\left\{\frac{1}{2N}\sum_z l_{wz}l_{zx} + \left(1-\frac{1}{2N}\right)\sum_z l_{wz}l_{zx}[\varphi_0(0)] + \sum_{z\neq x}\varphi_0(z, x)\sum_y l_{wy}l_{yz}\right\} \tag{5.51}$$

(Epperson 1999a). For greater temporal lags such equations become rather complex. However, other useful equations can be generated, in particular those relating space–time probabilities of IBD in terms of those for the next shortest time lag, $b - 1$, e.g., without mutation:

$$\varphi_{n,b}(x, w) = \sum_z l_{wz}\varphi_{n-1,b-1}(x, z) \tag{5.52}$$

for $b > 1$. With mutation, at equilibrium,

$$\varphi_b(x, w) = (1-k)\sum_z l_{wz}\varphi_{b-1}(x, z) \tag{5.53}$$

for $b > 1$ (Epperson 1999a). Equations 5.52 and 5.53 are also linear. Most importantly, together equations 5.53 (or 5.52) and 5.49 (5.48) provide a means for calculating all space–time probabilities of identity by descent from the spatial ones.

In the case of isotropic "homogeneous" migration, in terms of spatial lags, we have

$$\varphi_1(y) = (1-k)\left[\frac{1}{2N}l(y) + \left(1-\frac{1}{2N}\right)l(y)\varphi_0(0) + \sum_{z\neq y} l(z)\varphi_0(y-z)\right] \tag{5.54}$$

or

$$\varphi_1(y) = (1-k)\left[\frac{1}{2N}l(y) + \left(1-\frac{1}{2N}\right)l(y)\varphi_0(0) + \sum_{z\neq 0} l(y-z)\varphi_0(z)\right] \tag{5.55}$$

or

$$\varphi_1(y) = (1-k)\left\{\frac{l(y)}{2N}[1-\varphi_0(0)] + \sum_{z} l(y-z)\varphi_0(z)\right\} \tag{5.56}$$

Similarly,

$$\varphi_b(y) = (1-k)\sum_{z} l(y-z)\varphi_{b-1}(z) \tag{5.57}$$

or

$$\varphi_b(y) = (1-k)\sum_{z} l(z)\varphi_{b-1}(y-z) \tag{5.58}$$

In the above equations $l(y)$ denotes the migration rate from a population at $w + y$ or $w - y$ to a population at w, and $\varphi_b(y)$ is the probability of IBD between Γ and Γ' at two different populations, separated by b generations in time and by spatial lags in the vector y in space (Epperson 1999a). This is convenient notation for developing the next equation, but elsewhere when examining isotropic systems we will use l_y to denote this migration rate. Using equation 5.54 once and iterating equation 5.57 allows us to find the space–time probabilities from the spatial ones. Alternatively, for the case of homogenous migration, the closed equations are not quite as complicated and can be written

$$\varphi_b(y) = (1-k)^b \sum_{y_{b-1}} l(y_b - y_{b-1}) \sum_{y_{b-2}} l(y_{b-1} - y_{b-2}) \cdots \sum_{y_2} l(y_3 - y_2)$$
$$\sum_z l(y_2 - z) \left\{ \frac{l(z)}{2N}[1 - \varphi_0(0)] + \sum_x l(x)\varphi_0(z-x) \right\} \tag{5.59}$$

Thus the space–time probabilities of identity by descent can be determined for any homogeneous migration case in which the spatial probabilities are known, by directly using equation 5.59.

The simplest case is that for one spatial dimension (equations 5.11 and 5.12). It is most useful to apply the Fourier transform to both sides of equation 5.54, to show that

$$K_1(\alpha) = (1-k)\left\{ \frac{L(\alpha)[1-\varphi_0(0)]}{2N} + L(\alpha)K_0(\alpha) \right\} \tag{5.60}$$

The Fourier transform $K_0(\alpha)$ is given by equation 5.7, thus

$$K_1(\alpha) = \frac{(1-k)L(\alpha)a}{1-(1-k)^2 L(\alpha)L(1/\alpha)} \tag{5.61}$$

where $a = [1 - \varphi_0(0)]/2N$ (Epperson 1999a). Similarly,

$$K_b(\alpha) = (1 - k)\ L(\alpha)K_{b-1}(\alpha) \tag{5.62}$$

and

$$K_b(\alpha) = \frac{(1-k)^b L^b(\alpha)a}{1-(1-k)^2 L(\alpha)L(1/\alpha)} \tag{5.63}$$

for $b > 1$. Analytical solutions using the inverse of the Fourier transform are very similar to those for the spatial case, because the denominators of the Fourier transforms are the same (compare equations 5.63 and 5.7), and thus the same roots are involved (Epperson 1999a). For the one-dimensional case at equilibrium,

$$\varphi_b(y) = \frac{(1-k)^b[1-\varphi_0(0)]\ L^b(\alpha_1)}{4N\sigma\sqrt{2k}} \left(1 - \frac{\sqrt{2k}}{\sigma}\right)^y \tag{5.64}$$

(Epperson 1999a). Various approximations can be made to simplify this further. For example, if k is very small compared to σ and $1/b$, then the following is a good approximation: $L^b(\alpha_1) = 1 + bk$. Also, when b is large the probability of identity by descent decreases essentially exponentially with the time lag as well as with distance of separation.

Several properties of space–time probabilities of identity by descent have been elucidated. The effect of the size of each population N is simple. N affects all the space–time probabilities in exactly the same way, at least in the one-dimension case: as N increases $\varphi_0(0)$ decreases linearly, and all of the space–time probabilities decrease by $[1 - \varphi_0(0)]/2N$. The relative values of the $\varphi_b(y)$ in space and time (for b and y not equal zero) are unaffected by N. It is also to be expected that variation of N among populations would have little effect, as is the case for spatial correlations of allele frequencies (Bodmer and Cavalli-Sforza 1968). The *relative* values are quite important because the probability that a gene in a particular population is descended from another particular population should be equal to the probability of identity by descent divided by the sum total for the entire system, assuming that the system is at equilibrium, and accounting for mutation. This provides a probability basis for the relative likelihood of populations of origin of polymorphisms.

Higher amounts of migration generally cause the purely spatial probabilities of identity by descent to be smaller for populations separated by short distances (Malécot 1975). Systematic pressures such as mutation generally decrease spatial correlations, but also cause the rate of decrease with distance to become sharper. The effects of migration rates on the space–time correlations are more complex. The behavior of space–time probabilities can be quite complex from an analytical viewpoint. For example, for short time lags the probabilities for spatial lags 0 or ± 1 may actually be greater than $\varphi_0(0)$, especially if the mutation rate is small and the migration rates are large. Similarly, the probabilities of the type $\varphi_b(0)$ tend to decrease rapidly as migration rates increase (Epperson 1999a). This can also happen at larger spatial and temporal orders. This complex behavior would not necessarily be expected from consideration of purely spatial patterns.

Perhaps the most revealing results from analysis of space–time probabilities of identity by descent is the remarkable degree of "flattening" of the probabilities function on spatial distance at long temporal lags. The probabilities of identity by descent, and hence by extension the relative likelihoods of populations of origin of genetic variants, becomes very uniform, especially when rates of mutation and migration are both high. For longer time lags b, there can be a high degree of "flatness," even when there are sharp declines in the purely spatial probabilities of identity by descent (as distance increases), which occurs particularly when mutation rates are high and migration rates are low. That is, looking at ancient generations, identity by descent decreases much more slowly with geographic or spatial distance. Such effects are greatest when migration rates are high, as would be expected. The same also occurs in space–time correlations in allele frequencies (Epperson 1993a,b). This means that if we consider a gene at the present generation at a particular geographic location, its ancestors are essentially just as likely to have been from populations within such a range of distances as from that location itself. A specific example of this was discussed in chapter 3 (figure 3.1). With a migration rate m of 10 percent and a per-sequence mutation rate k of 10^{-6}, an ancient gene from a population located 100 times more distant than the average distance between adjacent populations is approximately 82 percent as likely to share identity as is an ancient gene in the same population (location) itself, for genes from 10,000 generations ago. With $m = 0.10$ and $k = 10^{-4}$, at the present generation the value at spatial lag 100, $\varphi_0(100)$ (0.0114) is only 4 percent as large as at the origin, $\varphi_0(0)$ (0.2833), whereas for 10,000 generations ago, $\varphi_{10,000}(100)$ (0.0101) is 23 percent as large as $\varphi_{10,000}(0)$ (0.0441). It may be expected that a system of populations existing in two dimensions would show even greater flatness, as is the case for space–time correlations of gene frequencies (Epperson 1993b). Moreover, it may be that anisotropic migration could cause distant populations to be *more* likely than the same (or a nearby) population to be the ancestral source of present variants (Epperson 1993b). Theoretical analyses of space–time probabilities of identity by descent thus can also provide a means for determining the degree of certainty that may be placed on the historic geographic origins of molecular genetic polymorphisms.

SPACE–TIME COALESCENCE PROBABILITIES

The natural definition of a space–time coalescence is the probability that the two genes, one, Γ, randomly selected from a population at generation $n - b$ at location x, and the other, Γ', from a population at generation n and located at w coalesced s generations prior (time forward) to $n - b$ (Epperson 1999a). This probability can be denoted $\pi_{n,b,s}(x, w)$. This means that, looking backward from generation $n - b$, generation $n - b - s$ ($s > 0$) saw the "first" gene that is a direct ancestor of the two genes Γ and Γ'.

For the space–time coalescence probabilities, the n is not necessary, as long as $b + s$ does not exceed n, and equilibrium is sufficient but not necessary in this regard. Equations analogous to equations 5.57 and 5.54 for φ, respectively, are (for $s > 1$)

$$\pi_{b,s}(y) = \sum_z \sum_x l_z l_x \pi_{b,s-1}(y + z - x) - \sum_z l_z l_{y+z} \pi_{b,s-1}(0)/2N \quad (5.65)$$

and for $s = 1$,

$$\pi_{b,1}(y) = \sum_z l_z l_{y+z}/2N \quad (5.66)$$

where l_y denotes the migration rate from a population at $w + y$ or $w - y$ to a population at w. The same equations are found for the coalescence probabilities for two sampled genes separated in space but not time (Malécot 1975), so that the equation holds for all $b \geq 0$. However, for $s = 0$ the value is not zero as it is by definition in the purely spatial case. Because we can choose any b, we can iterate these equations from the initial time lag of 0. Thus it is possible to determine the theoretical spatial and space–time coalescence times for general systems, and to conduct coalescence analysis of data collected from different time periods. It can also be shown that

$$\varphi_b(y) = \sum_{s=0}^{n-b} (1 - k)^{2s} \pi_{b,s}(y) \quad (5.67)$$

Equation 5.67 is very similar to equation 5.4 but note that the summation includes $s = 0$ (probability that Γ' is a direct descendent of Γ). Theory for purely spatial probabilities of identity by descent has already been developed for equilibrium and nonequilibrium general models by Malécot (1948, 1972, 1973a,b, 1975) and others, and these can also be related to coalescence events between pairs of genes or gametic kinship chains (Malécot 1975).

HETEROZYGOSITY

Spatial–temporal patterns of genetic differentiation among populations are measured by either probabilities of identity by descent or genetic correlations, and the amount of variation maintained within each population is quantified by $\varphi(0)$ and similar measures. The amount of variation in the total system is also of interest. Particularly important is the temporal rate of decrease of effective heterozygosity in systems with two spatial dimensions, which can be characterized using Laplace transforms. Malécot (1975) showed that the rate of change is essentially the same as that in an unstructured system if there is more than a rather minimal amount of migration. The lower bound is $8N\sigma^2 = 10$, where N is the size of the populations and σ^2 is the variance in dispersal distances, which should be exceeded in most cases. In contrast, in the one-dimension case, the rate of decrease of heterozygosity can be considerably different (Malécot 1975). This contrast reflects the enhanced "percolation" of stochastic effects when a second dimension is added. The rate of substitution, the spread of a new mutant allele to fixation, throughout a system of populations behaves similarly to heterozygosity, and hence it too should rarely be affected by structure, at least in the two-dimension case, if migration is nonnegligible. This is particularly important when one considers the differences in usual time scales operated on by mutation versus migration. In other words, coalescence times between genes differing by a mutation are generally much longer than those caused by migration not being free throughout the system.

MUTATION MODELS

The nature of mutational processes can sometimes be critical to theoretical population genetics as well as to optimality of statistical methods. Under the infinite allele mutation model, the fundamental connection between probabilities of identity by descent and gametic kinship chains or coalescence probabilities has been derived and is exemplified by equations 5.4 and 5.67. The recursion equations for each also are closely related. Attempts to take greater advantage of modern molecular techniques have stimulated a surge in population genetics theory and statistical methods that use coalescence probabilities together with different mutation models. Moreover, kinship chains have attractive similarities to phylogenetic reconstruction probability theory. Most studies have been conducted on the expected *order* of *pairwise* kinship chains (Malécot's terminology), i.e., pairwise coalescence *times* (e.g, Hudson 1990), T. These are generally found by taking the expected value of coalescence events weighted by the number of generations p:

$$E(T) = \sum_{p=1}^{n} p\pi_p \tag{5.68}$$

There are yet few theoretical studies of coalescence in spatially explicit models of populations, although recently Nagylaki (2002) analyzed the one-dimension stepping stone model. Moreover, much of the "coalescence theory" is development of statistical methods of inference. There can be a certain temporal depth to modern molecular data, due principally to the accumulation of identifiable *multiple* mutations of some genes during the time to coalescence. Further research in theoretical models will be needed to fully exploit this.

The relevant features of the new kinds of data are exemplified by DNA sequence data and microsatellites. Although a DNA sequence dataset can be used in conjunction with the infinite alleles mutation model, it can also be analyzed under the relatively new infinite sites mutation model. Microsatellites are believed to undergo primarily stepwise mutations in fragment sizes and therefore the differences in allele fragment sizes could possibly carry added information. Moreover, microsatellite alleles frequently revert (to previously existing ones), which violates the infinite alleles mutation model. The "k-allele" models, with reversible mutations, can be used for microsatellites, but they do not use the information contained in differential fragment sizes. The k-allele models have been studied in theoretical models of structure, both in terms of probability of identity by descent and coalescence (Malécot 1973b).

The Infinite Sites Mutation Model

The infinite sites mutation model usually assumes that sites along a haplotype or DNA sequence mutate independently and infrequently, and that the probability that the same site mutates twice is infinitely small. The sites could be loci along a chromosomal haplotype (e.g., chloroplast DNA, mitochondrial DNA, and the human Y chromosome), or they could be base pairs in a DNA sequence. Although Kingman (1982a,b) developed models of coalescence for entire sets of sampled genes, they appear to be prohibitively complex in spatially explicit theoretical models. Here we are concerned only with models that either are explicitly or collapse to (e.g., Hudson 1990) pairwise coalescence probabilities, which are treated precisely as Malécot's pairwise gametic kinship chains (Epperson 1999b). If mutation rate μ for the sequence is constant and spread equally over all sites, and there is no recombination, then the number of differences S between sequences is distributed as a Poisson variable with mean $\mu E(T)$ (Hudson 1990).

The oft-cited issue of applicability of measures to modern molecular data, such as DNA sequence data, is largely irrelevant in the context of genetic structure within a single, isolated population. Ewens (1974) has shown that there is almost no additional information in the infinite sites mutation model (e.g., DNA sequence data) compared to infinite alleles models (allele frequency data) when the product $4N\mu$, or four times the mutation rate and the population size, is less than 1.0. For normal genes, most reported populations have values an order of magnitude smaller or yet less (Ewens 1974). This result is for a single, unstructured population, but it seems unlikely that structure would change this conclusion, given the apparent rarity of effects of structure on rates of substitution. In a system of r populations, it is the total number of individuals that should be counted (population size N times r), suggesting that $4Nr\mu$ would have to be greater than 1.0. This reasoning suggests that large systems of populations may have more information (in the infinite sites model compared to the infinite alleles model) than single isolated populations. Nonetheless, at least for neutral models, it is unclear that there is any added information in haplotypes (or DNA sequences) with respect to migration processes, which operates on a shorter timescale than mutation. It may be that most information about migration is contained in the allelic classes and hence in the probabilities of identity by descent or spatial correlations of gene frequencies, as is the case for isolation by distance within large populations (Barton and Wilson

1995). In other words, the infinite sites model can be collapsed into the infinite alleles model with little or no loss of information, at least for selectively neutral loci.

In cases where the spatial pattern is ignored, as in Wright's island model, the infinite sites mutation model has been compared to the infinite alleles model (Hudson et al. 1992). The method first builds a gene genealogy using the coalescence probabilities. Then using the most likely genealogy, the *minimum* number of migrants can be identified, because genes that may be in different populations presently but nonetheless coalesce must have had ancestors in the same population sometime in the past. The minimum number provides a lower bound on the rate of migration m among all populations. Slatkin and Voelm (1991) analyzed the coalescences in the hierarchical model of Sawyer and Felsenstein (1983), and described the F_{ST} at the different hierarchical levels.

Little is known about the spatial patterns of polymorphisms of (infinite) sites for DNA sequence data. Various two-population models, for which there cannot be a spatial pattern in any meaningful sense, have demonstrated the degree of differentiation that can result from mutation. For example, Nordborg (1997) utilized differences in timescales of processes (Takahata 1991) to partition gene genealogies (again using coalescence) into classes. These methods also take advantage of the fact that two genes must be in the same population (class) to coalescence, and they do so on a short timescale. Mutations occur on a long timescale. Selection can be modeled by similarly forming classes.

Microsatellites

Recurrent mutation models can be included in the both probabilities of identity by descent models and coalescence models. These models become complex if the probabilities of mutational changes are different for different alleles. In other words, the probabilities of transitions from one allelic class to another may depend on both the allele before mutation and the allele resulting from mutation. In general, infinite allele models are simpler and should usually be adequate as long as the probabilities of recreating an allele are small during the time in which that allele exists before being lost by drift. However, the variation in transition probabilities among allelic classes could under some circumstances provide additional information. Microsatellite DNA appears to have the strongest variations. Precisely, because microsatellites have numerous small tandem repeats (typically 1, 2, 3, or 4 base pairs [bp]), they tend to mutate at high rates, and frequently in a stepwise fashion. For example, an allele with 20 repeats of a 3-bp motif (i.e., 60-bp size length) tends to mutate to 19 repeats (57 bp) or 21 repeats (63 bp), and then, for example, the new 21 repeats allele tends to either revert to 20 or mutate to 22. The fidelity or strictness of such a stepwise mutation model is still being determined, and there is ample evidence that mutations can occur other than ± one repeat motif.

Slatkin (1995) showed that if the probability μ that an allele mutates and the distribution of the size changes it mutates to (assumed to have mean zero and variance σ^2) are the same for all alleles, then the expected difference in the size of two genes is equal to $2\mu T\sigma^2$, where T is the expected coalescence time. The average squared length difference among pairs of genes may be defined for genes within a population S_W as well as for pairs of genes each from different populations S_B. In principle, the latter could be defined for pairs of populations in a spatially explicit, genetic isolation by distance model, but efforts have been directed toward Wright's island model, for which

$$\frac{E(S_B) - E(S_W)}{E(S_B)} = \frac{E(T) - E(T_0)}{E(T)} \tag{5.69}$$

where E is the expected value, T is the average coalescence time for two genes from any two populations, and T_0 is the average coalescence time for two genes from the same population (Slatkin 1995). Various estimators have been developed based on such results. Michalakis and Excoffier (1996) have examined the relationships of these to other estimators of average population differentiation.

CLINES

Real examples of clines shown in chapter 2 and exemplified by "rules" of phenotypic variations along latitude can require very little theoretical modeling when migration rates are low and selection is strong. Then the spatial pattern of genetic variation essentially reflects the pattern of environmental factors. Nonetheless, there are a few additional considerations that arise from the mode of genic action (Endler 1977). There are a number of theoretical reasons, to some degree grounded in observations, that under some circumstances heterozygotes may be more adaptable to varied environments and as a result have relatively high fitness over a wider range of environments than do the homozygotes. If the advantage of the heterozygotes is great, the cline may not depend much on the level of gene flow (Endler 1977). Heterozygotes may maintain their advantage even at the ends of the gradient, and overall there is balanced polymorphism. By contrast, if the heterozygote has intermediate fitness (additive or partial dominance gene action), clinal selection may cause overdominance only near the centerpoint of the gradient (Endler 1977). The steepness or conversely width of a cline is critical. In many cases, the scale and slope of the selection gradients may be normalized with respect to the distance and rates of migration. For example, let l be the standard deviation of dispersal distances and s a measure of the steepness of the selection gradient, then the ratio $l_c = l/\sqrt{s}$ captures many of the relevant considerations (Endler 1977). Other types of cline models include frequency-dependent selection and gradients that shift over time.

Sokal et al. (1989b) simulated a variety of spatially explicit models of environmental selection combined with significant gene flow. Although these simulations were based on spatially distributed individuals within population, apparently the general principles should also apply to spatially distributed populations. They modeled smooth gradients, where the strength of selection varied, but always directional against one allele. At short distances, there was little change from the large autocorrelation values that occur for neutral loci; however, the correlations for long distances had much larger negative values (Sokal et al. 1989b). More dramatic effects might be expected where the direction of selection is opposite on opposite ends of the gradient (Endler 1977).

Fix (1994) examined the spatial structures under the combined effects of clinal selection and kin-structured stochastic migration, in sets of simulations designed to mimic anthropological genetic processes. Several models with relatively small numbers of populations, small population sizes, and relatively high migration rates were simulated in systems of populations with one spatial dimension. The results indicate that, at least for one-dimensional stepping stone migration models (for which localized stochastic fluctuations percolate more slowly compared to two-dimensional processes), such conditions can cause considerable stochastic variations. Although loci under strong clinal selection generally produce the expected correlograms, some will not, simply by chance; in particular, the correlations for intermediate to long distances were not always highly negative (Fix 1994).

In addition to theory for clines, other important forms of environmental selection include the models of environmental patches (Sokal et al. 1989b). One cautionary result is that in some cases the spatial autocorrelations caused by selection can be nearly the same as for neutral loci, if the size of the environmental patches are similar to the genetic patches expected for neutral loci. In addition, the effects of the relative size of environmental patches can be deduced in some extreme cases. For example, if environmental patches are spatially scaled much smaller than both the distances among nearest-neighbor populations and the typical distances of migration, then selection should act such that the overall fitness of genotypes in the system may be more or less independent of the spatial pattern of environments, and dependent only on the relative total areas of the different types of environment patches (Epperson 1992). Another extreme type of case is where environmental patches are scaled much larger than the distances among populations and distances of migrations. In such cases, the spatial pattern of genetic diversity should evolve to that dictated by the fitness of genotypes in each environmental patch. The shape found in spatial patterns of genetic variation should correspond to the shapes of spatial distributions of the operative environmental factor. In this context, it is particularly illustrative to mention the properties of autocorrelation statistics calculated for patterns artificially generated in studies by Sokal and Oden (1978a) and Sokal (1979a). They found that when there are several patches of variable size, the *X* intercepts (estimate of patch size) of *I* correlograms are nearly equal to the average diameter. Rectangular or irregularly shaped patches have *X* intercepts closer to the smaller dimension. Helical and ridge-shaped clines result in large positive correlations at short distances and negative correlations at long distances, similar to the unidirectional clines ("inclined plane" Oden and Sokal 1986). Small patches may cause localized autocorrelations, and may be detected using local indicators of spatial association, LISAs (Sokal et al. 1998a,b).

Another important type of selection is where environmental variations have no consistent effect on fitness of genotypes, yet the genotypes are under natural selection. In other words, the fitness of genotypes in populations does not depend on where the populations are. Such selection can be included in the classic models of Malécot, and enter into his "recall coefficient" k. The stronger the selection, modeled with larger values of k, the more the spatial correlations are reduced, compared to neutral loci (e.g., Malécot 1948). The same is true for Moran's I statistics and other spatial autocorrelation statistics (Epperson 1990a). The reduction is especially strong when the fitness differential is on the order of 0.1 (Epperson 1990a).

POPULATION EXPANSIONS

Rapid expansions of populations in the past, as well as intermittent population contractions, are particularly important temporal events that may affect inferences about space–time population genetic processes. Models of expansions and contractions are generally based on single-population models, and are not spatially explicit. Rogers and Harpending (1992) showed that the shape of the distribution of the numbers of pairwise site differences among genes, under the infinite sites mutation model, changes following an expansion. If x_i is the number of pairs of genes that differ by i mutations (i site differences), and if there is no recombination, then the expected distribution *at equilibrium* is approximately $\bar{x}_i = \bar{x}_0\,(1 - \bar{x}_0)^i$, where $\bar{x}_0$ is the equilibrium frequency of pairs of genes that are identical (Watterson 1975). Note that the equilibrium distribution smoothly decreases (geometrically) with the number of differences i. Li (1977) showed how the distribution $x_i(\tau)$ changes over time, measured in units of $\tau = 2\mu t$ (t = number of generations; μ = rate of mutation), from a posited initial distribution $x_i(0)$:

$$x_i(\tau) = \bar{x}_i + e^{-\tau(1+1/\theta)} \sum_{j=0}^{i} \frac{T^j}{j!} [x_{i-j}(0) - \bar{x}_{i-j}] \qquad (5.70)$$

(adapted from equation 4 of Rogers and Harpending [1992]). Here θ is $2N\mu$, where N is either the population size for haplotypes or the number of diploid genes, and θ is also the mean value of pairwise differences. Immediately following a sudden and very large increase in population size (i.e., θ) there are modes in the distribution, rather than a smooth decrease with i. These are called "waves" because they tend to be centered on increasingly large values of i as time proceeds (Rogers and Harpending 1992). The amounts of stochasticity in the distributions appears to be rather high, but many real data sets exhibit such modes. It would be of considerable interest to model the effects of migration from other populations on the frequency distributions of pairwise differences, and to allow for recombination.

MULTILOCUS PROCESSES

The differentiation of multilocus genotypes among populations has a number of important implications. Wright's (1943) shifting balance theory requires stages where there are locally concentrated genes at multiple loci. Such concentrations favor the combination of "new" multilocus genotypes that otherwise would be unlikely to form. For example, a multilocus genotype homozygous for a number of alleles that are rare in the system on the whole is more likely to form in a population that happens to have locally higher frequencies of these alleles. Newly formed multilocus genotypes may have a local and global adaptive advantages, and accordingly could spread throughout the system. Wright's scenario thus involves some degree of isolation among the populations of a structured system, as well as inbreeding within populations relative to the system. In such scenarios, the genes may be effectively neutral, until the favorable genotypes are formed, and in any case neutral models of multilocus genotypes reveal how much multilocus differentiation is expected in structured systems. From another viewpoint, the correlations of genes at different loci across populations, or by one measure gametic phase disequilibrium D, are important to our understanding genomic associations with traits of interest when surveying across populations or in recently admixed populations. For example, population structure and admixture can cause correlations between a marker and a disease gene, even if the two are not linked. Special tests procedures, such as the transmission disequilibrium test (Spielman et al. 1993), must be used to remove from consideration markers that are not closely linked yet are in linkage disequilibrium.

The main issues for multilocus structure of neutral genes involve how much linkage disequilibrium can be produced and, from the viewpoint of explicitly spatial or geographic patterns in spatial–temporal processes, how linkage disequilibrium depends on spatial scale. Prout (1973) showed how linkage disequilibrium can arise from overlapping spatial structures for two loci. Linkage disequilibrium D_T in a system of r populations can be generally expressed as

$$D_T = \mathrm{Cov}(p, q) + \sum_{i=1}^{r} w_i D_i \tag{5.71}$$

where $\mathrm{Cov}(p, q)$ is the (possibly weighted) covariance in the frequencies p and q for two genes respectively for different loci, and the second term is a (possibly weighted) average of the linkage disequilibrium coefficients D_i within each population i. The values of the coefficients D_i should be very small among two unlinked neutral genes in large outbreeding populations, but the covariance term may not be depending if dispersal is limited. Similar results were found by Christiansen and Feldman (1975).

Extensive results for spatially explicit systems come from a simulation study (Epperson 1995b), and although the model was a lattice model of *individual* genotypes within a population, the relevant principles should be the same. The results showed, perhaps unexpectedly, that when dispersal is highly limited there develops strong local concentrations of double homozygotes within certain-sized regions, whether the loci are linked or not. Different regions may have different homozygotes, or concentrations of different alleles. The results can be used to illustrate how spatial structure-caused linkage disequilibrium depends on the spatial scale of the sampled area. If a sample covers a small area containing mostly two patches of opposite double homozygotes (e.g., one of *aabb*, the other of *AABB*), there is near maximal linkage disequilibrium in the sample. At the other extreme, in the large simulations of 10,000 lattice points, values of linkage disequilibrium in the total system populations are very near zero for unlinked loci, and they are very small even for loci that are tightly linked (Epperson 1995b). The simulated system is large enough that patches (and concentrations) of all combinations of genotypes for each locus are likely present in approximately equal numbers, thus the correlations are of both signs and hence cancel out. The same principle should occur in systems of discrete populations. We should expect that the disequilibrium among a small number of neighboring populations will exhibit a wide variety of amounts and signs, but as the number or regional range of populations included in the total sample is increased the disequilibrium in the total approaches zero. It also appears that the recombination rate is largely unimportant, although further studies need to be done.

THEORETICAL DISPERSAL CURVES

An understanding of how mechanical factors combine to determine the probabilities of migration and the distances of dispersal could aid parameterization of migration processes. Theoretical mechanistic models of dispersal can be complicated, but one of the unifying principles is the set of conditions that give rise to specific forms, the two most important of which are the approximate Gaussian and the strict stepping stone migration model (migration only among nearest neighbor populations). The latter is the focus of this section, and because mechanistic Gaussian and similar models tend to involve short distances in continuous space, they are covered in more detail in discussions of the within-population models of dispersal in chapter 9.

It may be envisioned that movements of an individual between populations consist of either a single discrete jump movement or, more commonly, a large set of small step movements (where each step is much smaller than the distance between populations). Many factors may affect movement patterns. In some animals where movements are *active*, the factors may include innate patterns of "steps," such as the tendency to "walk" in a straight line. They may also have various "search behaviors." In many other animals and in plants transport may be "passive," depending on winds and turbulence, water runoffs, etc. Plant propagules that are transported by animals may have patterns of movements similar to those of the animals, with possibly additional complications. In addition to movement, once a potential population is encountered, there may be other factors determining whether the individual is deposited there, or moves on and either is lost or succeeds to reach another population, and so on.

One relatively simple migration rule may occur when animals are so adept at finding populations that they always do so, and they always know about neighboring populations, hence once they decide to migrate, they move first to one of the adjacent populations. If they choose to stay in this population, then this behavior equates with the strict stepping stone model of migration. Alternatively, they may have some propensity to continue to move, and as a result the migration probability distribution might fit a geometric progression. Other models of animal dispersal are based on random walk models, where individuals make series of short steps, before reaching a potential recipient population. The simplest case is the "drunkards" walk, where each step has equal unit distance, random direction, and is independent of previous ones. If the size λ and length of time t taken for each step are such that the limit of λ^2/t as λ and t go to zero exists (say equal to $2D$ [Okubo 1980]), then the discrete space and time model can be approximated by a diffusion process with coefficient $D = \frac{1}{2}\delta\sigma^2/\delta t$. Other modifications may incorporate patchy environment and population size growth and be extended to multiple spatial dimensions (Okubo 1980). On the other hand, if the time T during which small steps are correlated is much larger than t, i.e., the correlation extends over times during which an animal is very likely to "pass" by many locations, then a wave equation results.

CHAPTER 6

Synthesis: Tying Spatial Patterns among Populations to Space–Time Processes

> Frequently . . . the observed patterns are the results of an historical process for which records are generally unavailable. For this reason the methods to be applied have to be inferential and the processes arrived at by means of deduction from the available evidence. . . . The nature of the inferences will differ depending on the model that underlies the studied variables.
>
> —Robert Sokal (1986)

The subjects of chapters 2–5, observed spatial patterns, spatial statistics, and space–time process theory, are interrelated and several lines of reason can be traced. Several conclusions can be made using the introduced technical terms. First and foremost, it is clear that many important subjects in evolutionary genetics, ecological genetics, conservation genetics, medical, and other aspects of human genetics imply spatial–temporal context. We may generically view studies of these subjects as attempts to parse off especially interesting aspects of space–time processes. The set of conditions under which such parsing may be done validly is nearly as complex as space–time processes themselves. Precisely how a subject of population genetics attempts to do this will drive data considerations, choices of statistical methods, and appropriate theoretical models. Moreover, we need to consider that there may be available information that is ancillary to spatial or spatial–temporal genetic data. It is also clear that issues of spatial and temporal scale and averaging over space, time or genetic loci, are critical.

Some typical population genetic questions are

What is the distribution of genetic variation, within vs. among populations?

What is the spatial pattern of genetic variation?

What are the averaged migration rates among pairs of populations?

What do patterns reveal about natural selection, particularly environmental selection?

What do patterns tell reveal about ancient events?

Statistical issues include methodology for measuring spatial patterns and estimation of the relevant parameters of appropriate models. Modeling issues include the circumstances and timescales over which more complex migration processes average out in the sense of producing standardizable genetic isolation by distance (drift and migration) patterns during the same time period. Another key modeling issue is whether the system is in equilibrium, i.e., temporally stationary. It is also sometimes important to consider the promise of space–time data, which in principle resolve many statistical and modeling problems, but which are difficult and sometimes impossible to obtain.

Perhaps the first issue is whether it is even necessary to consider space, i.e., to work with spatially explicit models. If not, a great deal of complexity is obviated at all levels—experimental, statistical, and theoretical. Non-spatially explicit theoretical models, for example, those derivative of Wright's island model, are entirely appropriate when all populations exchange migrants with one another at nearly the same rate. They can still be complicated, for example, where populations are changing sizes or there are extinctions and colonizations, such as in metapopulation models. Standard measures of genetic diversity within and among populations are sufficient. Either the variance in gene frequencies or such measures as F_{ST}, averaged among all populations, may be used, although issues surrounding the use of the denominator of F_{ST} remain. Theoretical F_{ST} can be expressed in terms of probabilities of identity by descent. However, it is unclear what various F_{ST} estimators measure, and their relationships to theoretical parameters in models such as Wright's island models are not exact. For example, the standard estimator of differentiation, F_{ST}, is biased, in that estimated values tend to exceed the input parameters (Weir and Cockerham 1984). The distributions of the estimators F_{ST} and modified estimators are difficult to obtain for two reasons. First, the theoretical F_{ST} includes terms with the a priori expected gene frequency, which is generally unknown, and second the expected values involve ratios of two terms that have random variables, and the two terms are not independent. The θ estimator of Weir and Cockerham (1984) appears to be much less biased. If spatial context cannot be ignored, then spatially explicit methods must be used, and usually these are based in some way on measures of genetic similarity or shared ancestry among *pairs* of populations in a manner that reflects the spatial locations of each population.

Equally important is timescale. In some cases it may be valid to ignore time altogether. One important set of circumstances is where migration rates and distances are small relative to other evolutionary forces, in particular natural selection, and the system is at equilibrium. Specifically, environmental selection, where different populations have different fitness values for genotypes, depending on the environment, can determine spatial patterns of selected genes, solely so if migration is small relative to fitness differentials. Measures of spatial patterns may be directly related to the spatial distribution of fitness values. Classically, clines of gene frequencies follow environmental gradients, often with a known north–south orientation. Such patterns may be measured using standard measures of variation among populations, standard Moran I statistics and Mantel statistics, directional Moran and Mantel statistics, or trend surface analysis. Other patterns can be created by environmental selection, for example, when it occurs in the form of patchy environments. Various spatial statistics, including those mentioned above, as well as local indicators of spatial association (LISAs) and wombling, can be used in a straightforward manner to quantify the environmental patches. Theoretical models may be deterministic, and hence the connection of spatial measures to the parameters of selection may be fairly straightforward.

When population genetic questions center on change over short timescales the migration matrix approach is suitable, especially in cases where migration rates differ widely and in a highly irregular fashion. If the migration rates are known, then it is simple to create recursion equations for the spatial–temporal dynamics. In contrast, the utility of migration matrices for longer terms is questionable. If there is great spatial irregularity of migration rates, it seems unlikely that these rates would be constant over many generations. Highly irregular migration matrices and drift are of limited utility as theoretical models of the general effects of migration, because of their very specificity. Migration matrices also provide an obvious means for estimating migration rates if genetic data are available across a small number of generations (space–time genetic data). However, in practice, estimation may be difficult if populations receive migrants from many other populations, or if genetic drift or other stochastic inputs (in particular, stochastic migration) are substantial.

The temporal dimension of evolutionary genetic studies can be further considered in terms of two diametrically opposed situations. One of these explicitly relies on some sort of averaging over time, if not temporal stationarity, equilibrium, or quasi-stationarity; the other is explicitly temporally nonstationary. Much of the subject of genetic isolation by distance falls into the first category. We should not generally expect that migration rates are really fixed and constant over time. Generally viewed as a long-term process, isolation by distance is studied under the assumption that migration and drift or other stochastic inputs have averaged out spatially and temporally. Isolation by distance produces spatial patterns that are controlled by the types of stochastic inputs, and the averaged rates and patterns of migration. If the populations are distributed in a fairly regular fashion, and the pattern of migration rates among populations is primarily a function of distance, then a fairly standardizable form of isolation by distance is produced. In the simplest form, it assumes that the only stochastic input is genetic drift within populations. The spatial pattern can be rather fully characterized by how pairwise measures of similarity (or ancestry) decrease with distances separating populations. Standard forms of Moran's I statistics and Mantel statistics may be used. Naturally, in real systems migration rates may themselves be functions of complex processes, as, for example, in fusion/fission cycles of the Yanomama. However, it appears that such complexities often average out, and the spatial patterns fit standard forms of isolation by distance. The regularity of standard forms of isolation by distance also avoids problems concerning spatial stationarity.

Directionality in two-dimensional spatial patterns is often observed in nature and can be caused by various processes, including clinal selection and directionality in the rates of migration among populations. Directionality can be detected and measured using distance/direction autocorrelation statistics (Oden and Sokal 1986). If the distribution of populations is fairly regular or uniform, and the form of the directionality is simple, then available theoretical models can easily incorporate directional migration. Spatial stationarity may be maintained if spatial proximity is expressed in terms of lags in each dimension, and higher order spatial lags can be defined (Hooper and Hewings 1981). If the distribution of populations is strongly nonuniform and migration rates are complex functions of distance and direction, it is difficult to compare observed patterns to theoretical models of genetic isolation by distance, particularly because it may be difficult if not impossible to define higher order lags. This is also a strong argument that only the "finest lag structure" should be used, not hierarchical measures such as hierarchical F_{ST}, unless migration actually acts in a hierarchical fashion.

Spatial stationarity in a space–time stochastic process implies that the expected values of the gene frequencies and the correlations for the various degrees of spatial proximity between pairs of populations are the same throughout the system. For example, a pair of populations located at **a** and **b** (**a** and **b** are vectors containing the coordinates) would have correlations that depend only on some measure of the difference between **a** and **b**, for any **a** and **b**. Populations that do not directly exchange migrants are still generally correlated, because effects of migrations are mediated through spatially intervening populations. The correlations are compounded functions of the steps (migration between intervening populations) of a path of migration effects. Even if Gabriel-connectedness schemes are used, it is very unlikely that spatial stationarity may be achieved, if the spatial distribution of populations is highly irregular. Apparently, migration rates would have to be the same for all edges of the Gabriel graph, which is unlikely because the distances typically vary among edges. Spatial stationarity is not required for the validity of the spatial autocorrelation test statistics of the null hypothesis of a random distribution. However, it is required in order to have reasonably simple alternate hypotheses, and nonstationarity can complicate inferences about the amount of spatial autocorrelation. Spatial stationarity is also an important experimental component, because it allows averaging many pairs of populations, to boost statistical power and the ability to make inferences. It also plays an important role in obtainment of analytical results in theoretical models.

It is not clear what kinds of weights should be chosen for weighted spatial autocorrelation statistics, but they generally will not be directly determined by rates of migration among populations. This problem is analogous to issues regarding the relationship of purely spatial statistical models versus the underlying space–time processes in autoregressive models and issues of the noncorrespondence of statistical interactions to process interactions (chapter 4).

Stochastic migration can be another important factor in spatial distributions, particularly for measures such as F_{ST} of the overall variance among populations. Standard forms of isolation by distance specify that genetic drift occurs only during population size regulation. This is appropriate when the absolute numbers of propagules that migrate (or not) are much larger than the size N of mature populations after population size regulation. Other stochastic changes in gene frequencies can occur during the migration process itself. For example, whenever it is the (post-population size regulation) adults themselves that migrate, after genetic drift has acted, then the number of individuals in migrant groups is smaller than N. The skewing of gene frequencies within migrant groups is greater than that caused by population size regulation (Rogers and Harpending 1986; Rogers 1988). In such cases, stochastic migration substantially increases the level of variation among populations. Stochastic migration generally causes difficulties for F_{ST} and similar estimators of migration rates, because the increase it causes in the variance of gene frequencies among populations is greater when migration rates are higher. This is opposite genetic drift, where higher migration rates decrease the variance. Many real systems will have substantial levels of both stochastic migration and genetic drift, and F_{ST} will be completely unreliable as an estimator of migration rates. If migration is kin structured the effects of stochastic migration are amplified (Fix 1978; Levin and Fix

1989). Stochastic migration can be investigated by sampling multiple life stages (Rogers and Eriksson 1988), although this has rarely be done in genetic survey studies.

In contrast, spatial correlations are unaffected by stochastic migration (even if it is kin structured) as long as the added stochastic effects are independent among populations. They are changed when the gene frequencies of emigrant groups are correlated, and hence the stochastic inputs are shared by multiple recipient populations. Shared stochastic migration effects may classified as "positively" shared or "negatively" shared. Positively shared stochastic migration is probably much more common than negatively shared, and if it is also kin structured, then it will generally increase the spatial autocorrelations at short distances. Negatively shared stochastic migration effects may be rare, but they can cause dramatic changes in spatial correlations, especially reductions in values for short distances. A possible example is the fusion/fissions of Yanomama villages (chapter 2), which may be expected to have a partial effect of negatively shared stochastic migration. Many real systems of migration may have a mixture of (1) positively shared stochastic migration effects, (2) unshared stochastic migration effects and (unshared) effects from drift during population size regulation, and (3) sometimes, negatively shared stochastic migration effects. The spatial correlations will be a weighted average of the three types of processes (Epperson 1994, chapter 5).

In principle, space–time data can resolve a number of difficulties. Statistical interactions (over time lags) can be directly related to process parameters primarily of migration, for example, in the space–time autoregressive (STAR) framework. STAR models are very general and can incorporate complex patterns of migration, similar to the general migration matrix approach. However, with respect to statistical power, the more complex or idiosyncratic the migration pattern and the greater the number of parameters, the lesser the extent of spatial replication, and hence lesser precision for estimates of migration rates, etc., although this could be ameliorated by using larger numbers of genetic loci. Naturally, the collection of sufficient space–time genetic data increases total sample size, but arguably it should be the *default* of genetic surveys. With the ever-increasing ease of assaying numerous polymorphic markers, the overall expansion of sample sizes required may often be justified. But, space–time data may be infeasible in the first place. In many species the period of observation may not allow multiple generations to be observed in contemporary populations, and in others, e.g., microbes, this does not pose an insurmountable problem. Moreover, in some cases it is possible to sample populations "posthumously," using museum specimens and other archived materials. For example, for many freshwater sport fish, extensive archives of fish scales have been collected (by fisheries' management officers) that often can be genotyped for DNA markers.

Processes of genetic drift and migration, in the absence of other strong forces, exhibit considerable stochasticity in spatial patterns, spatial correlations, and levels of genetic variation among populations. While single-locus data are usually adequate for testing the null hypothesis that the spatial distribution of genetic variation is random, it appears that in most cases data for multiple loci are required to precisely quantify spatial patterns of genetic variation, to reveal patterns of migration, and to precisely estimate migration rates and other parameters. This is true even in "well-behaved" systems, such as STAR processes of genetic drift and migration. For example, the correlograms in figure 2.1 show considerable variations, even though 400 populations were sampled. Some of the correlograms in figures 2.1A and 2.1B overlap, even though the simulations had a fourfold difference in migration rates. However, when averaged over five loci the correlograms easily distinguished the migration rates. Most genetic surveys assay multiple loci, and often the effort and expense of adding sample individuals are greater than those for adding assay loci. Multiple-locus data also appear to be generally required for using wombling to identify barriers to gene flow (Barbujani and Sokal 1991a,b), in part because stochasticity in the null hypothesis patterns produced under genetic isolation by distance, without barriers, occasionally gives artifacts. Correlograms are also quite variable for single loci subjected to kin-structured stochastic migration (Fix 1994). Such stochasticity (which is the result of a single realization of a stochastic process) should be distinguished from uncertainty via stochasticity in terms of expected probabilities, such as, for example, the space–time probabilities of identity by descent. The latter may be considered systemic, and may not be ameliorated by the use of multiple loci.

While the field of theoretical modeling of population genetic isolation by distance, exemplified by the pioneering works of Malécot and Wright, established the general effects of migration and drift on expected patterns and amounts of genetic variation among populations, it has proven difficult to precisely apply theoretical models to genetic surveys of geographical patterns. Many of the difficulties center around the a priori expected value, because it is usually unknowable and differs from the actual mean observed in any system of populations, no matter the size. This causes problems with estimators, variously called "kinship" coefficients or "conditional kinship" measures. Measures based on a posteriori expected values (which are conditioned on values in the preceding generation) involve simple recursion equations. They are generally not the same as theoretical a priori kinship measures, which are always relative to some prior generation. What they actually measure is unknown, and their statistical as well as stochastic distributions are also unknown. Like standard measures of F_{ST}, their expected values and/or distributions are unknown even under the null hypothesis of panmixia throughout the system. These problems are avoided by many spatial autocorrelation statistics, for example, Moran's I statistics. Indeed, it appears that the randomization hypothesis is what allows the distribution of Moran's I statistics, which are also genetic measures of kinship, to be derived. Under the randomization null hypothesis, the statistics are sample-based, and there is no need to know the a priori expected values.

Most of the above discussions are aimed toward studies of evolutionary questions that involved some sort of averaging over time. Other questions are driven by a diametrically opposed condition, in that they are explicitly nonstationary in time. The opposite of how fast things "average out" is how long before temporary effects dissipate. Prominent among these questions are inferences about "ancient" events, discussed in chapter 3, and, in particular, the geographical origination of genetic variants or the species itself. Most empirical studies parse off particular features of the past that have greatest interest, from the spatial–temporal context. In particular, they typically (1) use only genetic variation (not the geographic pattern of), sampled from the present; (2) conduct phylogenetic reconstruction to infer the gene genealogies, ignoring the spatial pattern of variation; and then (3) find among the sample locations the ones that are most like the inferred "ancestral" state and hence inferred as the location of ancestors. Such studies are a good example of the interplay of temporal and spatial dimensions, and illustrate the issues of when can we ignore one at one step and another at another step. Molecular data, typically DNA-sequence or other haplotype data, are used in steps one and two, because such data carries temporal information in the form of temporal sequences of mutation events. Ignoring spatial structure in the first two steps may not cause serious problems if the timescale on which coalescences occur within populations is much shorter than that for genes in different populations (Nordborg 1997). The third step has often been made without comment, but in fact its validity depends on how long it takes before the spatial patterns at the time of the ancient event are erased by time, being replaced through more recent events and processes.

Space–time probabilities of identity by descent provide one means for assessing the relationships of genes in present populations to ancestral genes by source. It has been shown, in relatively simple models that appear to yield general results, that the probabilities of identity by descent, and hence the relative likelihoods that various populations are the origin of genetic variants, are very flat or uniform functions of distance, when rates of mutation and migration are both high (Epperson 1999a). It appears that the critical range for migration is from 1 to 10 percent, and that for mutation is from 10^{-4} to 10^{-6}. When the curve is flat it means in essence that the spatial information from the past has been erased. The existence of a most recent common ancestor for a set of genes is a fact of life, and it reflects the effective size of a system of populations (i.e., the effective total number of individuals) averaged over the time since the ancient event. Among studies of ancient events, those events that occurred more recently are more often successfully identified and characterized.

In studies of ancient events the need for multiple locus data is clear. For example, with respect to the origins of anatomically modern humans, the realized outcome for a single (nonrecombining) type of variant, for example, mitochondrial haplotypes, appears to be too stochastic. Multiple replications of the process are needed, in the form of data for multiple loci or multiple variants. Data for mtDNA and Y chromosomes are probably inadequate, inherently so, and ultimately the answers are most likely to come from datasets with multiple nuclear markers. Space–time data, in the form of ancient DNA, may be inordinately valuable, but difficult to obtain in large numbers. Nonetheless, the general study of ancient events is a key "promise" of molecular marker data, because of the temporal information generated by multiple mutations.

It is less clear that the information in multiply mutated haplotypes are extra-valuable for studies of processes averaged over time, especially if equilibrium is implied. The relevant process and statistical issues are similar to those regarding the utility of multilocus vs. multiple-locus data. Most models of haplotypes or DNA sequences disallow recombination, so that linkage disequilibrium among polymorphic sites of a haplotype is complete (e.g., the infinite sites mutation model). In contrast, it appears that for nuclear genes there will seldom be much disequilibrium among loci, thus the information content of truly multilocus data converges to that for multiple-locus data. Ewens (1974) compared the information content of alleles under the infinite alleles versus infinite sites mutation models, for a model of a single population, and found that there is almost no additional information in the infinite sites mutation model (e.g., DNA sequence data) compared to infinite alleles models (allele frequency data) when the product $4N\mu$, or four times the mutation rate and the population size, is less than 1.0. Most reported populations have values an order of magnitude smaller or yet less (Ewens 1974). Explicit results for a genetically structured system of populations have not been obtained, but it seems likely that the result would hold approximately if N were taken as the total number of individuals in the system.

CHAPTER 7

Spatial Patterns Observed within Populations

> The effects of restricted dispersion on the genetic properties of a continuous population have been treated mathematically in a previous paper (Wright 1943a). . . . The mathematical treatment was based on the assumption of completely random union of gametes within each neighborhood and thus would rarely be strictly applicable to actual cases. . . . A term is needed to designate the local population of which the parents may be considered as representative. . . . "Panmictic unit" applies only if there is completely random mating locally and this is not sufficiently general. . . . An essential property of the population in question is that the individuals are neighbors in the sense that their gametes may come together. The term "neighborhood" is thus an appropriate one for this important unit.
>
> —Sewall Wright (1946)

Spatial structure of genetic variation largely determines the genetic dynamics and ultimate trajectories of evolution in natural populations. For example, sibling species sometimes occur sympatrically or parapatrically, and this implies that spatial differentiation of genetic variation likely is an important part of speciation. The frequent coexistence of well-defined ecotypes in close proximity indicates a complex interplay between genetic isolation and local adaptation. Spatial differentiation and isolation allow specialization while reducing the costs of reproductive isolating mechanisms. More fundamentally, spatial structure can be the primary determinant of the mating system, and together with dispersal can have a large effect on the frequencies of genotypes at both the single-locus and multilocus levels. Because spatial structure codetermines genotypic frequencies, it can affect selection and other population genetic and demographic processes over both the short term and the long term. In turn, spatial structure of genetic variation may be affected by many of these same processes. Indeed, process and structure may be inextricably linked. Spatial–temporal processes are not simple, and the mathematics quickly become prohibitively complex in genetic systems. This very complexity seems to be an important part of evolution. Spatial structures range widely among species, and perhaps are part and parcel with biologically determined levels of dispersal and population genetic dynamics, e.g., levels of biparental inbreeding and levels of inbreeding depression.

The important effects of spatial distributions of genetic variation in the field of population genetics can be placed in three categories. First, in many populations, various population genetic processes both influence and are influenced by spatial structure: in some cases process and structure may be inseparable, e.g., microenvironmental selection. Thus theoretical models of population genetics that include spatial context add to our understanding of short-term ecological and long-term evolutionary genetic processes. Second, because spatial distributions are sometimes distinctively altered by various evolutionary factors, we may expect to use spatial patterns to quantify different evolutionary factors. This includes means for obtaining indirect measures of gene flow or dispersal. Such an approach is especially powerful where spatial patterns for multiple genetic loci may be compared within the same population, in part because neutral loci are subject to the same dispersal and demographic parameters. For example, spatial patterns for neutral genes may differ from patterns for loci under natural selection. Third, spatial structure causes genetic correlations among the elements of spatial samples from natural populations. Samples from natural populations are almost always taken over space. Spatial correlations imply nonindependence among the elements of samples, and when they are strong enough they can cause serious biases in estimators of many important population genetic parameters, including the level of inbreeding within the sample. Spatial correlations violate a fundamental assumption of random samples (Epperson 1993a). The object of this chapter is spatial structure per se, and not this third category. It is simply noted that any sample design should include consideration of spatial structure, either by making corrections based on information about dispersal and breeding system, or by using knowledge of the spatial genetic structure itself to modify sampling schemes to avoid autocorrelations.

To begin to disentangle the complex interplay of spatial structure with evolutionary processes, it is important to characterize the spatial structure expected for neutral genes. Patterns under neutral theory form a starting point for inferences about other processes based on observed spatial structures. The proximal cause of spatial structure is limits to the distances of dispersal, either of individuals for most animals, or seed and pollen dispersal. However, the development of structure for neutral genes may also depend on various other demographic factors. In this chapter "standard" forms of genetic isolation by distance are distinguished from "nonstandard" forms and are presented first. Spatial distributions under standard forms of isolation by distance depend only on dispersal distances relative to density. There are well-developed theoretical models for the standard theory and spatial statistical models for its measurement. Analysis of standard forms is followed by review of experimental studies, and then an examination of how various demographic factors may cause violations of the assumptions of the standard theory, and sometimes result in differing types of spatial structures. Examination of various case studies leads to insights about several demographic factors, including clonal reproduction, age structure, and population growth rates. The final topic is changes in spatial patterns that can be caused by natural selection.

STANDARD FORMS OF GENETIC ISOLATION BY DISTANCE

Individual progeny of an animal typically do not disperse long distances away, nor do seed from their maternal parent. This is how the process of the buildup of genetic isolation by distance within populations starts. In the next generation, because of spatial proximity sibs have an increased chance of mating, beyond random mating probabilities. Even if they do not mate, their related progeny disperse limited distances and then these progeny are also more likely to mate than dictated by random mating. As this process proceeds, the degree of relatedness via spatial proximity builds. Spatial differentiation also builds up, even though the population is continuous. In the simplest models, it is assumed that there is no age structure, and that density remains constant and uniform over the spatial surface occupied by the population. The theories of Wright and Malécot, developed in the 1940s, considered single, large, even-aged populations in which individuals are effectively continuously or uniformly distributed (Wright 1943; Malécot 1948). In other words, densities were assumed uniform over space and fixed over time. Although such circumstances may often be reasonably approximated in nature, they imply constraints on various demographic processes, which are explored later in this chapter. In the standard theory, the process does not depend on details of the dispersal. The original standard models of Wright were in terms of male and female dispersal distances, and it was only the total of the two that determined the spatial structure (Wright 1943). In plants, dispersal is not male and female, rather seed and pollen; however, it appears that as long as neither is too extreme, spatial structure under the standard model depends only on a weighted average of the two (Crawford 1984). Thus, measures, such as Wright's neighborhood size (N_e), of the total amount of dispersal, standardized for density, capture essentially all effects on spatial structure, under the assumptions of the standard models. The intensity and physical scale of spatial genetic structure depends on measures of dispersal distances relative to density, which together determine Wright's neighborhood size. N_e determines the spatial scale in terms of numbers of individuals, and density must be factored in again to determine the physical scale of spatial structures. It is largely meaningless to compare physical scales of structures for different species, or different studies, without considering standardized dispersal and density.

There are various ways to describe the spatial structure generated by standard processes of isolation by distance. Originally, Wright (1943) showed that limited dispersal produces locally inbred demes. He showed how groups (demes) of individuals, spatially contiguous on a two-dimensional landscape, are inbred with respect to the total population. The amount of inbreeding depends only on Wright's neighborhood size. In a second paper, Wright (1946) did not assume that there is random mating within neighborhoods, which he considered to be a rarely fulfilled presumption (see the quotation heading this chapter). Moreover, unless there is random mating within neighborhoods, there is no basis for using neighborhood size as a spatial unit. Indeed, the general form of Wright's (1946) model provided no clear basis for choosing any block size or deme size, be it the size of a neighborhood or else. Wright used the inbreeding coefficient F_{IS} within blocks or demes to partially examine the genetic results of isolation by distance. However, measurement of the degree of theoretical genetic isolation by distance is more effectively done by the theoretical values of F_{ST} among demes for various block sizes. Generally, F_{ST} increases as rates of dispersal decrease, and it decreases as the block size increases. Several difficulties exist in applying this theory to data, and it has proven especially difficult to obtain unbiased estimators of F_{ST}, based on observable genotypic frequency data. Although Wright (1951, 1965) later developed F statistics as estimators, with the same labels, he also pointed out that these were not the same as the theoretical values.

In a classic study, Wright (e.g., 1978) analyzed the F statistics in a survey of enormous proportions conducted by Epling and colleagues (1960) on flower color polymorphism in *Linanthus parryae*. As many as a few hundred plants were assayed from each of 260 quadrats, with 10-foot quadrat lengths, for many years. The frequency of blue-flowered plants versus white-flowered plants ranged from about 1 percent to about 80 percent among quadrats separated by only 1000 feet. The values of F_{ST} for different size blocks of quadrats are shown for three of Wright's designated time periods (A, C, and E) in figure 7.1. The value shown for "zero" size (Wright 1978) is apparently the F_{ST} among individual quadrats. Values do not depend much on whether the block size is 10 or 40 quadrats, but the values are somewhat higher for individual quadrats. The values differ markedly among time periods. Interestingly, Wright (1978) also presented the correlation coefficients as functions of distance, also shown in figure 7.1. Adjacent quadrats have very high correlations, up to about 0.76, which tail off smoothly with increasing distance. Wright concluded that "the correlations suggest that it [the distances over which there is substantial panmixia] is of the order of 10 to 20 ft [corresponding to one to two quadrat lengths]." Wright (1978) plotted the correlations on log-distance, and it appears that the decrease with distance is

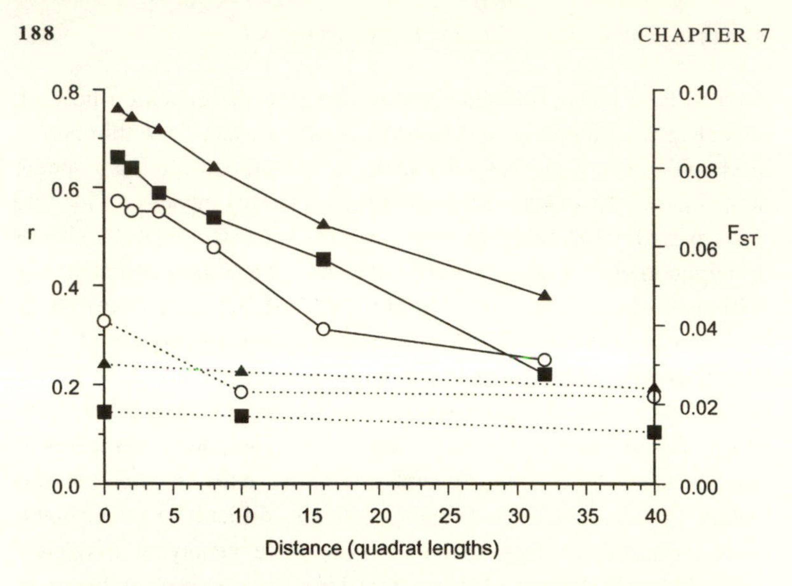

FIGURE 7.1. Values of F_{ST} and gene frequency correlations, modified from tables and figures in Wright's (1978) analysis of Epling and colleagues' (1960) data on the frequency of blue (versus white) flower types of *Linanthus parryae*, in quadrats along a transect. Values of F_{ST} (right side *Y* axis) were calculated for different-sized areas in terms of numbers (namely 1, 10, or 40 quadrats) of quadrats linearly arrayed along the transect, and the graphed values are given by dotted lines and symbols ▲, ○, and ■, for three different time periods *A*, *C*, and *E*, respectively. Graphs of values of the correlation coefficients *r* (left side *Y* axis) of frequencies among quadrats separated by various distances are given by solid lines and symbols ▲, ○, and ■, respectively, for the *A*, *C*, and *E* time periods. Distance units displayed are in quadrat lengths, and can be converted to feet by multiplying by ten.

rather faster than exponential. *Linanthus parryae* densities changed radically over the study years; thus, too, would the neighborhood size, and Wright concluded that it ranged from about 10 to about 100. Because of the changes in densities it is difficult to assess the true numbers of individuals per quadrat. Nonetheless, in general terms the observed correlations appear to be consistent with simulations of genetic isolation by distance (e.g., see values for quadrat size 25 in figure 8.3).

Inspired by Wright's work, Malécot (1948) also examined standard forms of genetic isolation by distance. Although Malécot reproduced with mathematical rigor Wright's results in terms of inbreeding coefficients, much of his work focused on consanguinity among pairs of individuals separated by different distances (Epperson 1999b). Doing so allowed him to obtain many analytical results on how the process results in adjacent or nearby individuals being kinship related, and precisely how the degree of kinship falls off with the distance of separation. In Malécot's analytic results the curve depends only on the total amount of dispersal (e.g., Wright's neighborhood size) and another parameter (the outside systematic pressure). Alternatively, Malécot showed how the probability that two genes (one randomly sampled from each of the two individuals) are identical by descent tails off with distance of separation. As with Wright's theory, it has proven difficult to obtain unbiased estimators of the consanguinity coefficients or the probabilities of identity by descent. As is discussed in chapter 8, such estimators of pairwise similarity as Loiselle et al. (1995) and other relative measures of similarity appear to be biased. These measures would be attractive if unbiased because they may allow pairwise estimates to be based on multilocus genotypes, thereby possibly increasing precision. These theoretical models, as well as those in terms of the conditional covariance of gene frequencies and heterozygosity, are discussed in detail in chapter 9. Pairwise measures of relatedness as a function of distance are closely related to spatial autocorrelation measures. There seems to be little reason to use hierarchical approaches in spatially explicit models, in population genetics, or other fields. Exceptions may include some animal populations where it may be difficult to assign individual locations, especially if they are social animals or if they have overlapping home ranges. Note that Malécot's methods could not obtain explicit analytical results for the especially relevant case of short distances of separation in systems with two spatial dimensions, because of the very nature of such a system.

Although analytic results are preferable, the existing theory focuses on stationary distributions that require mutation-drift (spatial) equilibrium and usually are in terms of parameters for which no unbiased estimators (from spatial genetic data) are available. While the former are useful for analyzing general trends caused by various factors, a great deal can also be gained from using a combination of (1) space–time Monte Carlo computer simulations, explicitly based on spatial distributions of genotypes; and (2) spatial autocorrelation statistics that are essentially distribution free (Cliff and Ord 1981). Moreover, simulations generate not only the average values of spatial correlations or parameters, but also the variances of these created by stochasticity inherent in the processes, as well as sampling errors. Particularly important are models of relatively short distances for populations existing in two-dimensional space. A series of simulation studies (e.g., Rohlf and Schnell 1971; Turner et al. 1982; Sokal and Wartenberg 1983) of the standard model have revealed that a dominant spatial feature is the growth of patches within 10–50 generations; where a patch is loosely defined as an area of concentration of an allele. In addition, although true equilibrium in models without recurrent mutation occur with fixation for one or another allele, in large populations a "quasistationary" state, in terms of spatial autocorrelations, obtains within about 50 generations and persists for hundreds if not thousands of generations (Sokal and Wartenberg 1983).

The nature of spatial structure in two-dimensional populations is most clear when neighborhood size N_e is small (less than about 25). The predominant spatial features are distinct genetic "patches," where each patch consists of a fairly large area (much larger than N_e) of several hundred individuals, most with the same homozygous genotype (Turner et al. 1982; Epperson 1990a) (figure 7.2). Heterozygotes occur primarily in the margins between patches of opposite types. If a movie were made of the process, it would show how a potential patch begins as a small "seed" area of a few homozygotes, and then either local stochastic events erase the seed or the seed grows. After reaching an intermediate size, its long-term persistence is almost assured, and then it grows further, perhaps combining with other seed patches of the same type, until it bumps up against patches of opposite types, whereupon it becomes rather stable. It may then slowly shift around, but maintains roughly its size, and this is at the quasistationary phase that persists for long periods.

In cases where dispersal is at higher levels ($25 < N_e < 200$) the homozygosity concentrations are weaker and heterozygosity is higher. Adjacent groups of individuals have moderately similar or autocorrelated gene frequencies compared to the rest of the population. The patches are not as obvious, but they are visible as gene concentrations. The boundary areas between genetic patches are less distinct, and the size of the patches are somewhat larger. This tendency con-

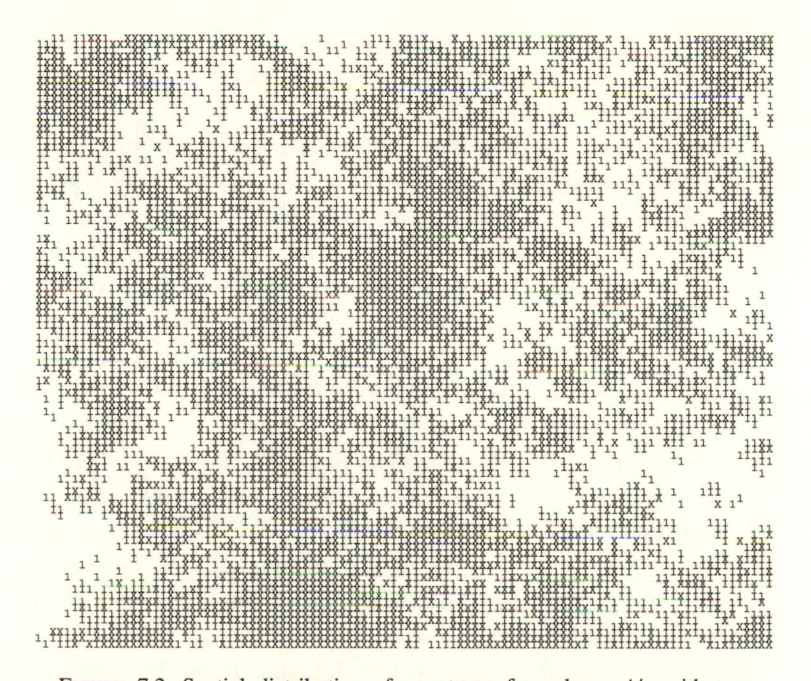

FIGURE 7.2. Spatial distribution of genotypes for a locus *A*/*a* with two alleles in equal frequency ($q = 0.5$) in a simulated population of 10,000 plants with nearest-neighbor pollination and no seed dispersal (Wright's neighborhood size 4.2). Individual genotypes *AA* are denoted with an X, *Aa* with an 1, and *aa* with a blank space. Note the large areas or patches containing mostly one homozygote, and that heterozygotes are found primarily along the edges of such patches.

tinues for even larger neighborhood sizes (>200), in which cases the patch structures are scarcely visible, yet there are measurable amounts of spatial autocorrelation.

Earlier Monte Carlo simulation studies by Rohlf and Schnell (1971), using *F* statistics and a variety of dispersal functions, supported Wright's theory in general terms, especially with respect to the general trend of decreasing structure with increasing neighborhood size. Interestingly, the form of the dispersal function on distance, whether normally distributed or leptokurtotic, had little effect. Rather, the variance of dispersal distances, as it enters the formula for Wright's neighborhood size, controlled the spatial structure. Nonetheless, Rholf and Schnell (1971) concluded that *F* statistics yield "relatively little" insight into actualized spatial patterns of genotypes at single loci. This is due in large measure to the stochastic effects that come in to play for genotypes at a locus.

Wright's (1943) fixation index has been measured in many species. In the absence of selfing, significant excesses of homozygotes caused by biparental inbreeding are often attributed to mating by proximity and hence spatial structure. However, there may be other causes of inbreeding that are not particularly spatial. For example, assortative mating in animals may in some cases draw from a large area. In plants, genetic correlations in flowering time may cause inbreeding even if there were widespread pollinations, by wind, for example. Selfing may further confound inbreeding. The use of F_{ST} as a measure of spatial structure is not as limited as F_{IT} or F_{IS}, because it measures inbreeding within blocks *relative* to total inbreeding in the population area. A good example of the use of F_{ST} for measuring localized spatial differentiation is the large study of *Liatris cylindracea* by Schaal (1975). She examined a large number of isozyme loci in quadrats set to sizes approximating the neighborhood size (apparently in the range of 130–150, based on the published range of the total numbers of individuals covered in the sample area). The mean value of F_{ST} among quadrats was about 0.069, although it ranged from about 0.009 to 0.224 among loci (Schaal 1975).

SPATIAL AUTOCORRELATION OBSERVED WITHIN POPULATIONS—MORAN'S *I* STATISTICS

Spatial genetic patch structure can be well characterized using spatial autocorrelation statistics, such as Moran's *I* statistics and join count statistics. As is described in chapter 8, Moran's *I* can be used to measure the degree of spatial autocorrelation among the gene frequencies of groups of individuals, in quadrats or other delineations of subsamples of the population. Alternatively, Moran's *I* and join count statistics can be used to measure the autocorrelation in spatial patterns of individual genotypes. Statistical methods based on Moran's *I* for individual genotypes proceed by considering only one allele, A_i, at a time. The genotypes are converted to frequencies q_i for this allele as follows: 1.0 for genotype A_iA_i, 0.5 for heterozygotes for A_i, and 0 for all other genotypes. This procedure, suggested by Dewey and Heywood (1988), has been widely adopted in experimental studies, and its stochastic and statistical properties are now well described. For loci with two alleles, the information contained in the spatial pattern is identical for each allele, and so are the Moran's I statistics. This is not the case for loci with more than two alleles.

Recall that Moran's I statistic is a pairwise measure that can be written

$$I_k = \frac{n\sum_i \sum_j w_{ij}(k) Z_i Z_j}{W_k \sum_i Z_i^2}$$

where $Z_i = q_i - q$, and q is the mean of the allele frequency in the sample. The $w_{ij}(k)$ can be general weights but in the more typical, unweighted, case they are binary, thus $w_{ij}(k) = 1$ if the pair of locations i and j is separated by a distance in class k, else $w_{ij}(k) = 0$, and then W_k equals twice the number of pairs of locations in the distance class k. Usually the classes are mutually exclusive and exhaustive, and based on ranges of physical distances separating pairs. Under the randomization hypothesis, I_k has expected value $u_1 = -1/(n - 1)$, where n is the number of locations. The variance, u_2, is given by equation 4.5. If the number of locations is fairly large, then the statistic $(I - u_1)/\sqrt{u_2}$ has an approximate standard normal distribution under the null hypothesis (Cliff and Ord 1981). A set of unweighted I statistics for mutually exclusive distance classes is known as an I correlogram. Thus I correlograms measure relative correlations in allele frequencies as a function of distance. I correlograms can be tested as a whole for significant deviation from the random hypothesis (Oden 1984), but exact tests are lacking for differences between correlograms from different data sets, for example, spatial distributions for different loci (Sokal and Wartenberg 1983). Other configurations for Moran's I statistics were described in chapter 4, and these include methods for classifying pairs in terms of direction as well as distance, other ways of calculating distance (e.g., using Gabriel connections), and weighted I statistics. There is a close but nonexact

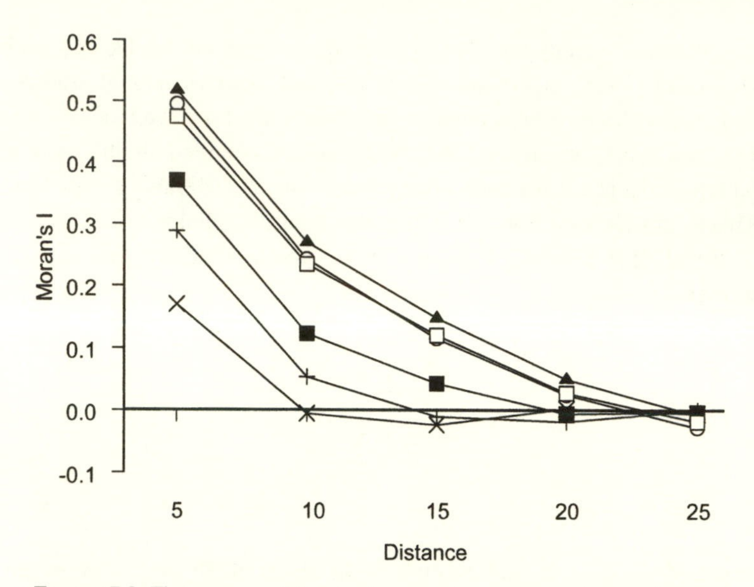

FIGURE 7.3. Time course of the development, in numbers of generations, of spatial autocorrelation, as measured by Moran's I statistics calculated on gene frequencies in quadrats of 25 (dimensions 5 by 5) individuals each. Distance is measured in units of the lattice of individuals. Results are summarized from Epperson (1990a) for a set of simulations of genotypes for a two-allele locus ($q = 0.5$). Simulated populations consisted of 10,000 plants, having near-neighbor pollination and no seed dispersal (Wright's neighborhood size 12.6). Generations 10, 30, 50, 100, 150, and 200 are denoted by symbols X, +, ■, ○, ▲, and □, respectively.

relationship of the standard I correlogram with theoretical correlations of allele frequencies, as is discussed in chapter 9.

Some general properties of unweighted Moran's I statistic are illustrated by gene frequencies in quadrat subgroups of 25 individuals each in computer simulations of isolation by distance. Here the Z_i are mean-adjusted gene frequencies in quadrats. Some typical correlograms that illustrate the accumulation of spatial structure over time are shown in figure 7.3. The structure and the spatial autocorrelations scarcely change after about 50–100 generations (Epperson 1990a). Similar properties are observed for I statistics based on converted individual genotypes. A typical I correlogram has high positive correlations between adjacent locations (figure 7.4), and the correlations drop off smoothly with distance, and typically become negative at

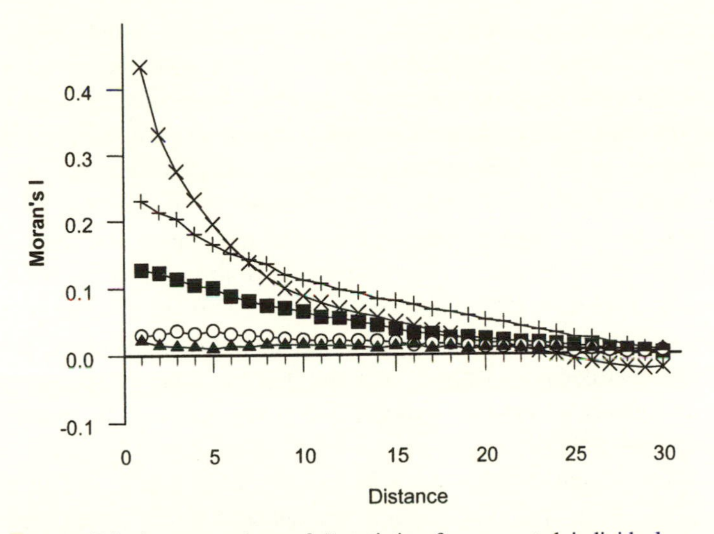

FIGURE 7.4. Average values of I statistics for converted individual genotypes for sets of 100 simulations each for the five different dispersal models outlined in table 8.1, at generation 200. Arbitrarily, only one allele of a three-allele model with equal allele frequencies was used. Symbols for sets are as follows: (X) set 1 (N_e = 4.2); (+) set 2 (N_e = 25.1); (■) set 3 (N_e = 50.2); (○) set 4 (N_e = 115.2); and (▲) set 5 (N_e = 632.4). Distance is measured in units of the lattice of individuals.

about 20–25, or more, times the distance between adjacent individuals. The distance at which negative autocorrelation occurs has been used as an operational estimate of the width of genetic patches. Patch size is a property that was not evident in the classical theories of Wright and Malécot, which were based on correlations relative to the population gene frequencies in the remote ancestors. It is usually much larger than the neighborhood size. The genetic patch size is a spatial unit, unlike the neighborhood size, which appears to form no basis for spatial units of genetic variation. Moreover, although patch size does increase with neighborhood size, the increase is not linear, and thus neighborhood size cannot be a unit of spatial genetic structure. Dispersal also reduces the deviations of Moran's I away from the expected value under the null hypothesis, at short and long distances. Again this reflects the decrease in spatial genetic structure as dispersal increases.

Despite having some similar general properties, correlograms of Moran's *I* statistics calculated for the converted gene frequencies differ from those for quadrat sizes of 25, which were used to characterize structure in early simulation studies (Sokal and Wartenberg 1983; Epperson 1990a). This difference caused considerable confusion in the many experimental studies that subsequently used Moran's *I* statistics. In fact, experimental studies that have used converted genotypes obtain the lower values expected from isolation by distance theory (Epperson 1995c; Epperson et al. 1999). Extensions of the above results apply to loci with multiple alleles (Epperson et al. 1999), multilocus genotypes, and the genetic value of quantitative traits, subject to some provisions, which are reviewed in chapter 8 (Epperson 1995b).

Under standard forms of isolation by distance (i.e., unless dispersal is great) spatial autocorrelation statistics should detect genetic structure for well-designed (Epperson 1992, 1993a) samples of a few hundred or less individual genotypes (Epperson et al. 1999). Moreover, in large populations the structure for any given locus is highly predictable; e.g., there is little stochastic variance among simulations. As sample sizes decrease statistical error may be added to stochastic variation. Allele frequency has little or no effect unless it is near 1.0 or 0.0. For the Moran's *I* statistics for converted individual genotypes, unlike those for some quadrat sizes, the *I* correlograms decrease monotonically with increases in dispersal, and overall the precise level of dispersal has a large effect (Epperson 1995a; Epperson et al. 1999).

In natural populations observed values of Moran's *I* statistics for converted genotypes are generally consistent with isolation by distance. Values for the shortest distance class are listed for plant species in table 7.1. These statistics are especially important, in part because they should have the largest values under isolation by distance. Moreover, it is difficult to imagine any other real spatial pattern that has autocorrelation at long distances but not at the shortest distance. This is also illustrated by the related concept of the interdependence of global autocorrelation and local autocorrelation (e.g., Sokal et al. 1998a,b). In addition, when the Bonferroni-type corrections for multiple significance tests are applied to spatial correlograms, they treat all distance classes equally, thus they are overly conservative. Finally, using the autocorrelation values for the shortest distance class is statistically powerful (Oden 1984) and robust to experimental variations in the size of the area sampled, whereas those for longer distances may not be. (However, the minimal scale of distances among sampled individuals remains critical for the values expected for the first distance class [Epperson et al. 1999].) The values given in table 7.1 were obtained from published information. It was possible to determine, at least approximately, the sampling densities. Some of the averages given in the table may not be exact, but they should be very close. Various degrees of information on mating system, dispersal, and sampling design were also available. The data are essentially all from isozyme polymorphisms.

The wide-ranging values in table 7.1 bear a close relationship with the pertinent features of the mating system and level of dispersal, and several classes emerge. The first class of patterns are those typical for plants that are insect-pollinated and have heavy seeds that usually fall within short distances of maternal parents and have little or no secondary dispersal. Plants may also be in this class due to high rates of self-fertilization, which may cause even wind-pollinated species to have small average distances of pollen dispersal. Plant species in this first class, such as most of the *Androcymbium* species and *Silene acaulis*, generally have Moran's I statistics for the shortest distance class in the range of 0.15 to about 0.40 (table 7.1). The value (0.06) for *Androcymbium pulchrum* is considerably lower. However, based on available information, the minimum spatial scale of sampling appears to be quite large, although it could not be precisely determined. One measure of minimum spatial scale is "porosity," the proportion of plants that are sampled in the sample area (this definition is slightly different from that used in simulation studies, e.g., Epperson et al. [1999]). Estimates of porosity in the table were in most instances calculated by the square of the distance for the upper bound on the first distance class times the density of individuals. Porosity is critical and when it is made much larger than one the spatial scale of the autocorrelation increases, and hence the predicted value of autocorrelation is reduced (e.g., figure 7.4). The precise effects of porosity on predicted values under isolation by distance are examined in detail in chapter 8. Overall, the plants in this class may be categorized as having highly limited dispersal of both pollen and seed, or

TABLE 7.1. Values of Moran's *I* statistic for the first distance class for various plants

Species	*Mating system/dispersal*	I	*UB*	*Po*	P	L	n	*Citation*
Androcymbium gramineum	Hermaphroditic, partial selfing, +clonal, pollen: bees and other insects, seed: indehiscent capsules, gravity (0.5–1 m), secondary by wind and surface water.	0.17	4.3	92	1	17	200	Caujape-Castells and Pedrola-Monfort (1997)
Androcymbium bellum	Hermaphroditic, preferential outcrosser, low selfing rate (s), +clonal, pollen: bees and other insects, seed: dehiscent capsules, gravity.	0.19	1	?	1	3	68	Caujape-Castells et al. (1999)
Androcymbium guttatum	Hermaphroditic, obligate outcrosser, +clonal, pollen: bees and other insects, seed: dehiscent capsules, gravity.	0.38	0.25	?	1	3	59	Caujape-Castells et al. (1999)
Androcymbium pulchrum	Hermaphroditic, preferential outcrosser, low selfing rate (s), +clonal, pollen: bees and other insects, seed: dehiscent capsules, gravity.	0.06	1	?	2	3	22, 37	Caujape-Castells et al. (1999)
Pteris multifida	Sexual and asexual. Sperm short lived, depends on temp. presence of water, highly limited distances. Asexually produced spores wind-dispersed perhaps long distances.	0.20	1	2.4–5.6	3	2	111 to 653	Murakami et al. (1997)
Silene acaulis	Gynodioecious, pollen: bumblebees and other insects, seeds: gravity and wind (no special structures).	0.18	1	0.1–2.9	5	5	83 to 473	Gehring and Delph (1999)

Rhus trichocarpa	Dioecious, pollen: bees, seed: gravity (short distance) and birds (perhaps long distances).	0.17	14	6.5	1	4	200	Chung et al. (1999)
Rhus javanica	Dioecious, +clonal (long-distance), pollen: bees, seed: gravity (short distance) and birds (perhaps long distances).	0.05	10	5.6	1	5	135	Chung et al. (2000)
Ipomopsis aggregata	Self-incompatible, pollen: hummingbirds, seed: gravity (no special dispersal features). Independent measures place neighborhood size in the range of 28–45.	0.10	1	0.5–0.9 fl. plts.	1	8	~200	Campbell and Dooley (1992)
Psychotria nervosa	Monoecious, distylous, pollen: insects, primarily nearest neighbor pollination, seed: gravity and possibly birds ("infrequently moved").	0.09	?	?	1	2	245	Dewey and Heywood (1988)
Eurya emarginata	Dioecious, +clonal (long-distance), pollen: insects, seed: gravity.	0.10	12	4.1	1	8	239	Chung and Epperson (2000)
Adenophora grandiflora	Monoecious (highly outcrossing), +clonal, pollen: bumblebees and other bees, seed: small (1 mm) wind-dispersed, plus secondary by surface water.	0.15 to 0.18	7–8	3.8 to 4.0 sexuals + clones	2	10	122–143	Chung and Epperson (1999)
Gleditsia triacanthos	"Essentially dioecious," pollen: insects, seed: gravity ("highly limited"), plus secondary by birds and other animals.	0.08	mean 4.3[a]	0.6[a]	1	16	258	Schnabel et al. (1991)
Maclura pomifera	Dioecious, pollen: wind, seed: gravity ("highly limited"), plus secondary by birds and other animals.	0.05	mean 7.9[a]	0.5[a]	1	3	207	Schnabel et al. (1991)

TABLE 7.1. (*Continued*)

Species	*Mating system/dispersal*	I	*UB*	*Po*	P	L	n	*Citation*
Calluna vulgaris	Pollen: insects plus wind, seed: gravity (most fall within a few meters).	0.06	5.2	> 1.2 (genets)	1	4	41	Mahy and Neve (1997)
Sclerolaena dicantha	Mixed mating ($s \sim 0.7$), pollen: wind, seed: gravity, with secondary by wind, flooding and most importantly by certain ants, when these are present.	0.10[b] 0.01	3	?	4	2	150–300	Peakall and Beattie (1995)
Acer saccharum	Pollen: wind and insects, seed (winged samara): wind.	0.04	14	2.5	3	5	100	Perry and Knowles (1991)
Fagus crenata	Mixed mating (*s* unknown), pollen: wind, seed: gravity plus secondary by small mammals	0.05	30	0.4	1	7	138	Kawano and Kitamura (1997)
Quercus petraea	Mixed mating (*s* small), pollen: wind, seed: gravity, birds and other animals.	0.07	10	0.4	1	7	190	Bacilieri et al. (1994)
Quercus robur	Mixed mating (*s* small), pollen: wind, seed: gravity, birds and other animals.	0.02	10	0.4	1	7	217	Bacilieri et al. (1994)
Quercus laevis	Mixed mating, pollen: wind, seed: gravity, birds and other animals.	0.07	1	0.2 (with DBH > 2.5 cm)	1	9	3400	Berg and Hamrick (1995)
Cymbidium goeringii	Perfect flowers, pollen: bumblebees, seed (small): wind dispersed probably long distances.	0.02	3 to 4	1.6–2.2	2	7 to 8	110–138	Chung et al. (1998)

Chionographis japonica var. *kurohimensis*	Gynodioecious (hermaphrodites primarily selfing).	0.04	1	1.5 (rosette forming plants)	1	3	302	Maki and Masuda (1993)
Hibiscus moscheutos	Obligate outcrosser, pollen: insects (low dispersal distances), seed (bouyant): water (perhaps long distances).	0.04	3	8.5–19.8	2	3	60 each	Kudoh and Whigham (1997)
Picea marianna	Primarily outcrossing, pollen: wind, seed: wind.	0.00 to 0.01	7.5–24	10.2–48.4	2	5 to 8	~500 each	Knowles (1991)
Pinus strobus	Primarily outcrossing, pollen: wind, seed: wind.	0.05	15	4.7	1	4	210	Epperson and Chung (2001)

[a]These values reflect the average distance among pairs in first distance class, not the upper bound.

[b]First value is for when ants are present, second for when they are not.

Note: Approximate averages are generally based on the published information and may be over alleles or loci. Other information listed is the approximate upper bound (in meters) on the first distance class (UB) (the lower bound can be taken as zero), the inferred porosity (Po), i.e., the percentage of total individuals in the sample area that were actually sampled, the number of populations in the study (*P*), the number of loci assayed (*L*), and the sample size (*n*).

in the case of *Pteris multifida*, highly limited dispersal of sperm and asexually reproduced spores.

A second class consists of species in which either one or the other, but not both, of dispersal of pollen or seed is less limited than in the first class, and total dispersal in this category would be considered moderate. One important feature is primary and secondary dispersal of seed by birds. Values of Moran's *I* statistics may range from about 0.14 down to about 0.07. Naturally, any effects would depend on how frequently birds disperse seeds and how far they carry them before deposition. In some species, such as *Rhus trichocarpa*, they are perching birds that favor *R. trichocarpa* for perching; hence, much of the seed dispersal remains local. In contrast, secondary dispersal by birds and other animals may have substantially increased dispersal and decreased autocorrelation in *Gleditsia triacanthos* and *Maclura pomifera* populations (Schnabel et al. 1991). Bird dispersal of pollen is generally expected to result in greater pollen dispersal distances than that from insect dispersal. A good example of this is *Ipomopsis aggregata*, which in a study population (Campbell and Dooley 1992) had considerably lower values of Moran's *I* for distance class one (mean ca. 0.10) than do species with highly restricted dispersal. Detailed, independent, direct measures of dispersal in *Ipomopsis aggregata* (Waser and Price 1983; Campbell 1991) indicate that pollen is dispersed considerable distances, as are seeds. In sum, the experimental studies place Wright's neighborhood size in the range of 28–45 plants. The observed values of *I* fit well within the range expected based on computer simulation studies. For simulated neighborhoods in the range of ca. 25–50 the Moran's *I* statistic ranges from 0.20 down to 0.12 for full census of a large population, and 0.11 down to 0.07 for samples like those in the study (Epperson et al. 1999).

The role of clonal reproduction (for a review, see Ellstrand and Roose [1987]) can also be examined. Clonal reproduction is generally considered to contribute to spatial structure. In such species as *Adenophora grandiflora*, clones "disperse," i.e., form, short distances from progenitors, and they do indeed contribute to structure. The average values of Moran's *I* for the total population (0.15–0.18) are relatively high (Chung and Epperson 1999), given the other characteristics of the species, whereas spatial autocorrelation among sexually reproduced plants only is considerably lower (0.10–0.11). The clones are highly spatially autocorrelated in small localized areas

(Chung and Epperson 1999). In contrast, clones of *Eurya emarginata* (Chung and Epperson 2000) and *Rhus javanica* (Chung et al. 2000) form relatively long distances away from progenitors, and clones of the same genotype (which might represent several sequential clonal events) are found long distances apart, and indeed were essentially randomly distributed in the population. Hence in these cases clonal reproduction may have actually increased dispersal and the population shows relatively low autocorrelations (I = 0.05–0.10).

A third category may be considered, wherein species have long-distance dispersal of either seed or pollen, but not both, and where total dispersal may be considered high. These include species with wind-dispersed pollen (*Fagus crenata* and the *Quercus* species), or seed that either are small (e.g., *Cymbidium goeringii* and *Adenophora grandiflora*) or have special structural characteristics (e.g., *Acer saccharum*) that decrease settling velocity. Other species may have substantial wind dispersal of pollen in addition to insect pollination (e.g., *Calluna vulgaris*) or substantial secondary movement of seed by surface water (*Sclerolaena dicantha*). Such species may have Moran's I statistics averaging down to about 0.03. One of the largest studies to date was conducted on ca. 3400 individuals in a population of *Quercus laevis* (Berg and Hamrick 1995), where the average value observed was ca. 0.10 for juveniles and 0.05 for adults. The latter would correspond to a neighborhood size of ca. 125 or more, which is in reasonably close agreement with the authors' conclusion of 440 (based on observed and simulated F_{ST} values). In an interesting counterexample, secondary movement by ants apparently *increases* spatial structure when they are present in populations of *Sclerolaena dicantha* (Peakall and Beattie 1995), probably because they carry related seed to their mounds. Similar effects are observed with Clark's nutcracker in collecting whitebark pine cones (Furnier et al. 1987). As is discussed in more detail later in this chapter, limited seed dispersal may cause a temporary structure in young age classes in species with high pollen dispersal, and the ultimate effects of this on isolation by distance may or may not be important, depending in large part on demographic considerations (Epperson 2000b). Moreover, this situation is analogous to metapopulation concepts of population extinctions and recolonizations, and the results depend on population size growth rate and regulation.

A final category could be formed for those species with high dispersal of both pollen and seed (most frequently, wind dispersal of both), and thus total dispersal may be termed very high. In such species the spatial distribution may be very close to random, and the Moran's I statistics near zero (or more precisely $-1/(n-1)$). The exact level depends primarily on the terminal velocities and plant heights, and hence horizontal distance of dispersal of each, and equally on the density of reproductive individuals. Many of the coniferous tree species fall into this category (Wright 1976; Govindaraju 1988). However, the amount of autocorrelation may still vary within this class, especially because of differences in density (and hence neighborhood size). Conifers with relatively low density (such as the study populations of *Pinus strobus*, ca. 120/hectare) have weakly autocorrelated ($I = 0.05$) yet nonrandom distributions (Epperson and Chung 2001), whereas those with high densities have nearly random distributions. For example, the two study populations of *Picea marianna* (Knowles 1991) had densities of ca. 1800/hectare and 847/hectare respectively, and had very low values of I (0.008 and -0.001). Two study populations of *Pinus contorta* had densities of approximately 2500/hectare and had essentially random spatial distributions (Epperson and Allard 1989). In such populations of *Pinus contorta*, Wright's neighborhood size must exceed 1000 and could exceed 10,000 (Epperson and Allard 1989).

The dispersal categories—highly limited, moderate, high, and very high—appear to correspond well with the observed values of Moran's I statistics for converted genotypes, for distance class one, of about 0.15–0.45, 0.07–0.14, 0.03–0.06, and near zero to 0.02, respectively. In simulation studies these values correspond to Wright's neighborhood sizes of about 4–25, 25–100, 100–250, and greater than 250. Generally speaking, these values are in the range of those expected for the various dispersal characteristics of the reviewed species by groups. This suggests that I statistics can often be successfully used to estimate dispersal. The match is all the more remarkable, considering there may be other unidentified factors at work, reflecting various demographic details. Perhaps most important are differences in ages of populations, in particular the number of generations or cycles of re-generations that have occurred since the founding of the population.

This use of autocorrelation statistics can be compared to various ways of measuring gene flow directly. Using the standard isolation by distance model and spatial autocorrelations, it is possible to measure the total amount of dispersal indirectly, as long as the process does not depend on details of the demographics. In animals dispersal is measured through a variety of marker/recapture and similar methods. In plants, various methods of measuring pollen dispersal include using dyed pollen deposition onto female structures, observing pollinator movements, using pollen traps, and parentage analysis. Seed dispersal may be measured using seed traps or by parentage. Parentage analysis is often preferred because it distinguishes effective or actual gene flow as opposed to animal or plant propagule movement. For example, for plants pollen transport to stigmas may not result in fertilization (because of, e.g., cross-incompatibilities, or outbreeding or inbreeding depression), and seed dispersal may not correspond to successful seedling establishment, for various reasons. It is important to note that the neighborhood size (and hence spatial genetic autocorrelations) may strongly depend on density, as generally would be the case for wind pollinated species, for example. This can confound the extrapolation of estimates of neighborhood size from one population to another. In contrast, in cases where insect pollinators, e.g., bees, usually sequentially visit nearest-neighbor plants, neighborhood size estimates may be fairly independent of densities (but, see Ellstrand et al. 1978). In animal vector systems pollen dispersal may depend on various details of foraging strategies (e.g., Zimmerman 1981; Levin 1981).

A number of the studies used relatively large upper bounds for the first distance class. For some studies, for example, those on the *Androcymbium* species (Caujape-Castells and Pedrola-Monfort 1997), these were the results of the initial sampling intensities or porosities. Large porosities were needed to avoid extensive clonality and hence multiple sampling of genets. In other cases, relatively large upper bounds resulted from use of equilibration of the numbers of pairs or joins among distance classes, which is a standard default in statistical computing programs (e.g., D. Wartenberg's SAAP program). One favorable feature of equilibration is that it can result in relatively smooth correlograms. However, typically when sample sizes are large, equilibration forces the numbers of pairs of joins in the first distance class to be much larger than necessary to reach appropriate levels of statistical power. Thus it unnecessarily combines longer distance joins with the shortest distance joins, which should show the highest correlations. An example of this is the study on *Picea marianna* (Knowles 1991), which had large sample sizes (ca. 500) collected from contiguous individuals (i.e., sample porosity was 1.0). As a result of using the distance class equalizing option, the first distance class had about 30,000 joins, which is far more than needed. Thus the area covered by the square of the upper bound contained about 10 and 50 trees, respectively, in the two studied populations. However, in this case the observed result of very small to near zero correlations might have remained true if smaller upper bounds were used. There may be very little finest-scale autocorrelation in *Picea marianna*, because of its high densities and wind-dispersed pollen and seed.

It appears that in most instances 200–300 joins in the first distance class are sufficient for obtaining robust estimates of mean values averaged over loci (Epperson 2000b). If sample sizes are in the range of 100–200 or more, a good rule of thumb for the upper bound of the first distance class is to set it equal to $i_d\sqrt{2}$, where i_d is the square root of the inverse of population or sample density (number of points per unit area). This has the appealing feature that it corresponds to inclusion of nearest neighbors (rectangular plus diagonal neighbors), as if the points were located on a regular lattice. Of course, unless intentional, this is usually not the case. Nonetheless, this rule of thumb seems to capture nearest neighbors in a sensible way and insures sufficient numbers of joins in the first distance class. It generally results in first distance class porosity in the range of 4–8, as is evident in the studies by Chung and Epperson listed in table 7.1. If the number of sampled locations is closer to 50, somewhat larger bounds should be used.

Some of the studies are large enough that the shape and *X* intercept of *I* correlograms for converted individual genotypes may be compared to isolation by distance models of large populations. The study of *Androcymbium gramineum* (Caujape-Castells and Pedrola-Monfort 1997) consisted of a transect of 70 meters, implying an two-dimensional area of 0.49 hectare, and the density was ca. 50,000 individuals (many of them clones) per hectare. Thus, the represented spatial area consisted of about 25,000 individuals, including probably many thousands of sexually reproduced individuals. The *X* intercept was about 15 m, implying a (circular) patch area of ca. 175 m^2, containing about 875 individuals, or hundreds of sexually reproduced individuals. This appears consistent with the patch sizes predicted for species with low to moderate dispersal, in the range of ca. 300–2500 individuals (Epperson 1993a). Moreover, the shape of the *I* correlograms closely follows that predicted from isolation by distance theory.

In most studies the entire sample area, or in some cases the entire population, is much smaller than any reasonably expected genetic patch size area under the isolation by distance model at quasistationarity. Nonetheless, the *I* correlograms typically become negative at relatively short distances within the sample area or population. However, this should not be compared with the patch size predicted under isolation by distance in a large population. In such study populations negative autocorrelations may occur at the longer distances, but these are relative to the sampled area. In other words, the autocorrelations at different distance classes are relative to the local mean, not to the mean of the entire population (unless the entire population area is sampled) nor to the expected value under isolation by distance in large populations. The problem of which mean should be used, and how it may be estimated, to compare observed values has persisted throughout the decades since the first attempts were made to compare experimental values to isolation by distance theories of Wright (in terms of *F* statistics) and Malécot (in terms of spatial covariances or correlations). Moreover, it is intuitive that Moran's *I* statistic very likely will be negative for some longer distance classes, if sizable positive values exist at shorter distance classes. However, this may not be mathematically necessary, because each distance class contains a range of distances, some of which might have positive, others negative, predicted correlations relative to the mean in the sample area. In addition to the above stochastic process considerations, there will generally be statistical error that can result in negative estimates of small positive correlations at one or another distance class. The patch sizes in the simulation studies should not be viewed as merely an arbitrary result of the simulated population size (typically 10,000), because the patches occur in large numbers (about 4–16) of times within the populations. In other words, the patches in the simulations are actual, spatially repeated units of structure.

Few studies on plant species have utilized spatial autocorrelations among quadrat subpopulations. This can be done by first breaking up the population into quadrats, mapping the locations of each quadrat, calculating the allele frequency in each, and calculating the distances between quadrats, placing each pair of quadrats into distance classes, and calculating Moran's I statistic for each distance class. A good experimental example of the effects of combining individual genotypes into quadrats is the extensive analysis by Berg and Hamrick (1995), who studied 3400 genotypes of *Quercus laevis*. They report that the correlogram for the quadrats becomes negative at distances at about 40 m, implying a circular patch size of ca. 5025 square meters, an area that would contain about 1100 individuals (with diameter at breast height [DBH] $>$ 2.5 centimeters). Again, this appears to be consistent with isolation by distance correlograms (Epperson and Li 1997).

The size of the quadrat (i.e., the average number of individuals in a quadrat sample) can be critical. The size of the quadrat affects the spatial scale, as well as the number of sample quadrats, for a given total sample size. Most importantly, when the quadrat size is around 25, the I correlograms scarcely vary over a fairly wide range of dispersal parameters, representing from low to moderate dispersal (Epperson and Li 1996). Indeed, within this range at first the curves become steeper, and further then begin to flatten out, as dispersal increases. This is unfortunate in the sense that in such cases I correlograms cannot provide precise estimates of dispersal; however, predictions for neutral loci are then robust to dispersal, and this could be useful in certain circumstances. For small quadrats (e.g., 4 individuals) I statistics for distance class one decrease monotonically with N_e, except for N_e in the range of 4–8.

Generally, there appears to be little reason for combining individuals into quadrat subsamples, although it could be needed for reasons of sampling feasibility. The most efficient use of data for individual genotypes is join counts or I statistics, rather than combining individuals into subsamples, as required for the usual estimates of F_{ST} or θ. To do so causes loss of information and usually statistical power. When it is necessary because of experimental feasibility, smaller quadrat sizes are preferable. Although large quadrat size should give a more precise estimate of the subsample area gene frequency, the loss of scale and information on joins within quadrats is severe. It has been shown that larger quadrat size corresponds to larger expected standard errors caused by the combined stochastic and statistical noise, for a given fixed total sample size (Epperson and Li 1996). However, the standard errors of I statistics are still fairly small as long the total sample size (the number of quadrats times the number of individuals per quadrat) times the number of loci is on the order of 2000 (e.g., 125 4-individual quadrats for four loci), which is well within practical experimental ranges (Epperson and Li 1996). Nonetheless, use of quadrats will also limit the minimum spatial scales that can be assessed.

The quadrat-based Moran's I statistics are calculated in the same way as for discrete populations. One example where it was applied to subpopulations is a study of *Helix aspersa* (Arnaud et al. 1999), where for each colony area the snails were repeatedly sampled, in order to increase sampling. The average Moran's statistic was ca. 0.18 for Gabriel-connected (a nearest neighbor criterion) colonies. Most of the loci showed a smooth increase of genetic isolation as distance increased.

OBSERVED JOIN-COUNT STATISTICS

The other of the two types of spatial autocorrelation statistics commonly used to characterize and quantify spatial structure is the join-count type of statistic. Join-count statistics are based on nominal data rather than numerical data (Cliff and Ord 1981), and for genetics these data are usually spatially located individual diploid genotypes. For the spatial distribution of three genotypes at a single locus with two alleles there are six different types of combinations in pairs, or "joins," of genotypes. For a locus with three alleles, there may be as many as six genotypes and then 21 types of joins. For a sample of size n, there are "n choose two", $n(n - 1)/2$, joins in total. As with Moran's I statistics, distance classes (ranges of distances that may separate pairs of individuals) are formed. The number $n_{ij}(D)$ of joins between genotypes i and j for each distance class of separation distance D is counted. For each type of join, for each distance class, the mean and variance are calculated under the null hypothesis of randomly sampling pairs of genotypes without replacement (Sokal and

Oden 1978a; Cliff and Ord 1981). For example, like joins ($i = j$) have expected value $u_{jj}(D) = W(D)n_j(n_j - 1)/2n(n - 1)$, where n_j is the number of individuals with genotype j, n is the total sample size, and $W(D)$ is the total number of all joins for distance class D. The standard deviation, $SD_{jj}(D)$, is given in chapter 8. A standard normal deviate test statistic, or SND, of this null hypothesis may be formed, $SND_{jj}(D) = [n_{jj}(D) - u_{jj}(D)]/SD_{jj}(D)$. Because this statistic has an approximate standard normal distribution, values greater than 1.96, or lower than -1.96, are statistically significant (Sokal and Oden 1978a; Cliff and Ord 1981). A set of SND statistics for the different distance classes is called an SND correlogram. In addition, SNDs can be calculated for the total number of unlike joins, i.e., the number of joins where the two genotypes are not identical. It is a useful summary inverse measure of spatial autocorrelation, especially where there are multiple alleles. A deficit of unlike joins at short distances generally indicates positive autocorrelations.

Join-count statistics provide a more detailed description of spatial distributions of genotypes, so let us briefly reexamine the distributions produced by isolation by distance. To fully quantify structure in detail in terms of the distribution of *diploid* genotypes, neither F statistics nor I statistics, separately or in combination, are sufficient for complete characterization. Gillois (1966) has demonstrated that a large number of additional measures are required. As a statistical matter, the situation is confounded by the fact that the two genes of a diploid individual may not be independent. Sets of join-count statistics provide more detailed measures of spatial distributions because they are expressed in terms of all the different combinations or pairs of diploid genotypes (in this respect they are analogous to the measures of Gillois). Explicit spatial distributions of diploid genotypes make theoretical models of two-dimensional systems further complicated and analytically intractable. Indeed, there are no analytical results explicitly for the spatial distributions of diploid genotypes separated by short distances within two-dimensional populations, and instead Monte Carlo simulations are utilized. Spatial autocorrelation analyses based on join-count statistics genotypes have successfully detected spatial structure (e.g., reviewed in Epperson 1993a; Shapcott 1995; Leonardi and Menozzi 1996; Leonardi et al. 1996), even in populations where structure is very weak, whereas, in the some of the same populations, analyses of F_{ST} often failed to detect significant spatial structure (e.g., Bacilieri et al. 1994). Although the methods of analysis and statistical computing are more complicated, join-count statistics may be considered especially sensitive detectors of spatial genetic structures.

For large simulated populations of genotypes at a locus (*A*/*a*) the SND values are very large and highly significant (figure 7.5). Most distinctively, for short distances, large excesses are observed for joins between neighbors that have the same homozygous genotype, either both *AA* (i.e., *AA* × *AA*) or both *aa* (*aa* × *aa*). Highly significant negative SNDs for *AA* × *aa* joins indicate deficits of joins between neighbors that have opposite homozygous genotypes. The excesses and deficits "drop off" as the distance between individuals increases, and typically at some distance they switch sign. This distance is important because it measures the diameter of patches. Typically, there are large patches, each consisting of 400–500 or more individuals.

Starting either from a random spatial distribution of genotypes for multiple alleles or from an initial population that is fixed for an allele, and allowing sufficient input of new alleles, the spatial structure builds rapidly. Large increases in spatial autocorrelations are evident within 10 generations. Within about 30 generations, the structure and statistics have nearly reached, and by 50–100 generations the populations have reached, a quasistationary distribution. The spatial autocorrelations scarcely change after about 50–100 generations (Sokal and Wartenberg 1983; Epperson 1990a,b). The correlations drop off with distance, and switch sign at about 20–25, or more, times the distance between adjacent plants. The SND correlograms for like homozygotes intersect zero at about the same distance as the *I* correlograms, indicating that both measure patch sizes in the simulations. The SNDs for like homozygotes have properties very similar to those for Moran's *I* statistics for allele frequencies converted from individual genotypes, and have similar statistical power. They decrease monotonically with increasing dispersal (figure 7.5).

For the SNDs, the precise level of dispersal has a large effect (Epperson 1995a; Epperson and Li 1997; Epperson et al. 1999). The most distinct feature is that increasing dispersal causes reductions in SNDs for joins between identical homozygotes for the shortest distance classes (figure 7.5). Complementary effects are seen for joins

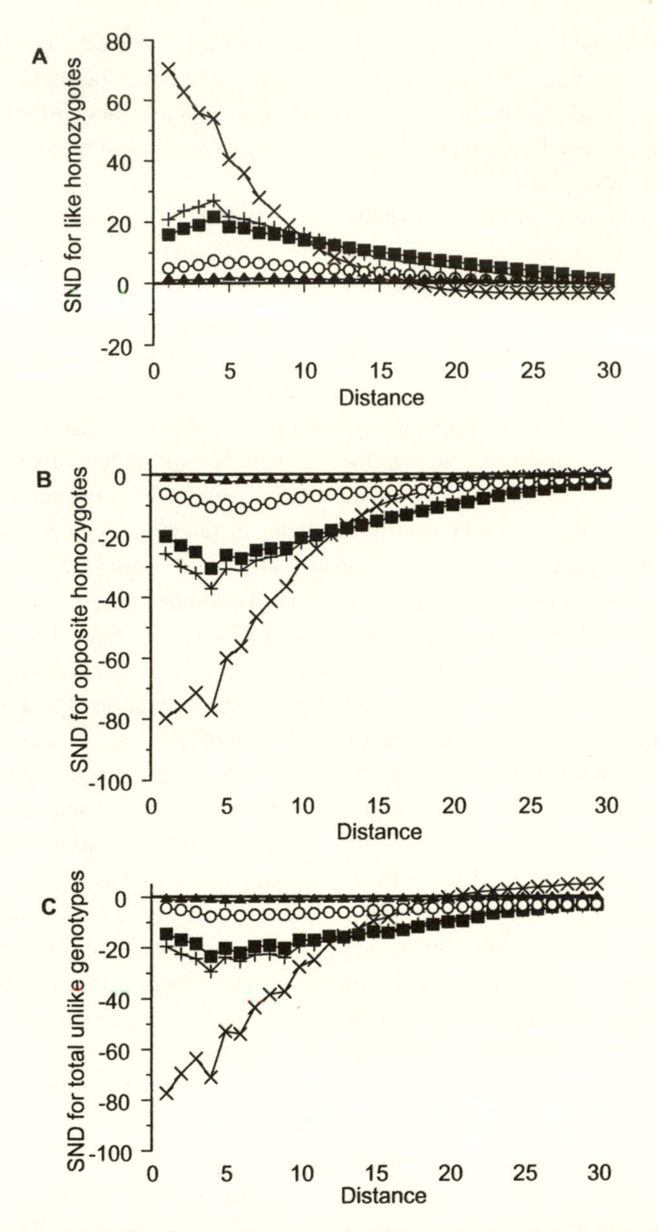

FIGURE 7.5. SNDs for simulations with different levels of dispersal, for joins: (A) like homozygotes (*AA* × *AA*); (B) opposite homozygotes (*AA* × *aa*); and (C) the total number of unlike genotypes. Average values are given for sets of 100 simulations each for the five different dispersal

between opposite homozygotes: greater dispersal causes lesser deficits of these joins. The same is seen for the total number of joins between unlike genotypes. The level of dispersal affects Moran's *I* statistics similarly, unless quadrat sizes are large (Epperson and Li 1996, 1997).

The first example of experimental studies is based on nine populations of the morning glory, *Ipomoea purpurea*, in the southeastern United States (Epperson and Clegg 1986), which contain a remarkable variety of flower color types. One locus controls pink versus blue color: two alleles segregate and the allele conferring pink is recessive to blue. Bumblebees account for more than 98 percent of visits to morning glory populations in the southeastern United States, and they cause a high degree of mating by proximity (Epperson and Clegg 1987). Morning glory seeds are large and do not disperse far. A series of direct experimental measures of gene flow places Wright's neighborhood size in the range of 5–15 (Ennos and Clegg 1982). The study populations are generally at least 50 or so generations old, and apparently stable in size. Based on the theoretical results reviewed above large, sharply defined patches should be present for neutral loci. Each population was sampled on a regularly spaced grid. The results clearly showed strong differentiation of pink homozygotes vs. the blue phenotypes, and these patterns are revealed in the join-count statistics. There were large excesses of joins between neighbors on the sample grids, where both are pink homozygotes, a similar excess of blue—blue joins, and a corresponding deficit of blue—pink joins. The distances at which the SND correlograms switch sign measured the diameters of the patches. Accounting for the overall density, which was also measured, the results indicated that the populations consisted of two types of patches, pink and blue, where each patch contained about 300–500 individuals, very similar to the simulated populations for neutral loci. The observed statistical values fit closely

---

models outlined in table 8.1, at generation 200. All were two-allele models with equal-frequency alleles. Symbols for dispersal sets are as follows: (X) set 1 (N_e = 4.2); (+) set 2 (N_e = 25.1); (■) set 3 (N_e = 50.2); (○) set 4 (N_e = 115.2); and ▲ set 5 (N_e = 632.4). Distance is measured in units of the lattice of individuals.

TABLE 7.2. Comparison of observed values of SNDs for like homozygotes in populations of *Ipomoea purpurea* and predicted values[a] from Epperson and Li (1997)

Population	*Sample Size*	*Porosity*	*Observed*	*Predicted* $N_e = 8.4$	*Predicted* $N_e = 12.6$
8201	166	9	2.0	3.6	2.6
8203(1984)	291	4	3.9	6.7	4.7
8312	52	10	4.0	2.0	1.5
8304	100	10	1.9	2.8	2.0
8205B1(1984)	366	3	4.0	7.5	5.2
8306	152	4	2.0	4.8	3.4
8303(1984)	318	4	2.4	7.0	4.9
8311	140	25	−0.9	2.7	2.0
		Average =	2.41	4.64	3.29

[a]Predicted using closest sampling schemes from tables in Epperson and Li (1997), and adjusting for actual sample size *n* by multiplying the table value by the square root of the ratio of actual *n* and the next higher simulation sample size.

those expected for the quasistationary distribution (table 7.2) for neutral loci in simulated standard isolation by distance models with Wright's neighborhood size in the range of ca. 10–25 (Epperson and Li 1997), which closely matches the direct measures of dispersal.

Schoen and Latta (1989) studied join-count statistics for loci controlling flower color and spotted corollas in *Impatiens pallida* and for waxy stems in *Impatiens capensis*. Both species have a high proportion of cleistogamous (selfing) flowers and gravity-dispersed seeds that disperse short distances. The sample sizes were several hundred for both species, and they had densities in the range of 80–300 per square meter. The SNDs for the joins between the same flower types, for the first distance class, were similar for the two traits in *I. pallida*, and averaged about 2.1, whereas those for unlike joins averaged −2.8. The porosity of the sample is not given precisely but apparently it was in the range of 4–9. As is shown in table 7.3, based on a sample size of 300 (extrapolating from tables in Epperson and Li [1997]) and sample porosities 4 and 9, the observed values for like joins are closest to those predicted (using linear interpolation between values for different simulated neighborhood sizes) for Wright's neighborhood size of 17 (if porosity is 9) to 45 (if porosity is 4). Similarly, values

TABLE 7.3. Comparison of observed values of SNDs for like joins and unlike joins in populations of *Impatiens pallida* and predicted values[a] from Epperson and Li (1997)

	Predicted				
Observed	N_e = 4.2	N_e = 8.4	N_e = 12.6	N_e = 25.1	N_e = 50.2
Like joins					
2.1					
Porosity 4	8.1	7.0	4.8	2.8	1.9
Porosity 9	3.8	3.3	2.4	1.6	1.1
Unlike joins					
−2.8					
Porosity 4	−10.0	−8.6	−6.1	−3.6	−2.4
Porosity 9	−4.9	−4.2	−3.0	−2.1	−1.4

[a]Predicted using closest sampling schemes from tables 8.3 and 8.4 and from tables in Epperson and Li (1997), and adjusting for actual sample size *n* by multiplying the table values by the square root of the ratio of actual *n* and the next higher simulation sample size.

based on extrapolated and interpolated values for the SNDs for unlike joins fit neighborhood sizes 15 (porosity 9) to 42 (porosity 4). Values for *Impatiens capensis* were similar, though it was sampled a little differently and it appears that the porosity was about 4. The mean value for like joins was ca. 3.6 and for unlike joins −4.5, fitting neighborhood sizes of 20 and 21, respectively. Correlograms for both species tended to intercept zero at distances of 2–4 meters, corresponding to genetic patch sizes in the range of 225–3800, based on the likely range of densities. These values are rough estimates, but nonetheless generally consistent with patch sizes on average ranging from about 289 for N_e = 4.2 to ca. 900 for N_e = 50.2 based on join-count correlograms in simulation studies (Epperson 1995a; Epperson and Li 1997).

Two studied natural populations of lodgepole pine, *Pinus contorta*, each continuous, large (with an excess of one million trees), and dense (with about 2500 trees per hectare), were analyzed using join-count statistics. The study populations had an estimated outcrossing rate of essentially 100 percent (Epperson and Allard 1984), and *P. contorta* has among the most highly dispersible pollen and seed even among conifers, which generally have fairly high dispersal. Wright's neighborhood size must be greater than 1000 and it could be ten times larger, thus no structure is expected in these populations. Each population was sampled on a grid, representing about one in every 50 trees in the sample area. Fourteen allozyme loci were assayed, many with multiple alleles and large numbers of genotypes. Out of the several thousand different join-count statistics that were computed, overall, about 5 percent of them were significant at the 5-percent level (Epperson and Allard 1989). The genotypes in these populations generally are extremely close to a truly random distribution, and this fits expectations because dispersal is so great. This is consistent with the isolation by distance model, and expected from the large amounts of dispersal, the stability of the population size, and the lack of age structure. We cannot be certain that there is no genetic structure at a finer scale than the distance between adjacent sampled trees. However, this seems unlikely because such structure should also be reflected at the sampled scales, under isolation by distance theory. It has been argued that genetically correlated (maternal half-sib in wind-pollinated species such as *P. contorta*) seed dispersal could cause fine scale structure in the absence of larger scale patterns. However, even this subtle structure is not clearly expected in species with even aged, stable population dynamics and high seed production and dispersal that results in highly overlapping "seed shadows."

A number of other species with very high dispersal rates have been analyzed using join counts. In a large isozyme study of *Picea abies*, Leonardi et al. (1996) found that overall, 7.3 percent of statistics were significant at the 5-percent significance level, although the percentage was higher (10.2%) in the shortest distance class. Some loci apparently exhibited some autocorrelation. Nonetheless, this case is reminiscent of the study on *Picea marianna* (Knowles 1991) and suggests that structure is very weak but not completely absent. Similar results (e.g., the SND for like haplotype joins for the shortest distance class was slightly negative and not significant) were found for genetic haplotypes for chloroplasts, which are paternally inherited in pines and thus dispersed through both pollen and seed, in *Pinus ponderosa* (Latta et al. 1998). However, the mitochondrial haplotypes, which are maternally inherited and thus dispersed only in the seed, did show significant autocorrelations (SND = 1.99). These results fit direct estimates of dispersal of pollen and seed, respectively

(Latta et al. 1998). Join counts have also been used in species with somewhat more restricted dispersal, such as *Fagus sylvatica*, where weak structure was found (Merzeau et al. 1994; Leonardi and Menozzi 1996), similar to other wind-pollinated species. Shapcott (1995) found many large SND statistics, particularly for joins between like genotypes, and a consistent patch structure, in the rainforest tree, *Atherosperma moschatum*. Join-count statistics have also been employed to identify genetic structure of nuclear DNA and mitochondrial DNA markers in microorganisms such as chestnut blight fungus, *Cryphonectria parasitica* (Migroom and Lipari 1995). They have more rarely been applied to animal species, but in species such as clonal soft coral, *Alcyonium* sp., significant autocorrelations have been found (McFadden and Aydin 1996).

In sum, spatial autocorrelation statistics are useful for describing spatial patterns and the kinship relations and genetic correlations caused by isolation by distance. They can also be used for forming a hypothesis-testing framework for detecting various forms of natural selection, especially by contrasting patterns for different loci. Moreover, they can be used to obtain estimates of the amounts of dispersal. In part, the power of the spatial approach is due to the fact that the spatial patterns may capture the cumulative effects of 50 or more generations.

VALUES FOR OTHER MEASURES

Studies using a variety of other statistics have similarly described genetic isolation by distance. Examples of geostatistical approaches, such as kriging, which are closely related to spatial autocorrelation techniques, especially Moran's *I* statistics, include the use of "second-order analysis" on the seaweed species *Pelvetia fastigata* (Williams and Di Fiori 1996). Other statistical methods center on the use of multilocus, often multiallelic data. Such methods may have somewhat different properties with respect to statistical power; assumptions under the null hypotheses; their relationships to models of stochastic processes; and the importance of ancillary estimates of local gene frequencies. For example, one statistic is based on the number of genes (alleles at loci, usually for multiple loci) shared in common by a pair of individuals, i.e., number of alleles in common, NAC (Loiselle et al. 1995). All of these statistics, as well as Moran's *I*, are closely related to measures of relatedness or kinship (e.g., Barbujani 1987; Hardy and Vekemans 1999), and thus bear connections with predicted levels of inbreeding caused by biparental mating by proximity (determined by the spatial distribution and dispersal function). Some have greater model dependence than do others. Although both spatial autocorrelation statistics and summary measures of relatedness are important, the stochastic and statistical properties of autocorrelation statistics are far better characterized. Little is known about the values of the other statistics, under isolation by distance, and their statistical properties. A number of fairly subtle statistical issues that distinguish these methods are discussed in detail in chapter 8, and only briefly here. Recently, Smouse and colleagues (Smouse and Peakall 1999) developed a measure that is closely related to NAC. Both measures are closely related to the haploid measure developed by Bertorelle and Barbujani (1995). The statistical properties, which often involve ways of averaging (e.g., over pairs of individuals first, then over loci in the case of spatial autocorrelations, and vice versa for these multilocus estimates), may also involve problems arising from nonindependence of the two genes of a diploid, problems regarding lack of knowledge about expected values for the local gene frequencies, as well as disequilibrium among loci. Interestingly, what few results there are on multilocus distributions under genetic isolation by distance strongly suggest that gametic phase linkage disequilibrium is negligible even in highly spatially structured but large populations (Epperson 1995b), even if loci are tightly linked. Recent work (discussed in detail in chapter 8) developed proper methods for constructing Moran's *I* statistic averages over alleles and loci. With these results there is every advantage to using Moran's *I* and the other measures derived from it. In addition, spatial autocorrelation statistics based on the total numbers of unlike joins among multilocus genotypes have been developed (Epperson and Allard 1989).

In the very large study of *Quercus laevis*, Berg and Hamrick (1995) computed both NAC and various Moran *I* statistics. There was an excess of NAC among adjacent individuals, compared to the average among all pairs of individuals, and the excess dropped off smoothly as distance increased, as presumably would be expected under genetic isolation by distance. At a distance of 28 meters the NAC became in deficit, and this apparently provides an estimate of patch size. Indeed, this estimate is similar to those based on *I* correlograms for quadrats (40–45 meters) calculated for the same data, or as based on visual inspection of homozygote concentrations (ca. 25 meters). Similarly, the measure developed by Smouse and Peakall (1999) suggested an intensity and scale of spatial autocorrelation consistent with correlograms for individual loci.

Multilocus join-count statistics, based on the total number of unlike joins between trees, in the previously mentioned study populations of *Pinus contorta*, were not statistically significant, in keeping with the single-locus measures. Moreover, 5-locus haploid genotypes among spatially located samples of individual pollen grains (actually the paternal contributions to seed in cones) also did not have a significant deficit of unlike joins. Under the conditions of sampling in this case we should expect this measure to be extremely sensitive to any heterogeneity in the pollen cloud of the population. Again this fits with the extreme dispersal and densities in the populations (Epperson and Allard 1989).

The spatial distribution of quantitative traits also appears to be predictive under isolation by distance. Note that technically the standard genetic isolation by distance theory requires the trait to be selectively neutral and this must be rare. However, it may be that the expectations are reasonably close to those for some types of selection (especially if not microenvironmental and/or not too strong). Simulations of neutral two-locus genotypes indicate that the dominant spatial features are patches of double homozygotes, in cases of low dispersal (Epperson 1995b). Processes with greater dispersal are perhaps better characterized as having patches of concentrations of specific alleles for both loci. Nonetheless, despite this implicit, very local (i.e., within patch) gametic phase disequilibrium, the alleles that are concentrated together vary among patches, and in large populations the disequilibrium of the population is near zero, even with tight linkage (e.g., recombination rate $r = 0.01$). In addition, it was shown that the Moran's *I* statistics for quadrat samples of two-locus genotypes, calculated after the genotypes were converted into additive genotypic values for a quantitative trait, are identical to those for the one-locus case (Epperson 1995b). This implies that the weightings of loci with respect to their contributions to the trait do not matter much if at all, and in the simulations neither did the recombination rate. Therefore, the results strongly suggest that for additive alleles and loci the expected *I* correlograms for the genotypic contributions to the trait generally are identical to the single-locus case. This is analogous to the identity of the predicted spatial correlations of quantitative trait models and single-locus models *among* discrete populations in stepping stone and similar systems (Rogers and Harpending 1983; Lande 1991). However, the stochastic variation of spatial correlations for quantitative traits might differ, and presumably is reduced, from that in single-locus models, because of averaging among loci. Environmental contributions to the trait might reduce correlations, although perhaps not in expected value. Among the few studies that have applied autocorrelation statistics to quantitative traits, there is a rather large study on *Impatiens capensis* (Argyres and Schmitt 1991).

BIPARENTAL INBREEDING CAUSED BY SPATIAL STRUCTURE AND ITS INTERACTION WITH INBREEDING DEPRESSION AND SELECTION

In plants, mating by proximity is one of the major determinants of inbreeding, and its importance is secondary only to the selfing rate. Most plants are either highly selfing or primarily outcrossing (Brown 1979). In highly selfing species of either plants or animals the inbreeding contributed by mating by proximity is usually minor in comparison. In highly outcrossing or obligate outcrossing plant species, biparental inbreeding is the major or only form of inbreeding, and it is usually determined by the spatial distribution of relatives and the pollen dispersal function on distance. Plants without selfing include those that are dioecious and other species that have strong self-incompatibility systems or strong mechanisms that prevent selfing. Although spatial structure and limited pollen dispersal are usually considered the main causal factors for inbreeding, there can be others, which may be unstudied in any given plant population. The other main category is temporal rather than spatial. For example, genetic correlations in dates of floral maturation may increase the likelihood of mates between relatives. However, in species with strong protandry or protogeny, such genetic correlations may *decrease* consanguineous matings by reducing the probability that females flowers are receptive at the same time related male flowers shed pollen. Such mechanisms may allow a population to limit gene flow, to better tract the microenvironment and increase local adaptiveness, and yet insure adequate pollination and limit inbreeding.

Many animal species exhibit complex mating system behaviors that can have quite powerful influences on levels of inbreeding. In many cases, for example, many forms of assortative mating, such behavior is not particularly spatial. However, behaviors that implicitly determine movement or migration patterns (Fix 1978) will affect the degree of mating by proximity and inbreeding. Nonetheless, even assortative mating may interact with spatial structure because usually the individuals considered as potential mates are within a limited area. Spatial structure, usually plays an important, if secondary, role in the mating system of animals at one spatial scale or another (Chesser and Baker 1996).

Levels of biparental inbreeding can have population effects ranging from disastrous to beneficial, depending on many factors. A low level of inbreeding may cause large populations (thus undergoing little inbreeding from finite population size per se) to maintain lower frequencies of recessive deleterious genes, which is beneficial in the long run. In contrast, in small populations it may trigger a process of meltdown, whereby the inbreeding makes the population less fit, the population size decreases, causing greater inbreeding, in a cycle that puts the population on a downward spiral to extinction (Lynch and Gabriel 1991). On the other hand, spatial structure (which itself is determined by dispersal of male and female animals, or plant seed and pollen) and effective gene dispersal itself may be changed by selection. For example, in *Chamaecrista fasciculata* biparental inbreeding apparently increases distances of effective pollen dispersal (Fenster 1991), because short-distance, more inbred pollinations produce progeny that are less fit and less likely to reproduce. In addition, selection in the form of inbreeding depression can directly cause changes in the spatial structure per se, probably in many cases reducing it. This is an important form of the interaction between selection, dispersal, and spatial structure (see also Epperson 1990a). At some (usually larger) spatial scales spatial structure sometimes has the opposite effect, in a phenomena termed outbreeding depression (Price and Waser 1979; Waser and Price 1989; Schmitt and Gamble 1990). In this process pollinations from individuals too far away produce less-fit offspring, presumably because those individuals carry genes that are less adapted to the local environment, and/or combinations of favorable gene complexes are less likely.

Complex interactions are expected between the selective deaths or reduced reproduction of inbred individuals and the spatial distribution of genotypes. By analogy, in nonspatial models the dynamics of genes that influence selfing rate interact strongly with those of genes that contribute to inbreeding depression (Holsinger 1988; Uyenoyama 1986). The dynamics may depend on the form of gene action such as overdominant selection versus deleterious recessive (Uyenoyama and Waller 1991). Doligez et al. (1998) examined some cases of inbreeding depression under a specific spatial form of selection. They found that heterozygote advantage at selected loci could change the Moran's *I* statistics for an unlinked neutral locus in a complex way, and, moreover, the effect apparently operates through the degree of spatial clumping (nonuniformity of density) in the population. However, this question warrants further study, as it would also address the fundamental issue of how selection acts on species that have no or limited movements, in relation to competition and population regulation. The situation becomes even more complex if we consider the dynamics of genes that directly influence dispersal, so-called dispersal genes (e.g., Olivieri et al. 1995).

The effect of spatial structure on inbreeding can be directly assessed by the convolution of the probability distributions of the spatial distances between relatives, using some measure of the intraclass correlation by distance, and the distribution of dispersal distances (Wright 1943). However, for plants, much of the evidence comes from measures of inbreeding coefficients in populations where either the selfing rate is known and accounted for or there is no selfing. A wide range of inbreeding coefficients are observed, generally reflecting the limits to dispersal (Brown 1979). Indeed, the distribution of these values across the landscape, either as a function of the pairwise values on distance (Malécot 1948) or within spatial groups of individuals (Wright 1943), forms the basis of genetic isolation by distance models. For plants, many other observations come from studies of the apparent population rate of self-fertilization. Typically, single-locus estimates of outcrossing rates are considerably lower than multilocus estimates, because the latter are usually far less biased by other forms of inbreeding (Shaw et al. 1981; Ritland 1985), and biparental inbreeding is generally attributed to spatial autocorrelation of relatives and limited pollen dispersal, i.e., mating by proximity. A rare exception occurred in my study of populations of *Pinus contorta*, presumably because there appears to be essentially no spatial structure, because of its extraordinary combination of densities and dispersal distances (Epperson and Allard 1989).

THE ROLES OF DEMOGRAPHIC FACTORS IN SPATIAL STRUCTURE

Standard models of genetic isolation by distance are framed in terms of either a lattice or a spatially continuous population, with discrete generations and stable population size and density. They generally ignore numerous demographic factors, such as population growth and regulation and age structure. Predictions of the standard models are robust to some deviations from these assumptions. Doligez et al. (1998) found that age structure tends to increase spatial structure, and clumping of individuals (i.e., nonuniform density) tends to increase short-distance spatial autocorrelations. However, in most cases the effects on the long-term ("quasistationary") spatial distribution of genetic variation were rather small (see also Epperson and Li 1996, 1997). Fairly marked effects were observed when there were extremely large clumps separated by large empty areas. This extreme form of clumping should occur rarely in nature, and only when there are specific ecologies. Formally, the model assumed that there were no effects of local density on survival and reproduction. In plants this requires that probabilities of successful reproduction within a very local area depend only on local seed sources (i.e., other plants are available in the area and seeds do not disperse very far), and that the success of the seedling is not harmed by the presence of other conspecifics. This is unlikely whenever the population size or density is being regulated. It could occur, for example, at some early stages after a pioneer species colonizes a new site at low initial density. In some such cases the presence of conspecifics may actually increase survival, for example, by stabilizing the soil against erosion or providing structural support or helping to retain soil moisture. Other causes of extreme clumping may simply reflect the inability of the species to live in certain areas of the population, although this would require a different model. In general, patches of available habitat could have a very wide variety of shapes, sizes, and distributions, and thus these might have various effects on the spatial genetic structure, although this remains unstudied. On the other hand, inclusion of stochastic sampling, in effect analogous to some very weak forms of clumping, had essentially no effect on spatial autocorrelations (Epperson and Li 1996, 1997). Spatial autocorrelation analysis should not be applied blindly in extreme cases of clumping, and often problems can be managed by substructuring the sampling and avoiding depauperate areas or adjusting density of sampling (e.g., Epperson and Clegg 1986).

The magnitudes and shapes of correlograms at the smallest distances may be particularly sensitive detectors of details of demographics. Combining genetic analyses with demographics may allow modeling and inferences of the processes that determine the genetic structure of populations. The amount and form of genetic variation maintained locally can be affected by demographics. Genetic differences among age classes may store genetic diversity. In some cases, structures generally considered to be short term, and therefore perhaps viewed as less interesting, may, in fact, be the dominant form of spatial structure for a species. Structure may depend on the age of a cohort or life stage and changes in population density. In plants, some key factors are the degrees to which limits to and correlations of seed dispersal produce and maintain substantial spatial structure of genetic variation through groups of kin-related (e.g., maternal half-sibs or full sibs) individuals. Spatial matrilineal groups may increase in numerical and spatial size if initial density is low, available habitat is abundant, the population size or density is well below local carrying capacity, and intraspecific competition is minimal. The critical point is how this structure is swamped out by long-distance pollen dispersal, or the length of time before the structure created by limited seed dispersal is replaced by more-standard isolation by distance curves. Analogous features exist for higher animals, in terms of how matrilineal, or less frequently patrilineal lineages, may occupy a spatial area, resulting in various consanguineous matings. An additional feature in many populations is clonal reproduction, which can be viewed as a short-term or temporary spatial structure. Clonal reproduction of sexually reproducing individuals may result in local inbreeding and may also affect the long-term genetic isolation by distance, but this depends on the distances that clones spread, the mating system (particularly dioecy vs. monoecy), and pollen dispersal distances.

It has been widely viewed that limited seed movement likely causes spatial clustering of matrilineal groups (maternal half-sibs, full sibs, or self-sibs, or some combination of these), and that this may be the *dominant* form of spatial structure in wind-pollinated (long pollen dispersal distances), outcrossing species (Linhart et al. 1981; Sakai 1985; Neale and Adams 1985). However, this may depend on density, population growth rates, and what can be termed correlated seed dispersal (typically, movement of matrilineal groups of seed together, often within a many-seeded fruit) or in animals kin-structured dispersal (see Fix 1978).

A good example of the role of population size regulation in the spatial structure of adults following formation of matrilineal groups of seedling is *Cecropia obtusifolia*, a dioecious tree that regenerates in gaps in the canopy of subtropical forests. It is wind pollinated and has many-seeded fruits, which are dispersed by bats and other animals. One population in southern Mexico was studied in detail by mapping and genotyping all individuals of all age classes within an area of several hectares (Alvarez-Buylla and Garay 1994). Both male and female adult trees, about 100 in number, were fairly evenly dispersed, displaying only slight clustering in the population. In contrast, the distribution of all 242 seedlings was very clumped, almost all occurred within six tight clumps, and each clump was located in a recent gap in the forest canopy. The distribution of the juveniles was similar to that of the adults.

Join-count statistics were calculated for eight isozyme loci, separately for the three age classes (Epperson and Alvarez-Buylla 1997). Strikingly nonrandom spatial distributions of genotypes for isozyme loci were observed among the seedlings. In addition to the usual distance class designations, it was noted that the first distance class (0–12.5 meters), contained no pairs separated into different gap clumps, and the second distance class, 12.5–37.5 meters, contained the remaining within gap pairs and many between gap clump pairs. For the first distance class, 31 of 61 (ca. 50 percent) SND join-count statistics were statistically significant at the 5-percent level. Many were also significant for distance class two. There was almost no autocorrelation at larger distance classes. In total, for distance classes one and two, there were many very large values, larger than had been previously observed for any other species, including *Ipomoea purpurea*. However, the statistics were quite different in form from those expected under standard isolation by distance theory. They indicate a very strong concentration of genotypes into gap clumps, and little relationship among gap clumps. The gaps contain far fewer individuals than do the patches of genotypes observed in simulations of the standard model. More importantly, while some types of joins fit the usual excess types expected under isolation by distance, others do not. For example, there are significant deficits of joins between some identical homozygotes and excesses of joins between some different heterozygous genotypes, for distance classes one and two.

In contrast, for the adults, 15 of 67 statistics (22 percent) and 11 of 79 (14 percent) were significant for distance classes one and two, respectively, and few statistics were significant at larger distances. The values were generally smaller than those for the seedlings, but more importantly almost all involved far fewer joins. Most of the significant values involve odd combinations of genotypes. For example, for locus Pgm-1 there was an excess of joins between the opposite homozygotes *22* and *33*, and this reflected concentrations of the rare *22* homozygotes into one of the patches. In addition, more standard associations were found. For locus Fe-2 the only two *13* heterozygotes are located next to each other, as are three of the four *22* homozygotes. For Fe-1 all three *23* heterozygotes are adjacent. There can be little doubt that such concentrations of genotypes are not statistical artifacts, and the type of structure differs from the standard model in form. Similar results were observed for the juveniles. For both adults and juveniles, what spatial nonrandomness of genotypes there is consists of clusters of few individual genotypes at a very fine spatial scale.

The structures observed at different age classes in *C. obtusifoli* are nonstandard, but make sense when we consider the demographic details of seed dispersal and population regulation. Because *C. obtusifolia* is dioecious and wind pollinated, most seeds on a tree should be maternal half-sibs and otherwise unrelated. Effective reproduction essentially requires the fruits, often carried long distances by birds and other animals, to be dispersed into recent gaps in the canopy, since the population size is stable and regeneration only occurs in gaps. Outside of gaps seeds are found, but not seedlings because they do not successfully germinate and grow (Alvarez-Buylla 1994). It is likely that animals may disperse one or a few fruits (groups of half-sib seeds) from only one tree, or a few trees, and deposit them at a particular locale. If this locale occurs in a gap the seeds may successfully germinate and grow. Thus the seeds that arrive in a gap may consist of a single half-sib array, or a few essentially unrelated (unrelated because there is little structure among adults) half-sib arrays. There are many low-frequency alleles among the eight loci, and hence it is likely that the fruits carry combinations of some low-frequency alleles, thus causing genotypic concentrations within one or another of the gaps, and statistically significant structure for the seedling population as a whole. As the seedlings grow, competition for the canopy gap removes all but one on average, but on occasion two, three, or four may survive to the juvenile or adult stages. This results in a nonrandom distribution with slight structure among the adults. The prediction of the standard isolation by distance model is essentially no structure, based on the high level of seed dispersal and wind pollination. Thus in contrast to the seed, the distribution of genetic variation in the adults, except for this relict, weak, and nonstandard form of structure, fits the standard model predictions. The structure among seedlings is strong, but transient, and essentially unimportant with respect to the buildup of isolation by distance over the long term of many cycles of regeneration.

Another, contrasting case where it may first appear that structure is temporary and unimportant is *Silene dioica*. *S. dioica* is a small dioecious perennial, which in certain regions of northern Sweden primarily inhabits islands. It is insect pollinated, primarily by bumblebees, in this region, and has no particular seed dispersal mechanism. Because the region has been undergoing rapid uplift (ca. 1

centimeter per year) since the last ice age and the landscape is rising from the sea from north to south, new islands are continually rising above sea level. Once above sea level, the islands undergo rapid ecological successional stages, starting as rocks frequently washed by sea, then having small plants and eventually mountain ash, alders, and conifers. Different successional stages are visible among islands, and also within an island as bands of vegetation types corresponding with topographic bands. *S. dioica* occupies an intermediate position in this succession (Ingvarsson and Giles 1999).

B. A. Giles has followed the precise demographics on a individual plant basis, since the initial foundings on different islands, including a population on Bigstone Island. Based on observations of demographics, Giles had previously described numerous spatial demographic patches (not to be confused with the spatial genetic patches), which appear to represent cohorts of seed from the same maternal plant. Moreover, three distinct subpopulations were identified, respectively on the north, west, and east sections of the island. Extensive isozyme genotyping was done for the entire population on Bigstone Island. Many of the main points from spatial autocorrelation analyses (Epperson 2000b) of genotypes for the entire island population are illustrated by statistics for the total numbers of unlike joins for genotypes for each locus. First, the average values of these statistics clearly indicated substantial autocorrelation in distance class one, zero to 1.0 meters, the upper bound of which is near the average diameter of the demographic patches. Second, there appeared be a "step" in the curve at about 5–6 meters, which is the distance separating the subpopulations. Normally, one would not make much out of such a "step," but it fits with other information indicating little autocorrelation among patches within subpopulations and substantial differentiation among subpopulations. Analysis of distinct types of joins revealed large excesses and deficits but in nonstandard types of joins, reminiscent of the seedling data for *Cecropia obtusifolia*.

Separate spatial analyses for each of the three subpopulations best reveal the dynamics during different population growth stages. The youngest section of the island population is in the north, where *S. dioica* occurs in low density and there is a relatively high percentage of isolated individuals and well-isolated demographic patches. Two separate spatial analyses were done for this subpopulation and also for the other two subpopulations: (1) an analysis that included one "distance" class for pairs that were within the same patches, as defined by the demographic studies, and all other pairs were classified according to distance of separation; and (2) an analysis based on the usual classification of all pairs into classes defined only by distance (Epperson 2000b). The results for the within-patch designations revealed many large correlations, and many of these clearly indicated odd excesses or deficits of joins that were of nonstandard IBD types. As a set, the results were like those observed in the distance-based spatial autocorrelation analysis of the total population. As for the total population, significant statistics were caused by concentrations of relatively-low frequency genotypes within one or another patch. For example, in the distribution of genotypes for locus Acn-1 all three *23* heterozygotes are concentrated into a single patch, which also has a concentration of *12* heterozygotes, suggesting that the founding matriarch was a *13* heterozygote (the most common genotype in the north is the *22* homozygote). There was no autocorrelation outside of patches, even for relatively short distances. These results indicate that the demographic patches do indeed represent genetic patches, precisely.

The western section of the island contains the second youngest subpopulation, and compared to the north, the west has higher density, larger patches, fewer isolated individuals, and less space between patches. Like the north, large correlations are observed and again these are often odd or nonstandard-IBD types, and they again represent concentrations of relatively low-frequency genotypes within one or another patch. There was little autocorrelation outside of patches, and although these values were small they were nonzero. Thus, in the west the demographic patches represent genetic patches not exactly but almost so.

The east is the oldest section of the island population. The autocorrelation results with patch designation revealed some large correlations, and again sometimes odd or nonstandard IBD types. As for the other subpopulations, the statistics represent collections of relatively low-frequency genotypes within one or another patch. However, unlike the north and west, there was also genetic relationships *among* adjacent demographic patches. Overall, most of the structure is within demographic patches, but not all. There is a fair amount of autocorrelation outside of patches, even for relatively short distances, although again no autocorrelation at larger distances (i.e., beyond distance class one). Otherwise, the amount and form of spatial genetic structure is very similar to younger north and west subpopulations and the spatial scale is the same. The matrilineal nature of patches remains strong and it remains the main form of spatial structure of genetic variation. Each maternal lineage must trace through seed, and because density has been high the recent dynamics apparently did not allow new seed much opportunity to grow, since available habitat had already been filled.

The kind of spatial genetic autocorrelation that is observed presently at all three population stages should persist long into the future, and this is likely even for the east. The demographics appear to follow a sequence. Once an isolated female plant becomes established, it produces seeds, which are heavy and generally fall within a circle with radius equal to the height of the inflorescence, ca. 20 centimeters, year after year (Giles, personal communication). After several years, some of the offspring become reproductively mature, and daughters can then begin to contribute to the local seed fall. Seedlings aggressively fill in available habitat within 10–30 years. Individuals can live for fairly long periods, 5–10 years, but even if a founding matriarch dies her legacy persists for the "life" of the patch. Within a few additional decades habitat in the patch is no longer suitable. *Silene dioica* must "move" elsewhere on the island (to a lower topographic zone in an earlier successional stage) or to another island (Giles and Goudet 1997). In the east, many of the adults are still relatively young and will live for many years, and because available habitat has already been largely filled, future pollen and seed flow (which is required for transition to a more standard isolation by distance) can occur only as these die. Moreover, it appears that the east region is at a late successional stage for *S. dioica*, and soon the available habitat will vanish. Thus, it will become increasingly likely that dead silene will be replaced by individuals of other species. It appears that the form and magnitude of genetic structure observed in all three subpopulations is in fact characteristic of the species on these islands. Moreover, the contributions of this population to elsewhere on the island or to new islands will be dominated by seed produced by this structure combined with matings expected based on pollinator movements. Because *S. dioica* is dioecious, this structure should be the main determinant of the mating system.

In both *Silene dioica* and *Cecropia obtusifolia* the patch "colonizers" are perennial maternal groups of seedlings, and the important difference is that *S. dioica* survive to maturity in high numbers, because the population density is increasing. It may be suggested that wherever available habitat first rapidly increases then rapidly decreases, which defines certain demographic parameters, nonstandard genetic structure exists on such a timescale that it is the main form of within-population genetic structure for the species. The structure exhibits extreme levels of spatial autocorrelation, predominantly matrilineal relationships, caused by highly limited seed and pollen flow at a specific spatial scale that corresponds to demographic patches, and these demographic patches are much smaller than the genetic patches expected in longer-standing populations.

Clonal reproduction also may have marked effects on the spatial genetic structure of plant populations. For monoecious, self-compatible species, clonal groups may increase effective selfing, whereas in dioecious plants any effects on the development of isolation by distance may be more subtle. Other factors may include the spatial distributions of clones and the timescale over which populations have existed. Moreover, the demographics of clonal structure may interact with the spatial–temporal demographics of sexual reproduction and the structure of genetic variation.

An example of clonal reproduction and spatial genetic structure is the study of *Adenophora grandiflora* mentioned earlier. Populations of *A. grandiflora* have a high degree of vegetative reproduction, and the species is a monoecious, herbaceous perennial, pollinated by bumblebees and possibly other pollinators, with reportedly wind-dispersed seed and high outcrossing rate (Chung and Epperson 1999). In addition to the analysis reported in table 7.1, separate analyses were done for the sexually reproduced plants. Clonal effects could be separated out, because plants with the same multilocus genotype must almost certainly be clonally related. Separation was achieved by randomly retaining only one individual of each multilocus genotype, and then conducting autocorrelation analysis on remainders, the so-called sexually reproduced individuals. Analysis of clonal distribution was conducted using statistics of the total number of joins between unlike (nonidentical and hence nonclonal) multilocus allozyme genotypes. These values indicated large spatial autocorrelation, concentrations contained within distances of ca. 15 meters (Chung and Epperson 1999), and this reflects clumping of clones. (This new way of analyzing spatial distribution of clones takes into account the spatial distribution of individuals, unlike simpler metrics based solely on the distances separating pairs of clones.) In both populations the autocorrelation among the sexually reproduced individuals corresponded to predicted values under the standard isolation by distance model for Wright's neighborhood size of about 50. The autocorrelation is rather larger than normally expected, in part because the seed are wind dispersed. It seems likely that additional consanguineous matings are caused by limited pollen flow combined with clonal structure, itself caused probably by both restricted seed flow and restricted clonal "dispersal" (distances at which clones form or are formed over time). Moreover, a substantial deficit of heterozygotes was observed, and this was probably caused by selfing and intracrossing among clones.

Intracrossing of clones cannot occur in dioecious species, and any effects of clonal reproduction on the buildup of genetic isolation by distance must be more subtle. Moreover, clones may develop long distances from their original progenitor, either immediately or after several cycles of clonal events. Both *Rhus tricocarpha* and *R. javanica* are dioecious, woody perennials, with insect pollination and seed dispersal by gravity and by birds. In two study populations of *R. tricocarpha* in Korea the frequency of clones was high (51 and 60 percent). However, contrary to typical expectations, clones were often found at long distances from each other, indicating long distances of either primary or secondary clonal reproduction, and there was no autocorrelation among clonal genotypes (Chung et al. 1999). Autocorrelation among sexually reproduced individuals was similar to that for the total population. Because in this case clonal reproduction does not cause spatial autocorrelation, it does not contribute to the buildup of isolation by distance; indeed, it may have the opposite effect, by increasing dispersal. Since there is no autocorrelation among clones, we cannot determine the influence of dioecy. However, we can note the effects of population age. Both populations are very young, according to demographic data, existing at the sites for only 26 and 12 years, respectively, and both have some spatial genetic structure. The genetic structure of sexually reproduced trees was greater for the older population, and it appears that both populations are in early stages of the buildup of isolation by distance, except that clonal reproduction should continue to randomize, to some extent, the structure at large spatial scales. Both *Rhus tricocarpha* and *R. javanica* in Korea are pioneer species that accelerate establishment of other tree species. Other tree species may replace *Rhus* within a few to several decades.

For two study populations of *Rhus javanica* (Chung et al. 2000), there was no autocorrelation among clonal genotypes for one, but there was slight autocorrelation within 4–6 meters for the other. For the first, there was no difference in the structures for the total versus sexually reproduced population, again because the distribution of clones was random. However, there was substantial structure among the sexually reproduced, presumably caused by limited seed and/or pollen flow. For the second population, there was an autocorrelated distribution of clonal genotypes, and this did contribute to total structure. In fact, clones contributed all of the structure, and there was essentially no structure among sexually reproduced individuals. In both populations, clones have not accelerated the build up of isolation by distance. However, it appears that limited pollen flow is in the process of creating a type of spatial structure similar to that for *R. tricocarpha*, because the more-structured first population is older than the second.

USING SPATIAL PATTERNS TO DETECT NATURAL SELECTION

In total, the results of numerous simulation studies, anchored by analytic results of mathematical models, lead to strong predictions for spatial correlations for neutral loci. In principle, these results can be used in two different ways in studies designed to detect or measure natural selection operating on individual loci. One is to use independent knowledge of the amount of dispersal, together with information on density, to form predicted spatial correlations for neutral loci, and compare to these spatial patterns observed for specific loci hypothesized to be subject to selection. The other is to examine spatial correlations for multiple loci and detect contrasts among loci. Loci under the same influences should have similar spatial correlations, and neutral loci are generally subject to the same dispersal and mating in a population.

Spatial patterns and spatial autocorrelations for neutral loci may depend on a combination of demographic factors as well as overall dispersal rates. However, one broad set of conditions that may apply to many species leads to a rather robust set of predictions. Specifically, the following situation may be particularly common: (1) the population has dispersal in a low to moderate range, say $N_e < 100$; (2) the population is at least 30 or more generations old; (3) the sample is taken over an appropriate spatial scale; and (4) the allele frequencies are not too close to zero. In this situation, all neutral loci should have large patch structure, and have the characteristic SND correlograms and *I* correlograms. In most cases, if sample sizes are large, the SND correlograms also are large and reflect the patch structure.

To predict spatial structure for neutral loci, it is often relatively easy to determine if a study population has dispersal within certain fairly broad ranges. In contrast, it is notoriously difficult to directly estimate dispersal with great precision. In such cases one might prefer to use quadrat sampling since it is less sensitive to precise level of dispersal within reasonable ranges, whereas values of Moran *I* statistics calculated on genotypes converted to gene frequencies are quite sensitive to dispersal. The size of quadrats is important. It is also important to note that values of SNDs for join-count statistics also depend on sample sizes, and tables that allow adjustments for sample sizes have been published (Epperson and Li 1997; Epperson et al. 1999). Predictions for neutral loci are quite robust to allele frequencies. Moran's *I* statistic is generally unaffected even if frequencies are as low as 0.05, and only marginally affected at still smaller frequencies. Join-count statistics are somewhat more sensitive, when frequencies decrease below about 0.1, but fairly unaffected otherwise. The spatial patterns observed for loci under various selection schemes may stand in contrast to those predicted for neutral loci. For example, they may have markedly reduced patch structure and greatly reduced autocorrelations (Epperson 1990a). It should also be noted that although contrasts are viable approaches in principle, there remain a number of statistical problems of comparing observed autocorrelations to hypothetical ones. Most of these arise from correlations of autocorrelations at different distance classes.

In most studies, numerous genes are studied and often it is expected that at least some of them are neutral. It is usually pretty easy, and getting easier all the time with molecular advances, to obtain data for many loci. Multiple-locus data can be used in two ways. One is to use observed values of SNDs or Moran's I, averaged over several apparently neutral loci, to precisely estimate dispersal (Epperson and Li 1997; Epperson et al. 1999). Spatial autocorrelations for the *first* distance class are very sensitive to dispersal level. Using them separately (rather than entire correlograms) to estimate dispersal greatly simplifies statistical procedures. Then these estimates can be used to find the predicted autocorrelations based on models of genetic isolation by distance and neutral theory. Second, the values for the first distance class can also be directly contrasted among loci, using standard statistical procedures for independent random variables.

The changes that selection causes in spatial patterns depend on how selection operates. Selection in a spatial context requires more precise specification as to how selective deaths or differentials in reproduction occur. Nonspatial selection models typically assume that selection acts independently among individuals. However, if there is variable microenvironmental selection differentials among genotypes, then the result of selection typically will depend on the individuals in each microenvironment. Even if selection is not microenvironmental, it will often act through localized competition among individuals, so again the local fitness of an individual depends on its neighbors. It seems clear that selection generally strongly interacts with spatial structure. Relatively little is known about how selection affects spatial genetic structure, and even less is known on how genetic structure affects the efficacy of selection or even its long-term direction.

Selection is not variable microenvironmentally when the factors that influence marginal fitness are not spatially distributed. Nonetheless, for most plants and many animals extensions of the standard selection model can still raise some fundamental issues. For example, plants generally cannot move to respond to the environment. Unless fitness differentials are purely genetic, and the environmental component is truly spatially constant, or random, and there are no genotype by environmental interactions, the spatial distribution of environments will factor in. However, when the spatial heterogeneity of environmental factors occurs on a scale smaller than the typical distance among individuals, selection may collapse to a nonmicroenvironmental form (Epperson 1992).

The effects of nonmicroenvironmental selection on spatial genetic structure have been evaluated using Monte Carlo simulations, where the probabilities of selective removal depend only on the genotype, not its spatial location (Epperson 1990a). However, in the simulations the removed individuals were replaced by one of their near neighbors, consistent with one form of competition among neighboring individuals. Remarkably and perhaps surprisingly, this form of selection can greatly decrease spatial genetic structure from levels produced by isolation by distance. Naturally, the reductions are most striking when dispersal is lowest, and the level of spatial autocorrelations is highest. The I correlograms for a locus with directional selection against one allele a, with additive fitnesses 1, $1 - s/2$, s for genotypes AA, Aa, aa, and with strong selection, $s = 0.1$, are much lower than those for selectively neutral loci (figure 7.6). With weak selection, $s \leq 0.01$, the effects are negligible even for populations with low dispersal. Differences in spatial patterns and size of genetic patches, both reflected in reduced values of spatial autocorrelations, are distinct within 30–50 generations (Epperson 1990a). In these simulations the spatial scale of local competition did not substantially influence the effects of selection.

Different forms of selection, in terms of genic action—dominant, recessive or overdominant—should have similar effects in many respects. However, it might be expected that patches of deleterious recessive genotypes (such as may be typical of inbreeding depression) would be somewhat larger for a given selection intensity. This expectation arises from the fact that heterozygotes, having full fitness, should continually contribute new homozygotes to the patches. Multilocus selection might exhibit more complex effects because it acts on overlapping spatial distributions of individual loci. Finally, the possible roles of biparental inbreeding caused by spatial structure, and that of spatial structure itself, in the mutational meltdown theory of demise of small populations (e.g., Lynch and Gabriel 1991) are unknown. More research is needed in this area.

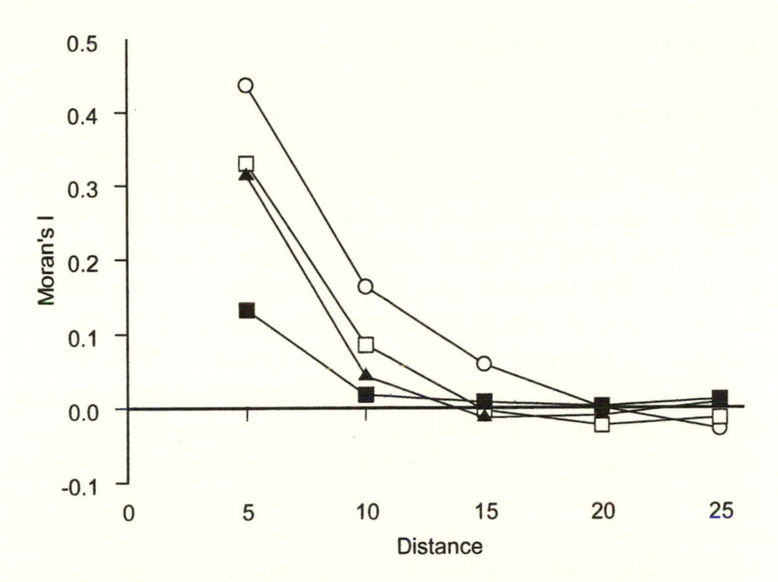

FIGURE 7.6. Effect of nonspatial natural selection on the spatial autocorrelation of genetic variation. Shown are Moran's I statistics calculated on gene frequencies in quadrats of 25 (dimensions 5 by 5) individuals each, for different sets of simulations at generation 200, summarized from Epperson (1990a). All models are two-allele models where additive directional selection acts against one allele. Simulated populations of 10,000 plants have nearest neighbor pollination and no seed dispersal (Wright's neighborhood size 4.2), at generation 200. Distance is measured in units of the lattice of individuals. Essentially, fitnesses are 1, $1 - s/2$, and s for genotypes *AA*, *Aa*, and *aa*, with a fitness differential s. Some sets also have a random replacement parameter (with rate μ) that could represent mutation, immigration, or long-distance (effectively uniform) migration within the population. Set 1 had $s = 0$, $\mu = 0$ (pure drift); set 2 had $s = 0$, $\mu = 0.001$ (neutral with systematic pressure); set 3 had $s = 0.01$, $\mu = 0.001$ (weak selection with equilibrium); and set 4 had $s = 0.1$, $\mu = 0.001$ (strong selection at equilibrium). Symbols for sets are as follows: (○) set 1; (▲) set 2; (□) set 3; and (■) set 4.

Flower color polymorphism in the morning glory, *Ipomoea purpurea*, is a good example of reduction in spatial structure caused by selection. Genes that cause white flowers are almost randomly distributed in populations (Epperson and Clegg 1986), because white flower genes are subject to selection through their effects on the mating system. Recall that the apparently neutral genetic locus controlling pink versus blue color exhibits genetic patch structures consistent with that expected for a selectively neutral gene. Bumblebees, which account for over 98 percent of pollinator visits, visit white flowers less frequently, but they visit blue and pink equally. Undervisitation causes white flowered plants to have rates of self-fertilization that are about 0.2 higher than the blue and pink types (Brown and Clegg 1984; Epperson and Clegg 1987). Genes that increase selfing by a value Δs confer a selective advantage of $\Delta s/2$, due to a transmission advantage through self-pollen (Fisher 1941), if all else is equal. However, the decreased visitation can also cause lower seed set (Epperson and Clegg 1987). Other factors may include transmission through outcross pollen (Rausher et al. 1993) and other, pleiotropic effects (Subramanian and Rausher 2000). Thus there must be very strong selection factors acting on the white flower genes, and it is very unlikely that the factors balance precisely. Although it is conceivable that the combination of factors could confer a selective advantage to white-flowered plants, it appears that most of the time white is at a selective disadvantage (Epperson and Clegg 1986, 1987). White flower genes are in low frequency in wild populations (Epperson and Clegg 1986; Subramanian and Rausher 2000).

Microenvironmental selection has predictable effects on spatial structure when (1) selection is strong, and (2) the microenvironmental heterogeneity occurs as distinct types, as in the case of microhabitat types, or along smooth gradients, and the spatial scale is much larger than the distances of dispersal. In such cases the genotypic frequencies follow directly from those expected for single populations within each microhabitat, and the gene frequency surfaces follow the pattern of the microhabitats. In other words, the spatial distribution of the environment determines the spatial genetic structures. The detection of selection depends on how differently these patterns compare to the genetic patches expected under the neutral genetic isolation by distance model. Obviously, there can be a wide range of structural aspects for populations on a two-dimensional landscape. Some aspects of how various (artificial) patches of various shapes and sizes affect the spatial autocorrelation statistics were explored by Sokal (1979b). Various clines that occur with one-dimensional smooth or step gradients were thoroughly examined by Endler

(1977). Such gradients are well known on large spatial scales (chapter 2), and some types of traits so commonly differ between animal populations that they are reflected in such "rules" regarding body size and appendage length along latitudinal gradients. Analogous traits in plants include cold tolerance, seedling emergence dates, various photoperiodic responses and flowering date. Clines may also exist on much smaller scales within populations, but still with the proviso that dispersal is negligible with respect to the spatial scale of environmental gradients. Important metrics of the spatial genetic distribution include various measures of the "width" of the cline (e.g., the length wherein the gene frequency changes from less than 0.2 to exceeding 0.8 [Endler 1977]). A number of studies have pointed out that it should be possible to scale the selection gradient to the amount of dispersal, for example, $l_c = l/\sqrt{s}$, where l is the standard deviation of dispersal distances and s is a measure of the steepness of the selection gradient (Endler 1977). The types of spatial distributions of genetic variation may also depend on the mode of genic action.

Some of the strongest and best examples of populations that exhibit striking spatial differentiation despite significant dispersal, in response to different microenvironments, involve genetic variation for metal tolerances in plants. In some cases, the spatial scale of differentiation is well within the scale of local gene flow, yet there are strong spatial differences in genotypes on mine tailings (see review by Bradshaw 1984).

Explicit models of microenvionmental selection in two-dimensional models with significant dispersal within populations have been examined in simulations by Sokal and colleagues (Sokal et al. 1989b; Sokal and Jacquez 1991; Sokal et al. 1997a). They simulated some populations with low to moderate size neighborhoods with several different spatial arrangements of micro-habitat patches. They found that the autocorrelations changed only slightly (Sokal et al. 1989b) in these particular cases, in part because the microhabitats were similar in size to the genetic patches expected for neutral loci. Smooth gradients of strengths of directional selection against an allele cause little change in the (large) autocorrelation values for short distance but correlations for long distances have much larger negative values (Sokal et al. 1989b). More dramatic effects might be expected if the direction of selection were opposite on diametric ends of the gradient

(Endler 1977). More theoretical and experimental work is needed in this area.

Although the null hypothesis of random distributions used in tests of statistical significance of spatial autocorrelation does not require the assumption of spatial stationarity (chapter 4), the spatial distribution produced for neutral genes under the standard isolation by distance model does. Thus far in this chapter we have examined only what may be termed global spatial autocorrelation (Sokal et al. 1998a). Spatial nonstationarities may be reflected in autocorrelations that vary over the landscape, deviating from the standard model, and perhaps caused by natural selection. If the overall or global autocorrelation is small, i.e., if spatial structure is generally weak, we can gainfully employ a new statistical method known as local indicator of spatial association (LISA) (Anselin 1995; Sokal et al. 1998a,b). LISAs are defined for each spatial sample point and sum to a proportion of the global autocorrelation measure (Anselin 1995). Sokal and colleagues have examined their general properties in detail in various simulated processes, and applied them to several biological examples (Sokal et al. 1998a). Among several LISA variations of various spatial autocorrelation measures, the local Moran coefficient at each sampling locality i is defined by

$$I_i = \frac{nZ_i \sum_j w_{ij} Z_j}{\sum_i Z_i^2}$$

(Sokal et al. 1998a,b). The w_{ij} are the weights (possibly binary) and the expected value under the null hypothesis is $-w_i/(n - 1)$, where w_i is the sum of the weights to locality i. Sokal and colleagues were able to detect simulated "hotspots" of autocorrelation, localized areas where there was greater dispersal. Also detected were "coldspots," where there were lower and often negative values, because of added randomly distributed "noise" (Sokal et al. 1998a,b). They noted that there could be a variety of other causes of hotspots, e.g., demographic expansions or microenvironmental selection on the spatial scale of the area occupied by a few individuals. Applications of LISAs can be complicated. In particular, it was found that significance tests were biased whenever there were large global autocorrelations; however, even here the method can be used on an exploratory basis. In addition, it appears that such examinations of hotspots and coldspots could be augmented with auxiliary information on spatial distributions of microenvironments or microhabitats. Application to Schaal's (1975) data on *Liatris cylindracea* resulted in additional evidence of spatial structure for some loci (Sokal et al. 1998a), beyond that suggested by an earlier global spatial autocorrelation analysis (Sokal and Oden 1978b). The results also point out the severe multiple-test corrections that are required (Sokal et al. 1998b) when testing many individual sampling points separately.

An alternative approach for revealing microenvironmental selection is to calculate correlations between the spatial distribution of genotypes and the spatial distribution of the microenvironment that may be examined. However, standard correlation coefficients will always be confounded with whatever patch structure overlays are produced and supported by limited dispersal per se (Epperson 1992). In other words, even neutral loci could show correlations with environmental variations, if dispersal is limited. There are spatial statistics that might be used to measure such "cross-correlations" (see, e.g., Cliff and Ord 1981) and that properly account for the autocorrelation in both environment and genetic distributions; however, to the author's knowledge these have not been used.

From an evolutionary perspective, microenvironmental selection and the spatial differentiation it causes can be viewed in terms of Wright's shifting balance theory (e.g., Wright 1978). After new gene combinations have been formed and selectively favored, the new combinations may spread spatially and numerically. On a more intermediate timescale, the effect of microenvironmental selection on population-wide genetics may depend on how the subpopulations in different microhabitats are represented at the population level, and on the nature of population size regulation. For example, some microhabitats may have subpopulations that are denser than others (equating to a form of spatial clumping). If so, then the distances over which individual animals or plant seed and pollen disperse may affect the total fitness of genotypes. This issue is analogous to the difference between hard versus soft selection (Levene 1953; Christiansen 1975). In addition to the issues of competition and level of population size regulation (within versus among patches), the spatial distributions and dispersal distances should be considered in spatially explicit processes. Little is known about hard and soft selection in an explicitly spatial context, because theoretical models have usually assumed that the subpopulations follow the island model of Wright. The situation becomes even more complex in instances where plasticity may be evolving (e.g., Zhivotovsky et al. 1996).

A good example of genetic differentiation caused by an environmental gradient within a local population is the spatial distribution of needle-number types, one of the key characteristics that distinguishes the sympatric taxa *P. ponderosa* (with three needles) and *P. arizonica* (with five) of the *Pinus* subsection *Ponderosae* species complex, in a transition zone in the southwestern United States (Rehfeldt 1999; Epperson et al. 2001). The spatial distribution in a belt transect that covers a transition zone from nearly pure three-needle types at the top of Mount Lemmon to five-needle types downslope, in the Santa Catalina Mountains, Arizona, was inconsistent with there being both free-interbreeding among types and selective neutrality of types. Indeed, it demonstrated one of the largest autocorrelations ever recorded. Needle number has high heritability (Rehfeldt et al. 1996), and both species have highly dispersible pollen and seed. Thus, either there is little or no hybridization or gene flow between the subspecies (or ecotypes) and there is very strong selection acting interspecifically, or there is some gene flow and there is strong selection acting either on needle number per se or on other traits of the ecotype with which it may be in linkage disequilibrium. Trees with intermediate types, having combinations of 3, 4, and 5 needles and/or mean numbers of needles between 3.0 and 5.0, are spatially concentrated in the middle of the transition zone (Epperson et al. 2001). The spatial distribution supports the occurrence of hybridization and introgression, and this is consistent with reported cross-abilities of the types (Conkle and Critchfield 1988). Isozymes show much less spatial structure than did needle number (Epperson et al. 2003), which further suggested that the ecotypes have been exchanging genes at a low rate for a long time, and there is advanced introgression. Further evidence has been obtained using chloroplast microsatellite markers (unpublished data), which are paternally inherited in *Pinus* (Neale and Sederoff 1989), and thus dispersed through both pollen and seed. These data again show much stronger spatial differentiation than expected for a freely interbreeding population. Yet the data also show some sharing of ancestries in opposite type trees located along the transect and miles away, in that haplotypes differing by several mutations are found in both ecotypes.

Spatial distributions have been examined in detail in various hybrid zones, which seem to provide a powerful framework for detecting natural selection. Wagner et al. (1991) used join-count statistics to discover strong fine-scale population genetic structure of chloroplast markers in a contact zone of *Pinus contorta* and *P. banksiana*, as well as for various cone characteristics. There is evidence of advanced introgression yet limited dispersal and hybridization in the contact zone. Moreover, the structure is in stark contrast to the nearly random distribution found in pure populations of *P. contorta* (Epperson and Allard 1989). A combination of mechanisms for partial reproductive isolation and selection forces appears to be responsible.

CHAPTER 8

Statistical Methods for Spatial Structure within Populations

> [The a priori expected values of gene frequencies may] vary from place to place in some region R, perhaps in an unknown way: so, we often obtain only the "pseudocovariances" evaluated from some "regional mean" (Morton) or some "working mean" . . . (Malécot, C. Smith). These may be considerably different from the a priori covariances, and will be negative at large distances if there is a cline.
>
> —Gustave Malécot (1973a)

Nearly all of the theory developments on spatial structure of populations are based ultimately on pairwise measures. Although a spatial pattern contains a great deal of information, it appears that much of this information is captured in pairwise measures of correlation as a function of spatial proximity, usually simply in terms of distance. Malécot's theoretical models of isolation by distance were almost always expressed in terms of probabilities that two genes are identical by descent (e.g., Malécot 1948, 1972). Wright's (1943, 1946) theory of isolation by distance within populations can be translated into averaged probabilities of identity by descent of pairs of genes within a local group of individuals, relative to the entire population. Presumably Wright chose this approach because of his earlier creation of inbreeding coefficients (Wright 1921). Statistical measures of spatial patterns in real populations are also based on pairwise measures, and although it is possible to define triplet measures, etc., these measures are in most instances inordinately complex and have undemonstrated utility. This chapter begins with a description of pairs of spatially located genotypes and their probability theory. Next, the distributions are derived for spatial autocorrelation statistics, including join-count statistics and Moran's I statistics for converted genotypes. Although many aspects of Moran's I statistics were described in chapter 4, those were based on samples from populations and did not emphasize their relationship to pairs of genotypes. Also examined are several features of their statistical properties. The next topic is a review of various ways that pairs can be averaged, including aspects of averaging the two genes within an individual, within locally contiguous groups (F statistics), or within groups of pairs separated similarly in space (spatial autocorrelations). This is followed by an examination of methods for averaging over alleles within a locus and across loci.

DISTRIBUTION THEORY FOR JOIN COUNTS

In general, there may be r different genotypes on a surface, representing a sample from or realization of a stochastic process with probabilities p_b. At each sampling station (locality) the genotype sampled is a realization of this process, and the probability of picking genotype b is p_b. All pairs of sampling points can be classified is some way, usually in terms of a range of Euclidean distances between pairs of points, and (distance) classes k are formed. Weighted join-count statistics can be generated, similarly as for weighted Moran's I statistics (Cliff and Ord 1981), but for simplification of presentation the discussions below focus on unweighted join counts. To develop a similar but different approach to that of Cliff and Ord (1981), let x_{ib} be an indicator variable such that $x_{ib} = 1$ if genotype b is found at location i, zero if not (Epperson, n.d.(a)). We are interested in the value of $n_{bc}(k)$, the number of joins between genotypes b and c for distance class k (i.e., the number of pairs of points that have genotype b at one point and c at the other and are separated by distances that fall within distance class k). For example, each distance class k may contain all pairs of sample points separated by d lattice units, where $k - 0.5 < d < k + 0.5$. Let $\delta_{ij}(k)$ be another indicator variable such that when the pair of locations i and j are separated by a distance in class k, it equals one, zero elsewhere. Then:

$$n_{bb}(k) = \frac{1}{2} \sum_{(2)} \delta_{ij}(k) x_{ib} x_{jb}$$

$$n_{bc}(k) = \sum_{(2)} \delta_{ij}(k) x_{ib} x_{jc} \tag{8.1}$$

for $b \neq c$. The double summation is twice over all pairs of locations (excluding where the same location is indicated twice, i.e., $\delta_{ii}(k) = 0$) without respect to order of locations. By the definition of a join $\delta_{ij}(k) = \delta_{ji}(k)$, and the factor ½ corrects for the double counting of like joins, $b = c$ (Cliff and Ord 1981). Let W_k be twice the total number of pairs or joins for distance class k. There are two ways to configure the probability distribution of $n_{bc}(k)$ under the hypothesis that there is no spatial structure, i.e., that the pairs of genotypes are independent of their spatial locations, and here it is useful to introduce notation that can be easily extended to Moran's I and to weighted autocorrelation statistics. The first way is appropriate if the genotypic probabilities (p_b) are known from other information, and then each location may be viewed as independent draw of a ball from an infinite urn model. Under this "free sampling" hypothesis, the expected values $\mu_{bb}(k)$ and variances $\sigma^2_{bb}(k)$ for joins between two individuals with identical genotype b are

$$\mu_{bb}(k) = \frac{S_0(k)}{2} p_b^2 \tag{8.2}$$

$$\sigma^2_{bb}(k) = \tfrac{1}{4}\{S_1(k)p_b^2 + [S_2(k) - 2S_1(k)]p_b^3 + [S_1(k) - S_2(k)]p_b^4\} \tag{8.3}$$

where

$$S_0(k) = \sum_{(2)} \delta_{ij}(k) \tag{8.4}$$

or W_k, twice the number of joins total for distance class k,

$$S_1(k) = \frac{1}{2} \sum_{(2)} [\delta_{ij}(k) + \delta_{ji}(k)]^2 \tag{8.5}$$

which is four times the number of joins ($2W_k$) and,

$$S_2(k) = \sum_{i=1}^{n}\left[\sum_{j=1}^{n}\delta_{ij}(k) + \sum_{j=1}^{n}\delta_{ji}(k)\right]^2 \tag{8.6}$$

(Cliff and Ord 1981). For joins between different genotypes b and c ($b \neq c$):

$$\mu_{bc}(k) = S_0(k)p_b p_c \tag{8.7}$$

$$\begin{aligned}\sigma_{bc}^2(k) = \tfrac{1}{4}\{2S_1(k)p_b p_c &+ [S_2(k) - 2S_1(k)]p_b p_c(p_b + p_c) \\ &+ 4[S_1(k) - S_2(k)]p_b^2 p_c^2\}\end{aligned} \tag{8.8}$$

Weighted join-count statistics, analogous to the weighted Moran's I statistics, can be configured by simply replacing the binary indicators $\delta_{ij}(k)$, with weights $w_{ij}(k)$ in the above expressions, although these are rarely used in practice.

As pointed out by Cliff and Ord (1981), normally the true p_b are expected values of some stochastic process that is usually unknown. Thus, because the means are usually unknowable, the second and preferred statistics are based on the null hypothesis H_o that the sampling distribution of the numbers of joins is "nonfree" random, i.e., that produced by randomly sampling each individual of a pair without replacement from a finite urn model of the total sample of genotypes. Effectively, H_o purports that the locations of sample genotypes are randomized. Under H_o the expected number of joins for any distance class k is $u_{bb} = W_k n_b(n_b - 1)/2n(n - 1)$ and $u_{bc} = W_k n_b n_c/n(n - 1)$ for $b \neq c$. Here n_b is the number of times that genotype b occurs in the sample of size n. The variances are

$$\begin{aligned}\sigma_{bb}^2(k) = \frac{1}{4}\Bigg\{&\frac{S_1(k)n_b^{(2)}}{n^{(2)}} + \frac{[S_2(k) - 2S_1(k)]n_b^{(3)}}{n^{(3)}} \\ &+ \frac{[S_0(k)^2 + S_1(k) - S_2(k)]n_b^{(4)}}{n^{(4)}}\Bigg\} - \mu_{bb}^2\end{aligned} \tag{8.9}$$

for joins between identical genotypes, and

$$\sigma_{bc}^2(k) = \frac{1}{4}\left\{\frac{2S_1(k)n_b n_c}{n^{(2)}} + \frac{[S_2(k) - 2S_1(k)]n_b n_c(n_b + n_c - 2)}{n^{(3)}} + \frac{4[S_0(k)^2 + S_1(k) - S_2(k)]n_b^{(2)} n_c^{(2)}}{n^{(4)}}\right\} - \mu_{bc}^2 \quad (8.10)$$

for joins between different genotypes, where $n^{(2)} = n(n - 1)$, $n^{(3)} = n(n - 1)(n - 2)$, $n^{(4)} = n(n - 1)(n - 2)(n - 3)$, and similarly for the $n_b^{(2)}$ etc. (Sokal and Oden 1978a; Cliff and Ord 1981).

The relationships among sets of joins are inherently complex. For example, consider the simplest case, where there are only two genotypes, B and W, which would be appropriate in the case of a polymorphic diallelic locus for haploids (e.g., pollen types, bacteria, or [haploid] plastid chromosomes). It may seem that any one of the numbers of types of joins, n_{BB}, n_{BW}, or n_{WW} contains all information, but this is not true. However, any one of these is redundant with respect to the other two since they sum to $W_k/2$. The values of n_{BB} and n_{BW} joins are not independent, even under the null hypothesis. The expected value of the product $n_{BB} \cdot n_{BW}$ is equal to

$$E[n_{BB}(k)n_{BW}(k)] = E\left\{\left[\frac{1}{2}\sum_{(2)}\delta_{ij}(k)x_{iB}x_{jB}\right]\left[\sum_{(2)}\delta_{kl}(k)x_{kB}x_{lW}\right]\right\} \quad (8.11)$$

$$= \frac{1}{2}\sum_{(2)}\sum_{(2)}\delta_{ij}(k)\delta_{kl}(k)\, E\, x_{iB}x_{jB}x_{kB}x_{lW} \quad (8.12)$$

$$= \frac{1}{2}E\left\{\sum_{(2)}\delta_{ij}(k)[\delta_{ij}(k) + \delta_{ji}(k)]x_{iB}x_{jB}x_{kB}x_{lw} + \sum_{(3)}[\delta_{ij}(k) + \delta_{ji}(k)][\delta_{ik}(k) + \delta_{ki}(k)]x_{iB}x_{jB}x_{kB}x_{lW} + \sum_{(4)}\delta_{ij}(k)\delta_{kl}(k)x_{iB}x_{jB}x_{kB}x_{lW}\right\} \quad (8.13)$$

(Epperson, n.d.(a)). The double summation in the first term of equation 8.13 requires that $k = i$, $j = l$ or $k = j$, $l = i$. Both give impossible assignments of both B and W to one location, hence this term has expected value zero. The second term sums over i, j, and k, with the constraint that $i \neq j \neq k$, but this requires that one of the following constraints applies: $i = k, j \neq l$; $i \neq l, j = k$; $i = l$; $j \neq k$; or $i \neq k, j = l$. The second two types are impossible, and give nothing to the expected value, whereas the first two have expectations $p_B^2 p_w$, for free sampling (Epperson, n.d.(a)). All of the permutations in the third term (summed over $i \neq j \neq k \neq l$) are possible and have expectation $p_B^3\, p_w$, for free sampling. Correct summing of the weights (converted to $S_0(k)$, $S_1(k)$, and $S_2(k)$) by expectations, and considerable rearranging, with the relatively simpler case of "free sampling," leads to an expected value of the product $n_{BB} \cdot n_{BW}$

$$\tfrac{1}{4}\, p_B^2\, p_W [S_2(k)(1-2p_B) - 2S_1(k)(1-p_B) + 2S_0(k)^2\, p_B] \quad (8.14)$$

The values for nonfree sampling can be found by expanding first to powers of p_B and p_W, and then substituting terms $n_B{}^{(i)}\; n_W{}^{(j)}/n^{(i+j)}$ for $p_B^i\; p_W^j$. Cliff and Ord (1973) used a different method to derive the same result, but only for the case where there are two types, whereas the above obtains in cases with arbitrary numbers of types, which is needed for virtually any surface of genotypes. One can combine this expression with those given above for the variances of n_{BB} and n_{BW} to see that the correlation of n_{BB} and n_{BW} involves $S_0(k)$ and $S_2(k)$, and hence the constellation of the array of sample points and designation of distance class. The situation can be expected to be more complex under alternate hypotheses. However, numerical evaluations (Epperson, n.d.(a)), which are discussed later in this chapter, demonstrate that the correlations are in fact effectively independent of $S_0(k)$ and $S_2(k)$, under both the null hypothesis and isolation by distance processes. Cliff and Ord (1973) also found that the n_{BW} join-count statistics performed better than n_{BB} (or n_{WW}) or combinations of n_{BB} (or n_{WW}) and n_{BW}. Similar relationships among various joins can be seen in the large formula for the expected number of the sum of joins between nonidentical types, given by Cliff and Ord (1981).

The complication of $S_2(k)$ arises in part because it is based on the numbers of joins to each location, which generally will be intimately connected to the actual distribution of sample points. However, something more can be studied when the sample is based on a regularly spaced grid or lattice. In such cases, we may define certain classes of joins, by analogy with one-step moves in the game of chess, for the short distance joins. The number of joins to each point, can be determined and expressed algebraically, simply from the dimensions (numbers of rows and columns) of the lattice. The various "single-step moves" include one-space rook's moves; one-space bishop's moves; and one-space queen's (rook's plus bishop's) moves. A number of two-dimensional rectangular sample lattices, with dimensions m and n, have been examined by Krishna-Iyer (1949, 1950). These formulations may be based on classifications of the types of lattice points, corner (4), border ($2m + 2n - 4$), and interior [$(n - 1)(m - 1)$]. For example, there are $2nm - (n + m)$ one-space rook's joins, $2nm - 2(n + m) + 2$ one-space bishop's moves (Epperson and Allard 1989), and $4nm - 3(n + m) + 2$ one-space queen's moves (Krishna-Iyer 1949). Similar logic can be used to find $S_2(k)$ (Krishna-Iyer 1949).

For larger distance classes in a lattice, the specifications become much more complex. The distributions of the numbers of (pairwise) joins depend on the distribution probabilities of various triplets and quadruplets, under the null hypothesis. Because ultimately we wish to examine entire sets of join-count statistics for various distance classes, hence, also, various triplets and quadruplets, there seems to be very little reason for examining the observed numbers of triplets or quadruplets. The complication of $S_2(k)$, hence the numbers of joins to locations, also creates complexity in the relationships of expected values among statistics for various different distance classes (or among various differently weighted statistics), even under the null hypothesis.

The methods used to obtain equation 8.14 lead to procedures for finding the expected value of all products of joins (Epperson, n.d. (a)). For example, the product $n_{BW}(k) \cdot n_{WW}(k)$, can be found simply by switching the B and W subscripts in equation 8.14. The expected values for all other products can be found by using appropriate changes in the subscripts B and W and substituting the appropriate forms of equation 8.1 into equation 8.11 (note that the ½ coefficients depend on whether the joins are like vs. unlike). The case of $n_{BB}(k) \cdot n_{WW}(k)$ can be derived by finding the appropriate form analogous to equation 8.13, and recognizing that only the quadruplets in the third term (not in the first or second) are possible, which leads to the result

$$E\, n_{BB}(k) \cdot n_{WW}(k) = \tfrac{1}{4}\, p_B^2\, p_W^2 [S_0(k)^2 - S_2(k) + S_1(k)] \quad (8.15)$$

in the case of free sampling (Epperson, n.d.(a)). In the multigenotypic situation we need to consider two additional genotypes, say *C* and *D*, and examine types of products of joins *BW BC*, *BW CC*, and *BW CD*. For any distance class *k*, the expected value of $n_{BW} \cdot n_{BC}$ is

$$\tfrac{1}{4}\, p_B p_B p_C \{S_2(k) - 2S_1(k) + 4[S_0(k)^2 - S_2(k) + S_1(k)]p_B\} \quad (8.16)$$

That for $n_{BW} \cdot n_{CC}$ is

$$\tfrac{1}{2}\, p_B p_W p_C^2 [S_0(k)^2 - S_2(k) + S_1(k)] \quad (8.17)$$

And that for $n_{BW} \cdot n_{CD}$ is

$$p_B p_W p_C p_D [S_0(k)^2 - S_2(k) + S_1(k)] \quad (8.18)$$

(Epperson, n.d.(a)). With the above equations, we can calculate the variances and covariances for any weighted sum of join counts for a given distance class. Like equation 8.14, equations 8.15–18 can be modified for non-free sampling by first expanding to powers of p_B and p_W, and then substituting terms $n_B^{(i)}\, n_W^{(j)}/n^{(i+j)}$ for $p_B^i\, p_W^j$, etc.

Test statistics of the null hypothesis can be formulated for each join count. The statistic $\text{SND}_{bc}(k) = [n_{bc}(k) - u_{bc}(k)]/\sigma_{bc}(k)$ has an asymptotic standard normal distribution (Cliff and Ord 1981). These significance tests generally have high statistical power (Cliff and Ord 1981), and they measure the excess (positive values) or deficit (negative values) of each type of join. A set of unweighted SND statistics for mutually exclusive distance classes (k) can be formed, and is known as a correlogram (Cliff and Ord 1981). The author freely distributes a computer program (JCSP) that calculates these (unweighted) statistics. It also calculates those for the total number of unlike joins, which for each distance class k, is the sum of joins for pairs of individuals that have different genotypes and which are separated by a distance in the distance class k. In general, the join-count test statistics may be based on either unweighted join counts or weighted counts by using binary indicators $\delta_{ij}(k)$ or weights $w_{ij}(k)$, respectively, although weighted join-count statistics are rarely used in practice.

MORAN'S I STATISTICS IN TERMS OF JOIN COUNTS

For spatial distributions of genotypic data, Moran's I statistics are written

$$I(k) = \frac{n\sum_i \sum_j w_{ij}(k) Z_i Z_j}{W(k)\sum_i Z_i^2} \tag{8.19}$$

where $Z_i = q_i - q$, q_i is the gene frequency converted from genotype at location i, and q is the mean of the allele frequencies. Moran's I statistics can be expressed as a function of the join counts, whether weighted (using $w_{ij}(k)$) or unweighted ($\delta_{ij}(k)$). For the sake of brevity, we will discuss this relationship as though there are unweighted counts of joins. The simplest case corresponds to a haploid system for a locus with two alleles, B and W, converted into values 1.0 and 0, respectively. Then q is either a theoretical mean (free sampling) or the observed frequency of B in the sample, n_B/n (non-free sampling). We may denote the I statistic $I_B(k)$. The values of Z_i Z_j are determined by the join counts. The number of terms in the numerator of $I_B(k)$ for which Z_i Z_j equals $(1 - q)(1 - q)$ is precisely $2n_{BB}$, those for which Z_i Z_j equals $(1 - q)(0 - q)$ number $2n_{BW}$, and those for q^2 number $2n_{WW}$. In the denominator, there are n_B terms of $Z_i^2 = (1 - q)^2$ and n_W terms of $Z_i^2 = q^2$. Thus the following identity exists:

$$I_B(k) = \frac{2n[n_{BB}(k)(1-q)^2 - n_{BW}(k)q(1-q) + n_{WW}(k)q^2]}{W_k(n_B(1-q)^2 + n_W q^2)} \tag{8.20}$$

This identity is useful for several reasons. Most importantly, it can be used to evaluate the expected value of the I statistics for converted genotypes and their higher moments. Doing so requires taking expected values of powers (1 for the mean, 2 for the raw second moment, etc.) of a ratio, which is usually problematic. Cliff and Ord (1981) have shown that under the assumption that the data are normally distributed from free sampling, and using a theorem by Pitman (1937) and Koopmans (1942), the expected values of such statistics equal the ratio of the expected values of the numerator and denominator. However, binary data are clearly nonnormal. Thus, we will focus on the data under the randomization (non-free sampling) hypothesis, wherein $q = n_B/n$. Then the value of q is fixed, and indeed the expectation is relative to all random permutations of pairs of the n_B and n_W observations. In fact, in this case the denominator in equation 8.20 is fixed. Thus we have

$$E[I_B(k)] = \frac{2n[(1-n_B/n)^2\, En_{BB}(k) - (n_B/n)(1-n_B/n)En_{BW}(k) + (n_B/n)^2\, En_{WW}]}{W_k[n_B(1-n_B/n)^2 + n_W(n_B/n)^2]} \tag{8.21}$$

Substituting the expected values of the join counts and simplifying this expression yields $E\,[I_B(k)] = -1/(n-1)$, as before (Epperson, n.d.(b)).

We can take expectations of products of $I_B(k)$, for different alleles B, to get the covariances of I statistics for different alleles (Epperson, n.d.(b)). Although the expressions are very long and not represented here, some numerical results for the covariances are given later in this chapter. This is a quite important issue in population genetics today, because of the popularity of using I statistics for converted genotypes and the fact that modern markers tend to yield greater numbers of alleles. Hitherto it has been unknown how to combine I statistics for alleles within a locus or how to calculate the standard error of a combined statistic, because the covariances were not known (except for the two-allele case in which the covariance is complete). However, we must first examine in some detail the case for diploid genotypes. For the case of a diallelic locus with three genotypes AA, Aa, aa (corresponding to types B, W, C) and letting q be the frequency of the a allele ($nq = n_B + n_C/2 = n_{aa} + n_{Aa}/2$), it has been shown (Epperson 1995c) that the Moran I statistic for allele a is

$$I_a(k) = \frac{2n[q^2 n_{AAAA} - q(0.5-q)n_{AAAa} - q(1-q)n_{AAaa} + (0.5-q)^2 n_{AaAa} + (0.5-q)(1-q)n_{Aaaa} + (1-q)^2 n_{aaaa}]}{W_k[q^2 n_{AA} + (0.5-q)^2 n_{Aa} + (1-q)^2 n_{aa}]} \tag{8.22}$$

where q is the average or total frequency of the a allele, and n_{AAAA}, n_{AAAa}, n_{AAaa}, n_{AaAa}, n_{Aaaa}, n_{aaaa} are the numbers of $AA \times AA$, $AA \times Aa$, $AA \times aa$, $Aa \times Aa$, $Aa \times aa$, $aa \times aa$ types of joins, respectively, for a given distance class.

For multiple alleles we must consider three types of joins: ones that involve aa, those that involve $\bar{a}a$ (where $\bar{a}$ means any allele that is not allele a), and those that involve neither (joins with $\bar{a}\bar{a}$ only). We can write the expression

$$I_a(k) = \frac{2n[q^2 \sum_{\bar{a}} n_{\bar{a}\bar{a}\bar{a}\bar{a}} - q(0.5-q) \sum_{\bar{a}} n_{a\bar{a}\bar{a}\bar{a}} - q(1-q) \sum_{\bar{a}} n_{aa\bar{a}\bar{a}} + (0.5-q)^2 \sum_{\bar{a}} n_{a\bar{a}a\bar{a}} + (0.5-q)(1-q) \sum_{\bar{a}} n_{aaa\bar{a}} + (1-q)^2 n_{aaaa}]}{W_k[q^2 \sum_{\bar{a}} n_{\bar{a}\bar{a}} + (0.5-q)^2 \sum_{\bar{a}} n_{a\bar{a}} + (1-q)^2 n_{aa}]} \tag{8.23}$$

where each summation is over all combinations of alleles that are not allele a (Epperson, n.d.(b)). As noted earlier, from this expression and those for the various join counts, we can provide the means for obtaining not only the variances of the $I_a(k)$ but also the covariances among I statistics for different alleles, under the null hypothesis of randomized pairs of genotypes (see Epperson, n.d.(b)). Note, for example, that some of the joins for alleles "not a" or "not b" will overlap for the expected value of product $I_a(k)\, I_b(k)$.

A different kind of covariance was described by Oden (1984), that for different distance classes for the same allele. The moments of two spatial autocorrelation statistics, one with weights w_{ij} and the other with y_{ij}, under either the normal or randomization hypothesis, depend only on a function of the weights. For example, the expected value of the product of two Moran's I statistics, I_W and I_Y (for the same allele but for different weights or distance classes), is given by

$$E[I_W I_Y] = \frac{n^2 S_{(WY)1} - nS_{(WY)2} + 3S_{(W)0}\, S_{(Y)0}}{S_{(W)0}\, S_{(Y)0}(n^2-1)} \tag{8.24}$$

which is analogous to the value for $E[I^2]$. $S_{(W)0}$ and $S_{(Y)0}$ have the usual interpretation and

$$S_{(WY)1} = \frac{1}{2} \sum_{(2)} (w_{ij} + w_{ji})(y_{ij} + y_{ji}) \tag{8.25}$$

and

$$S_{(WY)2} = \sum_{i=1}^{n} \left(\sum_{j=1}^{n} w_{ij} + \sum_{j=1}^{n} w_{ji} \right) \left(\sum_{j=1}^{n} y_{ij} + \sum_{j=1}^{n} y_{ji} \right) \tag{8.26}$$

In addition, as discussed in chapter 4, Oden (1984) also developed a test for the significance of correlograms under the null hypothesis.

STATISTICAL PROPERTIES OF JOIN-COUNT STATISTICS

A series of large simulation studies have examined many of the stochastic and statistical properties of spatial autocorrelation measures in population genetics (e.g., Sokal and Wartenberg 1983; Epperson 1995a; Epperson and Li 1996, 1997; Epperson, et al. 1999). Unlike most mathematical models, e.g., those based on kinship, these models are explicitly based on the spatial distributions of genotypes, and thus they can be directly compared to observed data. Importantly, they include the stochasticity inherent as individual genotypes, whether single- or multi-locus, disperse, mate, and reproduce. Appropriate models for genotypes in two-dimensional space are analytically intractable, for several reasons, and thus large (10,000 individuals), highly replicated space–time Monte Carlo simulations have been conducted (e.g., Sokal and Wartenberg 1983). It is more meaningful to replicate over entire simulations rather than over generations. The analyses have focused on stochastic realizations either without or with added statistical error (sampling from the realized genotypic surfaces in various ways), for the spatial structure of the "quasistationary" phase (e.g., Sokal and Wartenberg 1983). Because dispersal distances have such a large effect, a wide range of dispersal levels have been simulated (Epperson and Li 1997).

In the discussion that follows, a wide range of types of simulated samples from populations undergoing genetic isolation by distance (table 8.1) is examined. Dispersal levels were simulated by choosing either or both the female and male parents of an offspring at random

TABLE 8.1. Dispersal parameters in various sets of simulations

	Simulation Dispersal Model				
	1	*2*	*3*	*4*	*5*
N_f	1	25	49	1	625
N_m	9	25	49	225	625
N_e	4.2	25.1	50.2	115.2	632.4

Note: N_f and N_m are the numbers of nearest female and male individuals (including self) from which parents of an offspring are randomly chosen, and N_e is effectively Wright's neighborhood size.

TABLE 8.2. The sample sizes in terms of sampled individuals for various sampling schemes from the simulated populations

	Area[a]		
Porosity[b]	*10,000*	*5,000*	*2,500*
1	10,000	5,000	2,500
1/4	2,500	1,250	625
1/9	1,089	544	272
1/25	400	200	100

[a]The total number of individuals in the population that the entire sample lattice covered.

[b]Porosity: the proportion of total population individuals in the area that were sampled.

generally from one of the nearest N_f and N_m (respectively) neighbors including self. The range of sampling was intended to cover those likely to be pursued or contemplated by researchers. Key variations in sampling are (1) the total number of individuals in the population area sampled; and (2) the "porosity" of sampling, where porosity is best represented as the proportion of the individuals in a population area that were actually sampled (table 8.2). It is important to realize that porosity affects the spatial scale of sampling as well as the size of the sample. When all individuals in the sample area are sampled porosity equals 1.0.

The stochastic and statistical properties of join-count statistics are best illustrated by the three most important types of joins: joins between identical homozygotes, joins between opposite homozygotes in the two-allele case (Epperson 1990a, 1995a), and the total number of unlike joins. These three types of joins are slightly dependent on allele frequencies, but other types of joins (e.g., $Aa \times Aa$, $AA \times Aa$, and $aa \times Aa$) are strongly dependent on allele frequencies (figure 8.1). The SNDs for the full sampling case (n = 10,000, porosity = 1) include stochastic variation only. The large positive values observed for the $AA \times AA$ joins for short distances indicate positive autocorrelations and clustering of AA genotypes, and they generally decrease as the distance increases. They also decrease monotonically with increasing amounts of dispersal (Table 8.3), which may be standardized by what is essentially Wright's neighborhood size N_e. Values of SNDs for $AA \times aa$ joins for the shorter distances are highly negative, and the SNDs for the total number of unlike joins are very similar to values for $AA \times aa$ joins, in the two-allele case (Epperson and Li 1997). For complete sampling of a large population, for all but the highest amounts of dispersal ($N_e = 316.2$ or 632.4), the statistics for short distance classes (generally the most useful) have high statistical power and reject the null hypothesis either at or near 100 percent (tables 8.3, 8.4, and 8.5). Naturally, when the null hypothesis is essentially true, as in the random case, the observed type I error rate is not different from the expected 5-percent level and overall the SNDs are not statistically different from zero. For most of the sample schemes the statistical power of the SND test statistics remains high (tables 8.3, 8.4, and 8.5). Increasing porosity decreases the autocorrelations, as expected, because then sampling is on a larger spatial scale, where genetic autocorrelations are expected to be smaller, under the isolation by distance model.

In chapter 7, examples were given as to how experimental values can be compared to precise predicted values for the sampling scheme used to find the best-fitted dispersal level from tables 8.3, 8.4, and 8.5. This can be done by various interpolations among sample designs and dispersal levels. In general, the effect strictly of reducing sample size (i.e., keeping porosity fixed but reducing sample area) is to decrease the SNDs roughly by the square root of the ratio. For example, in the tables, for a given porosity, the values of SNDs for the quarter-population (an area containing 2500 individuals) are close to half of those for the total area. Generally, for diallelic loci, SNDs for $AA \times aa$ joins have the greatest statistical power, followed by those for the total number of unlike joins, and $AA \times AA$ joins. This is reminiscent of the finding by Cliff and Ord (1973) that *BW* joins cannot be improved upon in the two-color case. However, the differences in power are slight. Decreasing the sample size also changes the standard deviations of SNDs, but in a somewhat complex manner.

Importantly, greater amounts of dispersal generally tend to result in lower standard deviations of all spatial autocorrelation statistics. However, there appears to be a trend that suggests a noteworthy interaction. The effect of decreasing sample size on standard deviations of spatial statistics tends to diminish as dispersal increases. This appears to result simply from the fact that there is less autocorrelation

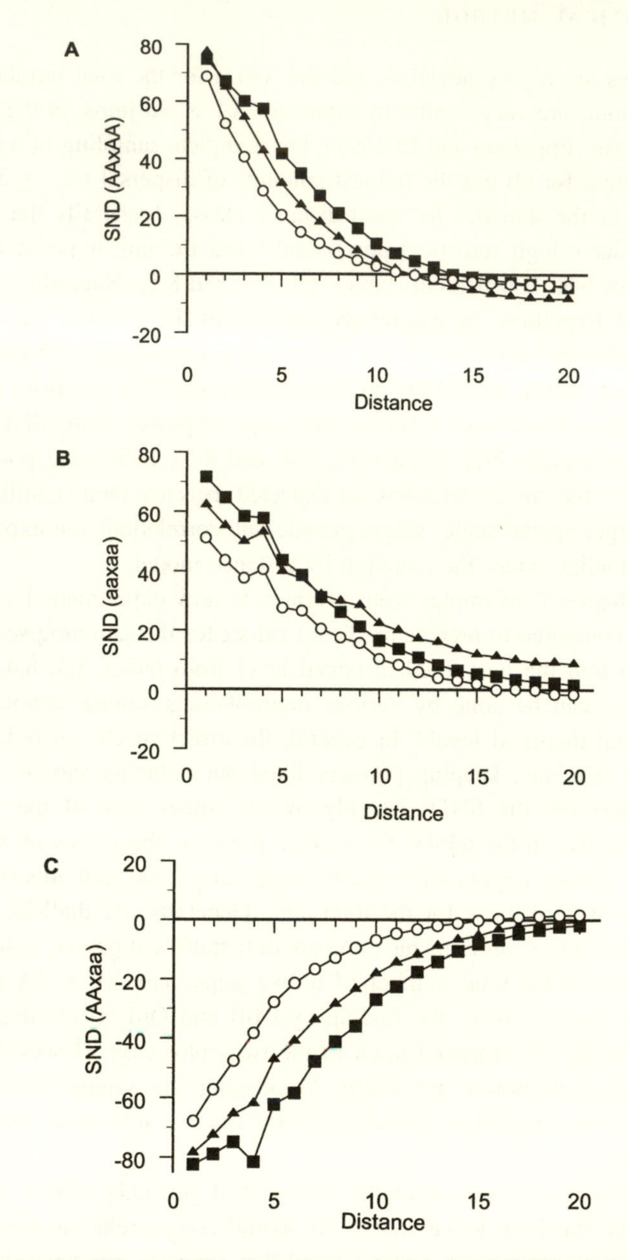

FIGURE 8.1. Effects of allele frequency on the join-count SND statistics for a locus (*A*/*a*) with two alleles. Averages are shown for different sets of five simulations each, and sets have different frequencies q of the a allele: for set 1 (■) $q = 0.5$, set 2 (▲) $q = 0.2$, and set 3 (○) $q = 0.1$. All simulated populations of 10,000 plants have nearest-neighbor pollination and no

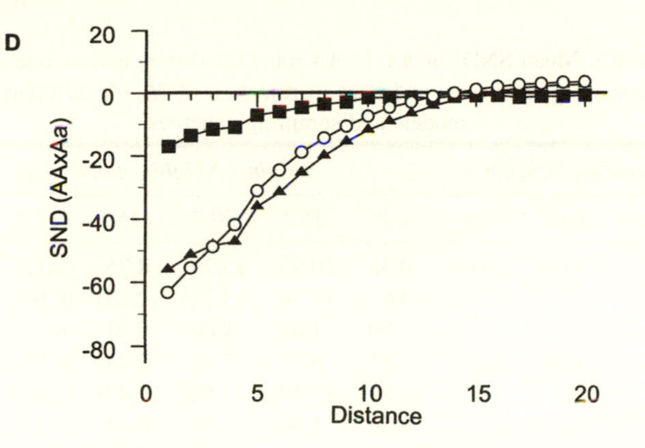

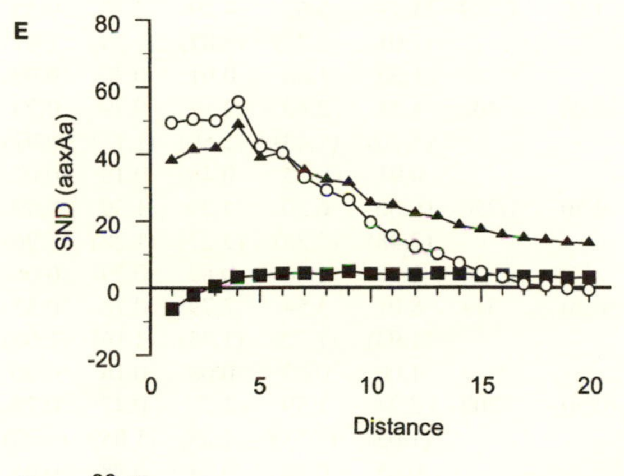

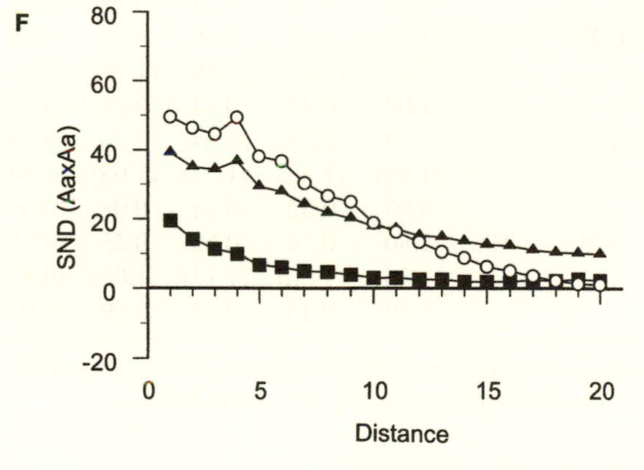

FIGURE 8.1 (*continued*) seed dispersal (Wright's neighborhood size 4.2), and results are shown for generation 200. Each subfigure shows results for a particular kind of join: (A) *AA* × *AA*; (B) *aa* × *aa*; (C) *AA* × *aa*; (D) *AA* × *Aa*; (E) *aa* × *Aa*; and (F) *Aa* × *Aa*. Figures adapted from results of Epperson (1995a).

TABLE 8.3. Mean SND for $AA \times AA$ joins for distance class one (standard deviations in parentheses) and rates of rejection of H_0 for different dispersal models and sampling schemes

Sampling Scheme			*Wright's Neighborhood Sizes*					
Porosity[a]	*Area*[b]	*Size*	*4.2*	*25.1*	*50.2*	*115.2*	*632.4*	*Random*
1	1.00	10,000	70.56	20.95	15.79	4.75	1.12	−0.01
			(4.62)	(4.56)	(5.21)	(2.07)	(0.96)	(1.03)
			1.00	1.00	1.00	0.91	0.17	0.04
4	1.00	2,500	24.47	9.63	7.42	2.44	0.55	0.04
			(2.87)	(2.49)	(2.80)	(1.40)	(1.00)	(0.99)
			1.00	1.00	0.99	0.64	0.11	0.04
9	1.00	1,089	11.79	5.62	4.59	1.37	0.35	−0.11
			(2.01)	(1.71)	(1.87)	(1.34)	(1.07)	(1.01)
			1.00	1.00	0.91	0.30	0.09	0.04
25	1.00	400	4.24	2.43	2.18	0.79	0.23	−0.07
			(1.70)	(1.43)	(1.59)	(1.17)	(0.97)	(1.02)
			0.91	0.55	0.48	0.19	0.02	0.07
4	0.50	1,250	16.66	6.50	4.39	1.39	0.49	0.02
			(2.48)	(2.25)	(2.27)	(1.28)	(0.96)	(0.99)
			1.00	1.00	0.89	0.29	0.06	0.03
9	0.50	544	8.01	3.54	2.79	0.78	0.33	−0.04
			(1.93)	(1.72)	(1.75)	(1.16)	(1.00)	(0.96)
			1.00	0.83	0.68	0.16	0.06	0.05
25	0.50	200	2.74	1.59	1.27	0.47	0.28	−0.04
			(1.63)	(1.53)	(1.28)	(1.05)	(1.00)	(0.96)
			0.65	0.37	0.24	0.08	0.05	0.06
4	0.25	625	11.42	4.01	2.66	0.74	0.23	−0.03
			(2.63)	(2.01)	(1.80)	(1.19)	(1.07)	(0.98)
			1.00	0.88	0.64	0.14	0.07	0.04
9	0.25	272	5.33	2.22	1.54	0.26	0.20	−0.04
			(1.97)	(1.46)	(1.44)	(1.05)	(1.04)	(1.01)
			0.95	0.52	0.41	0.08	0.09	0.05
25	0.25	100	1.80	0.79	0.65	0.25	0.17	0.01
			(1.36)	(1.30)	(1.17)	(1.06)	(0.97)	(1.04)
			0.49	0.15	0.14	0.09	0.03	0.06

[a]The inverse of the proportion of individuals sampled over the sample lattice area.
[b]Fraction of population area covered by the sample lattice.

TABLE 8.4. Mean SND for $AA \times aa$ joins for the first distance class (standard deviations in parentheses) and rates of rejection of H_0 for different dispersal models and sampling schemes

Sampling Scheme			*Wright's Neighborhood Sizes*					
Porosity[a]	*Area*[b]	*Size*	*4.2*	*25.1*	*50.2*	*115.2*	*632.4*	*Random*
1	1.00	10,000	−79.81	−25.74	−20.01	−6.31	−1.31	0.10
			(4.47)	(4.49)	(5.65)	(2.74)	(1.04)	(1.08)
			1.00	1.00	1.00	0.97	0.24	0.05
4	1.00	2,500	−29.76	−11.98	−9.43	−3.12	−0.47	−0.06
			(2.67)	(2.53)	(2.99)	(1.69)	(0.97)	(0.96)
			1.00	1.00	1.00	0.76	0.05	0.02
9	1.00	1,089	−15.02	−7.31	−5.55	−1.94	−0.30	0.09
			(2.05)	(1.75)	(2.00)	(1.26)	(1.08)	(1.11)
			1.00	1.00	0.98	0.43	0.06	0.06
25	1.00	400	−5.97	−3.23	−2.75	−1.07	−0.31	−0.08
			(1.61)	(1.40)	(1.64)	(1.09)	(1.08)	(1.02)
			1.00	0.82	0.62	0.22	0.08	0.07
4	0.50	1,250	−20.59	−8.13	−5.83	−1.82	−0.27	0.02
			(2.38)	(2.30)	(2.54)	(1.53)	(1.00)	(1.01)
			1.00	1.00	0.99	0.47	0.04	0.06
9	0.50	544	−10.31	−4.79	−3.55	−1.07	−0.12	0.14
			(2.09)	(1.82)	(1.71)	(1.20)	(1.05)	(1.12)
			1.00	0.98	0.83	0.20	0.07	0.08
25	0.50	200	−4.10	−2.29	−1.78	−0.51	−0.22	−0.17
			(1.59)	(1.38)	(1.43)	(1.00)	(1.15)	(1.14)
			0.94	0.59	0.42	0.07	0.10	0.08
4	0.25	625	−14.16	−5.14	−3.38	−0.86	−0.07	0.13
			(2.56)	(2.23)	(1.88)	(1.16)	(0.93)	(1.01)
			1.00	0.94	0.77	0.16	0.04	0.04
9	0.25	272	−6.93	−3.03	−1.99	−0.45	−0.02	0.08
			(2.14)	(1.64)	(1.39)	(0.93)	(1.15)	(0.95)
			1.00	0.71	0.48	0.05	0.09	0.02
25	0.25	100	−2.78	−1.32	−1.08	−0.19	−0.11	−0.17
			(1.50)	(1.19)	(1.32)	(1.06)	(1.06)	(1.11)
			0.64	0.27	0.26	0.06	0.07	0.06

[a]The inverse of the proportion of individuals sampled over the sample lattice area.
[b]Fraction of population area covered by the sample lattice.

or less distinct patch structure in cases of greater dispersal, thus decreasing the size of the sample area is less influenced by which patches happen to be contained. In other words, there is a tradeoff. For cases of moderate to high dispersal, it may be reasonable to sample all contiguous individuals, or this may actually be preferable over in-

TABLE 8.5. Mean SND for the total number of unlike joins for distance class one (standard deviations in parentheses) and rates of rejection of H_o for different dispersal models and sampling schemes

Sampling Scheme			*Wrights Neighborhood Sizes*					
Porosity[a]	*Area*[b]	*Size*	*4.2*	*25.1*	*50.2*	*115.2*	*632.4*	*Random*
1	1.00	10,000	−77.30	−19.54	−14.55	−4.36	−0.97	0.01
			(5.27)	(4.00)	(4.71)	(1.90)	(0.95)	(1.03)
			1.00	1.00	1.00	0.93	0.15	0.06
4	1.00	2,500	−26.95	−9.05	−6.91	−2.15	−0.56	0.09
			(3.17)	(2.16)	(2.51)	(1.32)	(0.98)	(0.88)
			1.00	1.00	1.00	0.51	0.12	0.05
9	1.00	1,089	−13.10	−5.41	−4.14	−1.35	−0.32	−0.05
			(2.06)	(1.58)	(1.78)	(1.26)	(0.98)	(1.05)
			1.00	1.00	0.90	0.31	0.07	0.05
25	1.00	400	−4.94	−2.42	−2.28	−0.70	−0.33	−0.01
			(1.47)	(1.32)	(1.50)	(1.00)	(0.91)	(1.01)
			0.99	0.63	0.55	0.13	0.05	0.02
4	0.50	1,250	−18.70	−6.09	−4.23	−1.27	−0.44	0.00
			(2.82)	(2.02)	(1.98)	(1.26)	(0.86)	(0.88)
			1.00	0.98	0.89	0.26	0.05	0.03
9	0.50	544	−9.24	−3.52	−2.66	−0.80	−0.31	0.01
			(2.05)	(1.71)	(1.62)	(1.27)	(1.01)	(0.94)
			1.00	0.79	0.65	0.20	0.09	0.05
25	0.50	200	−3.43	−1.57	−1.38	−0.51	−0.36	0.09
			(1.58)	(1.29)	(1.42)	(1.05)	(0.95)	(0.93)
			0.78	0.44	0.34	0.07	0.05	0.02
4	0.25	625	−12.90	−3.89	−2.55	−0.72	−0.19	−0.02
			(3.10)	(1.84)	(1.63)	(1.20)	(0.90)	(0.94)
			1.00	0.84	0.64	0.19	0.04	0.06
9	0.25	272	−6.33	−2.27	−1.53	−0.23	−0. 17	−0.02
			(2.06)	(1.45)	(1.38)	(1.23)	(1.10)	(1.01)
			0.98	0.51	0.40	0.11	0.11	0.04
25	0.25	100	−2.25	−0.84	−0.85	−0.38	−0.27	0.04
			(1.58)	(1.24)	(1.23)	(1.16)	(1.01)	(1.02)
			0.58	0.20	0.20	0.08	0.06	0.05

[a]The inverse of the proportion of individuals sampled over the sample lattice area.
[b]Fraction of population area covered by the sample lattice.

creasing the porosity, for a fixed total sample size. The loss of autocorrelation due to increasing spatial scale by increasing porosity may be considerably worse than the gain from stochastic spatial nonstationarity.

To reveal the effects of allele frequencies, and the potential direct

TABLE 8.6. Models of unequal allele frequencies for simulation sets with multiple alleles

Allele				
1	*2*	*3*	*4*	*5*
0.1	0.3	0.6	—	—
0.1	0.1	0.3	0.5	—
0.1	0.1	0.1	0.1	0.6

and indirect effects of numbers of alleles per locus, on the strength of observed spatial correlations, additional simulation studies have been conducted on a variety of sets with different numbers (three, four, or five) of alleles (Epperson et al. 1999). For each multiplicity, a set of simulations was conducted where all alleles had equal initial (and nearly equal end) frequencies. In addition, various sets with unequal frequencies of alleles were conducted (table 8.6). Together, the sets allowed contrasts that separate effects of allele frequency per se from possible effects of number of alleles.

The SNDs for joins between identical homozygotes are shown in table 8.7. The values for alleles with frequencies in the range of 0.3–

TABLE 8.7. Means SND values for joins between identical homozygotes, for various alleles, for different levels of dispersal (Wright's neighborhood size N_e)

Allele Model	*Allele*	*Allele Frequency*	*Dispersal Model*				
			4.2	*25.1*	*50.2*	*115.2*	*637.4*
2e	1	0.50	70.56	20.95	15.79	4.75	1.12
3e	1	0.33	63.19	23.51	11.97	3.33	0.68
3u	1	0.10	51.28	12.91	6.23	1.27	0.29
	2	0.30	62.00	20.94	11.56	2.97	0.70
	3	0.60	72.29	28.99	16.36	5.13	1.28
4e	1	0.25	59.21	20.44	9.83	3.02	0.45
4u	1	0.10	49.55	13.69	5.90	1.44	0.24
	3	0.30	61.18	21.80	12.02	3.32	0.78
	4	0.50	69.55	27.29	16.00	5.17	0.89
5e	1	0.20	57.13	18.37	9.17	2.36	0.58
5u	1	0.10	51.10	12.52	5.49	1.46	0.23
	5	0.60	73.04	30.04	17.19	5.26	0.93

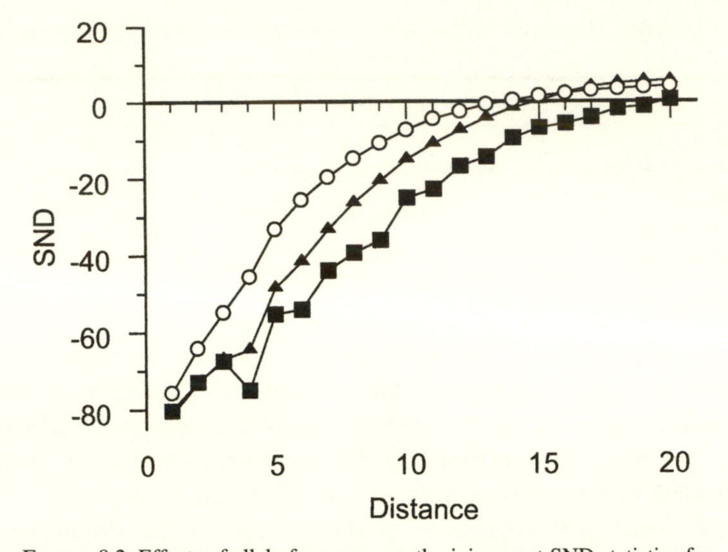

FIGURE 8.2. Effects of allele frequency on the join-count SND statistics for the total number of unlike joins for a locus (*A*/*a*) with two alleles. Averages are shown for different sets of five simulations each, and sets have different frequencies q of the a allele: for set 1 (■) $q = 0.5$, set 2 (▲) $q = 0.2$ and set 3 (○) $q = 0.1$. All simulated populations of 10,000 plants have nearest neighbor pollination and no seed dispersal (Wright's neighborhood size 4.2), and results are shown for generation 200. Figures adapted from results of Epperson (1995a). Distance is measured in units of the lattice of individuals.

0.6 are virtually identical, and this was also true for the two-allele case (figure 8.1 and table 8.3) (Epperson 1995a; Epperson and Li 1997). Skewing of allele frequencies has little effect unless they are reduced to about 0.2. A stronger effect obtains when allele frequencies are reduced further to 0.1 or less. Then the SNDs for the joins between like homozygotes for rare alleles are substantially decreased. Reductions of the same magnitude occurs in the two-allele case. Apparently, this simply reflects the fact that alleles with moderately low frequencies occur in fewer patches and patch sizes are unaffected. When the allele frequency is reduced to about 0.1 or lower, then the patch size must be reduced in populations of this size. In contrast, the SNDs for total number of unlike joins are nearly unaffected by allele frequencies per se (figure 8.2, table 8.8).

TABLE 8.8. Mean SND values for total unlike joins for various allele models for different levels of dispersal (Wright's neighborhood size N_e)

Allele Model[a]	*Dispersal (Wright's neighborhood size)*				
	4.2	*25.1*	*50.2*	*115.2*	*637.4*
2e	−77.30	−19.54	−14.55	−4.36	−0.97
3e	−108.23	−38.01	−19.99	−5.62	−1.11
3u	−95.51	−35.01	−19.51	−5.70	−1.44
4e	−133.32	−44.75	−23.18	−6.67	−1.23
4u	−110.60	−41.50	−23.25	−7.05	−1.23
5e	−152.11	−50.66	−26.11	−7.59	−1.53
5u	−123.27	−48.53	−27.26	−8.69	−1.65

[a]u denotes unequal allele frequency, e denotes equal.

The values for joins between like homozygotes do not depend directly on the number of alleles. The only major effect of increasing the number of alleles is indirect, in that it tends to skew allele frequencies away from 0.5. In addition, allele number per se has negligible effects on standard deviations (Epperson et al. 1999); however, allele frequency does have a small effect (Epperson and Li 1997; Epperson et al. 1999).

For sake of brevity, the statistical power in terms of the observed rates of rejection of the null hypothesis is not shown in tables 8.7 and 8.8. However, as is true for Moran's I statistics, these rates were very close to values calculated by the proportion of times that the estimated 95-percent confidence intervals overlap with zero. Values may be treated as normal deviates and the standard deviations were generally similar to those for the two-allele cases, which are shown in table 8.3. Thus the predicted rejection rates can be inferred from combining information in the tables. For example, mean values approximately one standard deviation from zero have rates of rejection near one-third.

The SNDs for the total number of unlike joins consistently take larger negative values as the number of alleles increases. This is in contrast to their near-independence of the allele frequencies (table 8.8). For almost every dispersal model, and every allele multiplicity, values for all unequal allele frequencies are very close to those for equal allele frequencies (table 8.8). The only substantial exception is the case of full sampling in the model with very low dispersal, $N_e = 4.2$, for which SNDs for equal allele frequencies are substantially more negative than for unequal allele frequencies. Standard deviations are also negligibly affected by allele number and frequencies, and they are virtually identical to those for two alleles with equal frequencies shown in table 8.5. Similarly, the statistical power or rates of rejection of the null hypothesis are virtually identical to those for the corresponding two-allele cases (Epperson and Li 1997).

In summary, the first-distance-class SNDs for joins for like homozygotes, opposite homozygotes, and the total numbers of unlike joins, for selectively neutral loci, change in monotonic and precisely predicted ways with changes in the amount of dispersal. This means that we can use analyses of survey data to accurately estimate dispersal. First we may make any needed adjustments for the allele numbers and frequencies. The values of the SNDs in experiments must also be adjusted by the using the square root of the ratio of experiment sample sizes relative to the closest one found in the tables. The standard deviations of the statistics must be similarly be interpolated by the square root of the ratio in sample sizes, and they must be adjusted for sample porosity.

Existing results also support several general recommendations for the design of samples. When dispersal is in the low to moderate range (say Wright's neighborhood size $N_e < 50$), the scale of the sampling lattice should cover an area that is expected to contain at least four to nine patch areas. This will avoid stochastic and statistical fluctuations arising from blind sampling of an area only within or between fewer strongly differentiated genetic patches that are initially invisible to the investigator. For low to moderate dispersal ($N_e < 50$), this translates to an area containing about 2000–5000 individuals. Qualitative information about the pollen and seed dispersal mechanisms in plants, or casual observations about the movements of animals, are often available and sufficient to determine whether dispersal is within the range of low to moderate rather than larger. Thus, in many cases experimentalists can use the above guidelines to structure samples with near optimal properties. On the other hand, it is notoriously difficult to obtain *precise* direct estimates of dispersal (see review by Levin 1981) for a variety of reasons. Thus estimators based on standing spatial patterns may be quite valuable, and under many conditions they may be more feasible than direct measurement of dispersal. They may also be more reliable in many circumstances. It has been suggested that ca. 10–20 sample points should fall within each patch on average (Epperson 1993a). If a genetic sample contains a few hundred to several hundred sampled individuals, it follows that the porosity should be within the range of about 4–16, or possibly 25 (resulting in percentages of sampled individuals over sample area of 1/4 or 1/9 to 1/25, respectively). The main point is that while the one extreme, complete censussing (porosity of one), is very inefficient, use of porosities greater than about 25 forces the shortest intersample distances (and hence sample distance classes) to be so large that the spatial autocorrelations are substantially decreased, as genetic isolation by distance theory predicts (e.g., figure 8.1). Regularly spaced lattices of sampling points have several beneficial properties (Epperson 1993a). For populations with higher amounts of dispersal, the increased stochastic and sampling errors caused by sampling of patches (i.e., technically spatial–nonstationarity) are not important because the patches are not as distinct and they are very large. Moreover, the genetic isolation by distance or spatial autocorrelation curves drop more slowly with distance. Hence it appears that in such populations a wide range of sampling schemes, including contiguous sampling, may work reasonably well.

STATISTICAL PROPERTIES OF MORAN'S *I* STATISTICS FOR QUADRATS

In chapter 7 it was noted that, although the use of quadrat samples generally reduces the statistical precision by a factor equal to the number of individuals in the quadrats (Epperson and Li 1996, 1997), sometimes it is not feasible to avoid quadrat sampling. An important issue is the size of quadrats. For small quadrats (e.g., 4 individuals) I statistics for distance class one monotonically decrease with N_e, except for N_e in the range of 4–8. Completely monotonic decrease is observed for a quadrat size of one individual (Epperson 1995c) (see figure 8.3). In contrast, when quadrats are of size 25, the decrease is monotonic only after N_e exceeds 50 (Sokal et al. 1989b; Epperson and Li 1997). The design of experiments with the aim of indirectly

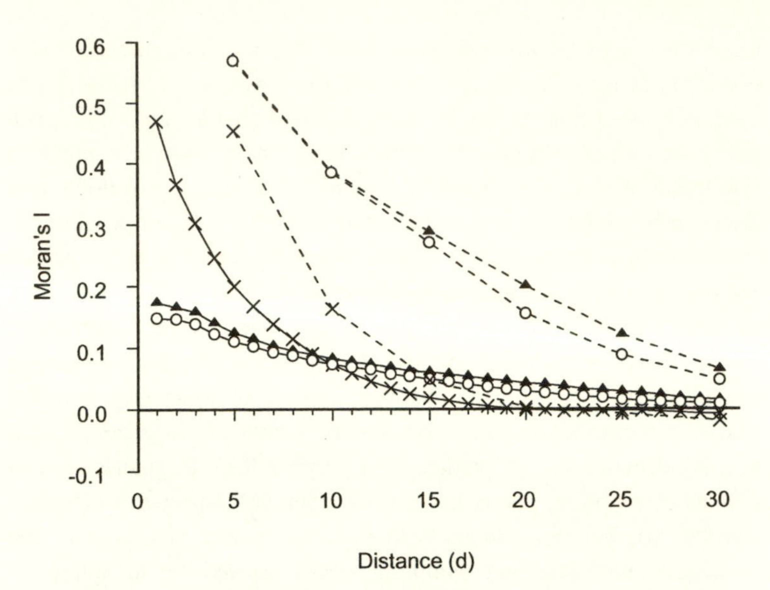

FIGURE 8.3. Moran's I statistics with different quadrat sizes for sets of simulations of populations with various levels of dispersal, at generation 200. Average correlograms are shown for converted individual genotypes (solid lines) and for local gene frequencies in quadrats of size 25 individuals (dimensions 5 × 5) (dashed lines). Set 1 (X) had Wright's neighborhood size $N_e = 4.2$, set 2 (▲) had $N_e = 25.1$, and set 3 (○) had $N_e = 50.2$. Distance is measured in units of the lattice of individuals. Note that autocorrelation measured by I statistics for converted genotypes decreases monotonically as dispersal increases, but that for quadrat allele frequencies does not.

measuring dispersal should take advantage of the convenience of large quadrats only when dispersal is believed to be in the moderate to high range ($N_e > 50$) (Epperson and Li 1996).

STATISTICAL PROPERTIES OF MORAN'S *I* STATISTICS FOR INDIVIDUAL GENOTYPES

Predicted values of Moran's I statistics for individual genotypes, as evidenced in the same types of simulated populations discussed above, depend on the sample porosity or distance class, but they are mostly unaffected by the sample size (table 8.9). Even when the sample size

TABLE 8.9. Mean values of Moran's I statistics for individual genotypes for distance class one (standard deviation in parentheses) for different dispersal models and sampling schemes; also shown are the rates of rejection of the null hypothesis

Sampling Scheme			*Wright's Neighborhood Sizes*					
Porosity[a]	*Area*[b]	*Size*	*4.2*	*25.1*	*50.2*	*115.2*	*632.4*	*Random*
1	1	10,000	0.454	0.199	0.115	0.036	0.008	−0.000
			(0.024)	(0.037)	(0.033)	(0.015)	(0.005)	(0.006)
			0.96	1.00	1.00	0.96	0.32	0.09
4	1	2,500	0.334	0.179	0.109	0.036	0.006	0.000
			(0.031)	(0.041)	(0.035)	(0.019)	(0.011)	(0.010)
			1.00	1.00	1.00	0.86	0.13	0.01
9	1	1,089	0.255	0.156	0.098	0.033	0.005	−0.002
			(0.034)	(0.045)	(0.035)	(0.022)	(0.017)	(0.017)
			1.00	1.00	0.98	0.49	0.11	0.07
25	1	400	0.163	0.117	0.082	0.029	0.006	−0.001
			(0.043)	(0.050)	(0.047)	(0.030)	(0.026)	(0.027)
			1.00	0.93	0.74	0.25	0.07	0.07
4	0.5	1,250	0.331	0.170	0.096	0.030	0.006	−0.001
			(0.039)	(0.054)	(0.040)	(0.024)	(0.014)	(0.013)
			1.00	1.00	0.99	0.50	0.06	0.05
9	0.5	544	0.256	0.152	0.090	0.026	0.003	−0.004
			(0.052)	(0.057)	(0.044)	(0.030)	(0.024)	(0.024)
			1.00	1.00	0.87	0.31	0.09	0.08
25	0.5	200	0.159	0.111	0.074	0.018	0.006	−0.000
			(0.063)	(0.067)	(0.058)	(0.040)	(0.037)	(0.041)
			0.95	0.73	0.51	0.12	0.05	0.11
4	0.25	625	0.325	0.148	0.080	0.020	0.002	−0.004
			(0.060)	(0.066)	(0.044)	(0.026)	(0.020)	(0.019)
			1.00	1.00	0.81	0.19	0.05	0.03
9	0.25	272	0.249	0.127	0.073	0.011	−0.002	−0.007
			(0.076)	(0.069)	(0.052)	(0.036)	(0.038)	(0.033)
			1.00	0.83	0.56	0.10	0.09	0.06
25	0.25	100	0.147	0.094	0.058	0.007	0.000	−0.001
			(0.066)	(0.093)	(0.075)	(0.056)	(0.053)	(0.058)
			0.70	0.42	0.29	0.08	0.06	0.06

[a]The inverse of the proportion of individuals sampled over the sample lattice area.
[b]Fraction of population area covered by the sample lattice.

is quite small (n = 100), only a small decrease is observed. Generally, the standard deviations are small for various distance classes, illustrating that the amount of stochastic and statistical variation is rather small. The effects of sample size on the standard deviation is simple and as for the other autocorrelation statistics, the stochastic

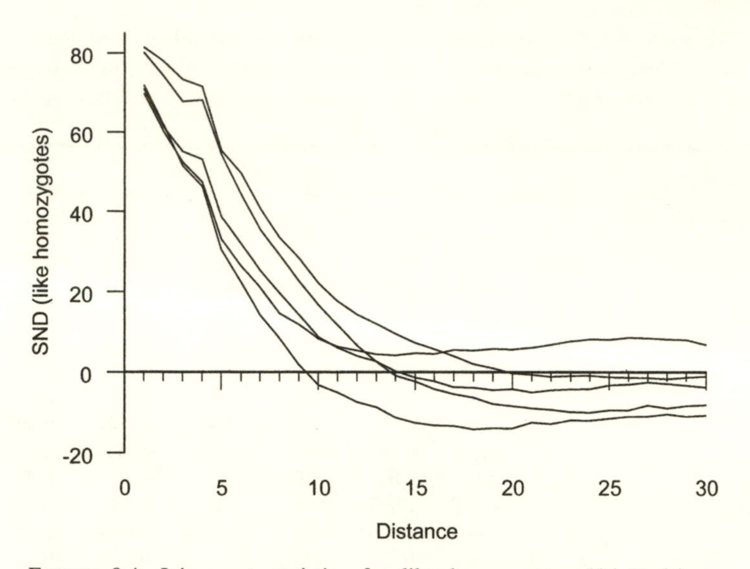

FIGURE 8.4. Join-count statistics for like homozygotes (*AA* × *AA* or *aa* × *aa*) for 5 individual simulation runs, at generation 200, of simulated populations of 10,000 plants that have nearest-neighbor pollination and no seed dispersal (Wright's neighborhood size 4.2), illustrating the stochastic spread of correlograms for individual loci. Allele model is two alleles with equal frequency. Adapted from results of Epperson (1995c).

variation is small (figure 8.4). When sample size (area) decreases, the standard deviation increases approximately by the square root of the ratio (Epperson et al. 1999). Moran's *I* statistics for individual genotypes generally have somewhat lower statistical power than do SNDs for the total number of unlike joins. These differences become large as the number of alleles increases (Epperson et al. 1999). The statistical power of *I* statistics is essentially equal to that of SNDs for like homozygotes. As for the join-count statistics, increasing dispersal has a strong effect on the mean values. In comparison, dispersal has relatively little effect on the standard deviations.

The frequency of an allele has very little effect on Moran's *I* statistics for individual genotypes (table 8.10), in contrast to most of the join-count statistics (figure 8.1, tables 8.7 and 8.8). Across the same array of allele models used to examine the join counts, there is virtually no effect. Changes in the number of alleles and the allele frequencies did not cause statistically significant differences in Moran's

TABLE 8.10. Mean Moran's I statistics for distance class one, for various multiple allele models with different levels of dispersal (Wright's neighborhood size N_e)

Allele Model	*Allele*	*Allele Frequency*	*Dispersal Model* 4.2	25.1	50.2	115.4	637.4
2e	1	0.50	0.454	0.199	0.115	0.036	0.008
3e	1	0.33	0.450	0.201	0.112	0.033	0.008
3u	1	0.10	0.455	0.191	0.114	0.035	0.007
	2	0.30	0.450	0.192	0.112	0.033	0.008
	3	0.60	0.449	0.190	0.109	0.035	0.008
4e	1	0.25	0.451	0.197	0.110	0.036	0.007
4u	1	0.10	0.447	0.197	0.113	0.036	0.007
	3	0.30	0.447	0.198	0.116	0.037	0.007
	4	0.50	0.448	0.193	0.117	0.039	0.008
5e	1	0.20	0.454	0.195	0.114	0.035	0.008
5u[1]	1	0.10	0.453	0.191	0.109	0.034	0.006
	5	0.60	0.453	0.197	0.113	0.036	0.007
5u[2]	1	0.02	0.438	0.179	0.102	0.032	0.007
			(0.066)	(0.054)	(0.041)	(0.005)	(0.007)
	3	0.05	0.444	0.184	0.107	0.036	0.007
			(0.037)	(0.040)	(0.032)	(0.005)	(0.007)
	5	0.86	0.450	0.186	0.113	0.037	0.006
			(0.030)	(0.032)	(0.030)	(0.005)	(0.006)

[1]Allele frequencies given in table 8.6.

[2]Allele frequencies given in text. Standard deviations for this model are shown in parentheses.

I statistics (or their standard deviations and rates of rejection of the null hypothesis) for individual genotypes (Epperson et al. 1999). Because of this, not all values for all sampling schemes (obtainable by request from the author) are displayed in the tables. However, there appeared to be a small effect of rare alleles, based on the values for $q = 0.1$ in multiallele models, and this called for a closer examination, with additional simulations with $q = 0.05$ and 0.02. The results for another five-allele model, for which there were two alleles with initial allele frequency 0.02, two with initial frequency 0.05, and the fifth allele with initial frequency 0.86 (Epperson, n.d.(b)), are included in table 8.10. These results show that there is still nearly negligible effect for alleles with frequency 0.05, but for even rarer alleles with $q = 0.02$, the values are consistently reduced. The reduction varies up to about 15 percent of the value for more common alleles. In addition, there is greater stochastic variance for very rare alleles. The standard deviations for very rare alleles, $q = 0.02$, are 0.066, 0.054, 0.041, 0.005, and 0.007 for the five dispersal models, and much larger than those for common alleles (table 8.9) when dispersal is low. This should be aggravated further by sampling error. Thus it appears that very rare alleles, which, for example, commonly occur for microsatellite loci, may be relatively poor for measuring dispersal when it is in the low to moderate range. Alleles with only slightly greater frequencies, i.e., $q \geq 0.05$, all had standard deviations very similar to the values for $q = 0.50$ shown in Table 8.9.

COVARIANCES OF MORAN'S *I* FOR CONVERTED INDIVIDUAL GENOTYPES OF MULTIALLELIC LOCI

Most empirical studies assay several to many loci and, especially with modern molecular markers, such as simple sequence repeats, studied loci often have large numbers of alleles. There have been outstanding issues on how best to utilize multiple-locus, multiallelic data. The issue of multiple-locus data is fairly straightforward, since unlinked loci are generally considered to be essentially stochastically and statistically independent. However, the question of what to do with multiple alleles has been more complicated. One approach is to use such statistics as the total number of unlike joins, which have some favorable properties, as described earlier. Yet the total number of unlike joins may present some problems in extreme cases where there are very large numbers of alleles, since it may become unlikely that *any* two individuals have the same diploid genotype in a small sample, regardless of the amount of genetic isolation by distance. The methods outlined below could be used for the other join-count statistics, but here the focus is on the popular method of using Moran's *I* statistic, I_a, for diploid genotypes converted to frequencies of a given allele *a*. Up until now, such studies have not been able to assess summary average values, because the values for different alleles of a locus are correlated, but the degree of correlation was unknown. Thus calculations of the standard errors of an average (weighted or unweighted) of values over alleles within a locus could not be obtained, either under the null hypothesis or else, thus also disallowing exact tests of significance for averaged values. The only known correlation is that for the two-allele case where the two values are completely correlated and only one value is kept. Experimental studies have taken a wide variety of approaches for analyzing multi-allelic spatial distributions, ranging from arbitrarily keeping only the value for one allele (discarding the rest), deleting the rarest allele (and claiming that the remaining values are independent), or keeping all values and not doing tests on the averages.

As was developed earlier in this chapter, Moran's I statistics for converted genotypes (and always for one allele at a time) can be expressed as sums of terms of the join counts, weighted by functions of the allele frequency (equation 8.23). Thus the covariances between I_a and I_b (for alleles a and b at a locus), and the correlations, can be found by summing terms of products of functions of allele frequencies and expected values of products of join counts, using equations similar to equations 8.14–8.18, only modified for non-free sampling (Epperson, n.d.(b)). Doing so generates completely general results on the expected correlations of Moran's I statistics, under the null hypothesis. We do not write the equations for the covariances here, because they involve large numbers of terms. Even for the smallest case of interest (three alleles), there are six genotypes, 21 types of joins, and hence $21^2 = 441$ terms of products of joins. In response to this problem the author wrote a computer program (available upon request) to calculate the values for any sample (any set of weights, $w_{ij}(k)$), for any number of alleles. The program takes advantage of the fact that regardless of the number of alleles, all products of joins can be placed into seven classes that determine the coefficient (multiplier of the allele frequencies function) and the powers of the genotypic frequency terms (Epperson, n.d.(b)).

The values of the covariances (although apparently not the correlations—see below) generally will depend on the sample array. Here we use as examples the covariances expected for a lattice population of 10,000 individuals, because this may be considered as the stochastic covariance for the null hypothesis for Moran's I in large populations. For such populations the values of $S_0(k)$, $S_1(k)$, and $S_2(k)$ for unweighted joins (i.e., to produce an unweighted Moran's I statistic) for the shortest distance class (single queen's move nearest neighbors) are 78,804, 157,608, and 2,497,968, respectively. For all allele models (those shown in table 8.6, plus the other five-allele case discussed above), the variances of I_a were identical (0.0000253) under H_o. It is particularly interesting that the allele frequency has little or no effect, indicating that the allele frequency terms cancel exactly or nearly completely. Moreover, as expected, the standard deviation (0.00503) is very close to that (0.006) observed for the simulated random mating ($N_e = 10{,}000$) cases (table 8.9), which also appeared to be invariant to allele frequencies. However, the variance under the alternative hypothesis of isolation by distance still might depend on the allele frequencies.

Under the null hypothesis, the covariances and correlations depend strongly on the allele model and the frequencies of the two alleles. For the allele models where all alleles have equal frequencies the correlations are 1.000, 0.250, 0.111, 0.062 for numbers of alleles 2 ($q = 0.50$), 3 (0.33), 4 (0.25), and 5 (0.20). This indicates that when allele frequencies are even, correlations are a significant issue only for loci with three or four alleles. This in turn suggests that the straight average has a standard error under H_o that is fairly close to the standard error for individual alleles (under H_o) divided by the square root of the number of alleles, when there are at least four alleles. The cases with unequal allele frequencies are summarized in Table 8.11. All of these models have one major or common allele ($q \geq 0.5$), some have another allele in moderate frequency, and all have one or more relatively rare alleles ($q \leq 0.1$). Generally, the most common alleles have the largest correlations with other alleles. This is not necessarily an intuitive result, although it fits with the idea that more of the spatial distribution is determined by the common alleles, whereas there is a "dilution factor" for the others. Moreover, by contrasting allele models with some of the same frequencies, it is clear that the number of alleles (as long as more than two) does not substantially affect the correlations directly, only indirectly if it is associated with changed frequencies. The general features indicate that much of the correlations can be removed by omitting I statistics for the most common allele, to form an approximate test statistic based on the straight average of the remaining alleles, especially if there are no other alleles with moderate frequency (say q in the range of 0.2–0.3), which are also the most correlated with common alleles.

The correlations involving rare alleles are generally very small, less than 0.05 in the three-allele case, and about 0.01 or less in the models with larger numbers of alleles. The correlations are quite similar, as expected, in sets of 100 simulations of random surfaces (table 8.11), especially given the high degree of stochasticity in the correlations within sets. Moreover, the correlations did not change detectably among a wide variety of other values of $S_0(k)$ and $S_2(k)$ in calculated cases where the ratio $S_2(k)/S_0(k)$ ranged widely from about 10 to 600 (Epperson, n.d.(b)). Thus the correlations are extremely insensitive to and, for all practical purposes, independent of the spatial distribution of sample locations and distance classes.

To calculate standard errors of estimated averaged Moran's I statistics over alleles within a locus beyond use in hypothesis tests, the correlations under isolation by distance must be characterized. The correlations calculated for sets of 100 replicate simulations of several dispersal models are also shown in table 8.11. There is a very large degree of stochastic variation in such correlations in sets of 100 simulations. The values in table 8.11 are averaged wherever there are redundant types of pairs of alleles (e.g., the pair of alleles 1 and 2 of model $5u^2$ have the same initial frequency, and so do the pair 3 and 4, thus values for pairs of types "1" and "3," are averaged among the values for 1×3, 1×4, 2×3, and 2×4). Illustration of the full degree of stochastic variation is made for two typical cases in table 8.12, which lists the actual correlations between all pairs of alleles for the allele model $5u^1$ in specific sets of 100 simulations. Even when some allele pairs are averaged there is still a high level of stochastic variation. This means that for any given locus, the values for different alleles are expected to be correlated, but the correlations are highly variable. This makes it somewhat difficult to assess whether the level of dispersal affects the correlations. However, because there are multiple comparisons of alleles it is clear, and perhaps very surprising, that there is no apparent systematic effect of dispersal on the correlations. Remarkably, this implies that *the degree of correlation does not depend on the amount of spatial autocorrelation*, at least not much on that created under isolation by distance in large populations. Moreover, as for the null hypothesis, the correlations did not change in any systematic way with distance class, nor did the number of alleles matter beyond its indirect effect through allele frequencies.

TABLE 8.11. Correlations of Moran's *I* statistics for converted genotypes for two alleles *a* and *b* within a locus for loci with unequal allele frequencies

Allele Model	*Alleles*[a]	*Allele Frequency*	*Dispersal Model* 4.2	25.1	50.2	115.2	637.4	Average	10,000	H_o
3u	1, 2	0.10, 0.30	0.159	0.051	0.280	0.063	0.052	0.121	−0.097	0.048
	1, 3	0.10, 0.60	0.291	0.306	0.175	0.234	0.220	0.245	0.050	0.166
	2, 3	0.30, 0.60	0.667	0.638	0.571	0.585	0.669	0.625	0.618	0.643
4u	1, 2	0.10, 0.10	0.095	0.139	−0.009	−0.067	0.150	0.062	−0.121	0.012
	"1," 3	0.10, 0.30	−0.133	0.081	−0.041	−0.035	−0.039	−0.033	−0.056	0.048
	"1," 4	0.10, 0.50	0.083	0.101	0.112	0.098	0.089	0.097	−0.015	0.111
	3, 4	0.30, 0.50	0.408	0.465	0.608	0.372	0.430	0.457	0.285	0.429
5u^1	"1," "2"	0.10, 0.10	−0.035	0.007	−0.096	0.019	0.176	0.014	−0.013	0.012
	"1," 5	0.10, 0.60	0.171	0.181	0.058	0.132	0.282	0.165	0.079	0.167
5u^2	1, 2	0.02, 0.02	0.102	−0.101	−0.065	−0.026	0.026	−0.013	−0.059	0.000
	"1," "3"	0.02, 0.05	−0.018	−0.002	−0.015	−0.081	0.074	−0.008	0.025	0.001
	"1," 5	0.02, 0.86	0.233	0.196	0.264	0.055	0.213	0.192	0.239	0.125
	3, 4	0.05, 0.05	−0.029	0.053	0.045	−0.041	−0.044	−0.003	−0.072	0.003
	"3," 5	0.05, 0.86	0.366	0.366	0.398	0.402	0.378	0.382	0.384	0.324

[a]Quotation marks mean either any one of alleles with same frequencies, in the case of the expected value under the null hypothesis, or the average of correlations found for alleles with the same initial frequencies, in the case of simulated processes.

TABLE 8.12. Correlations of Moran's *I* statistics for converted genotypes for two alleles *a* and *b* within a locus with five alleles with unequal allele frequencies (model $5u^1$), for dispersal with $N_e = 4.2$ (above diagonal) and $N_e = 115.2$ (below diagonal), as examples showing the full degree of stochasticity in sets of 100 simulations

	Allele				
Allele	*1*	*2*	*3*	*4*	*5*
1	1.000	−0.085	−0.063	−0.065	0.270
2	0.100	1.000	−0.164	−0.002	0.116
3	0.085	−0.111	1.000	0.172	0.171
4	−0.064	0.083	−0.019	1.000	0.128
5	0.104	0.102	0.003	0.320	1.000

Note: Alleles *1*, *2*, *3*, *4* have the same frequencies (0.1) and allele *5* is most common (0.6).

Similar features exist for the correlations for simulated isolation by distance models where all alleles have equal frequencies. When there is structure, the average values (i.e., averaged over nonrandom dispersal models—here, too, there were no apparent major trends for correlations to change systematically with level of dispersal) are 1.000, 0.278, 0.115, and 0.040 for numbers of alleles 2 ($q = 0.50$), 3 (0.33), 4 (0.25), and 5 (0.20). Again, the level of stochasticity was high, even among allele pairs within the same set of 100 simulations, and the averages are close to the values under the null hypothesis. The average values for the correlations for simulated allele models where all alleles have equal frequencies, for the random mating simulations, are 1.000, 0.154, 0.064, and −0.006, for numbers of alleles 2 ($q = 0.50$), 3 (0.33), 4 (0.25), and 5 (0.20), and these values are reasonably close to the expected values under the null hypothesis, given the levels of stochasticity.

In total, our knowledge about Moran's *I* statistics for converted genotypes for multiallelic loci allows certain recommendations to be made. They are completely robust, including with respect to the sampling array, the actual amount of spatial structure, the distance classes, and numbers of alleles. First, alleles that are not sufficiently informative should be discarded. Criteria may vary, but the simplest is to delete alleles that are not carried by at least two sampled individuals. Next, we may want to delete alleles that have very low frequencies by other criteria, but generally there is no increase in variances (available as standard deviations in tables 8.9 and 8.10) unless allele frequency is less than about 0.02. Moran's I statistics should probably be calculated for all remaining alleles (unless there are only two). Next, with respect to calculating averages either for hypothesis tests or estimation, there are two considerations. First, it appears unlikely that weighting of alleles is useful, because they almost always have very similar variances. This fits with results by Smouse and Peakall (1999), who found that weighted measures did not improve accuracy. Second, there are at least three choices for calculating the average: (1) ignore the correlations, if alleles all have moderate frequencies, and/or there are 4 or 5 alleles or more; (2) throw out the most common allele; or (3) use the correlations (together with the variances in tables 8.9 and 8.10) to correct the standard error of the average using the standard formula. The latter can be done using the predicted correlations and variances ($\sigma^2(I_a)$) to find the predicted covariances among alleles ($\mathrm{Cov}(I_a, I_b)$), and then use the standard formula for the (unweighted) sum of I among m alleles:

$$\sigma^2\left(\sum\right) = \sum_{a=1}^{m} \sigma^2(I_a) + 2\sum_{a,b} \mathrm{Cov}(I_a I_b)$$

where the second summation is over all alleles a and b such that $a < b$ (e.g., Feller 1957). The variance of the (unweighted) average σ^2 is equal to $\sigma^2(\Sigma)$ divided by m^2, or, for options (1) and (2), $\sigma^2(\Sigma)$ divided by $m(m - 1)$ (Epperson n.d.[b]). Similar methods could be used to find the variance for weighted averages. This could be applied to quadrat data or even data among populations (chapter 4). However, it still is based on all pairs of genotypes, and it appears that this confounds the average level of differentiation of allele frequencies among quadrats or populations with the spatial autocorrelation among quadrats or populations.

The role of rare alleles is particularly interesting. On the one hand, it appears that for moderate size samples (up to a few hundred genotypes) as the number of alleles becomes very large, there are diminishing returns, because many alleles will have small frequencies. This has considerable importance for the choice of molecular markers. On the other hand, in some cases rare alleles may contain special information. The above results on the variances in dispersal models assumed an initially random distribution, which is incidental if alleles have been present for long periods. However, if rare alleles are the result of recent mutations they may have more information, which may be especially important when dispersal is low. Thus, loci with very high mutation rates, such as microsatellites, may be exceptions. In addition, rare alleles might represent recent long-distance dispersal from outside the population, and this situation may be amenable to similar logic. The late Gustave Malécot considered rare alleles to be "very important" (personal communication). Moreover, the above logic seems to fit with Barton and Wilson (1995), who found that the most recent coalescence events determine much of the spatial structure caused by genetic isolation by distance.

MULTIPLE-LOCUS ESTIMATORS

The standard errors of spatial autocorrelation statistics generally are rather small, but for estimates based on k loci they will be $\sqrt{k}$ times smaller. Multiple-locus studies are an important strategy in experimental work. Using modern molecular methods a large number of multiallelic genetic loci may be easily assayed within natural populations. The results on join-count statistics and Moran's I statistics for the first distance class may be used to estimate the level of dispersal in a way that is more often feasible and more efficient than those based on direct observations of movements. These are based on predicted values, found by adjusting the values in the tables for experimental sample size and porosity (for this, approximate population density may be easily estimated and the scale of the sampling is generally known), for various levels of dispersal, and matching these with the values observed in the experiment. Some examples for data from real plant populations were given in chapter 7. Statistical power is high even when there is only a single locus, if sample sizes are a few hundred for low to moderate dispersal cases, or several hundred to one thousand for high dispersal cases (Epperson and Li 1997).

It has been often argued that true multilocus measures will give more accurate estimates of the kinship among any given pair of individuals, and therefore should provide better estimates of spatial structure than do single-locus measures (e.g., Loiselle et al. 1995).

Whereas the former is surely correct, the latter may not be if multiple single-locus measures are averaged. First, any additional information in multilocus data (as opposed to multiple locus data) depends on linkage disequilibrium. While relatively little is known about multilocus disequilibrium, two-locus disequilibrium is near zero in simulated populations, even if the two loci are tightly linked, with recombination rates as small as 0.01 (Epperson 1995b). This suggests that global disequilibrium is small in populations, although it seems possible to have substantial disequilibrium if only a small area of a population is sampled. Moreover, it appears that generally there is so much stochasticity in genetic isolation by distance, that the actual kinship among pairs of individuals is itself highly stochastic. Some measures, such as the number of alleles in common, first average (combine information) over loci in each pair of individuals, and then average among pairs of individuals. Use of Moran's I statistics, in the way prescribed above, involves averaging first among pairs of individuals, and then among alleles and loci. There seems to be no a priori reason to choose one method over the other. However, to date, almost nothing is known about either the stochastic or statistical properties (either under the random distribution null hypothesis or under any other hypothesis) of the multiallelic, multilocus pairwise measures of kinship such as in Loiselle et al. (1995). As is shown later in this chapter, these properties can be derived from the results on Moran's I statistics. In contrast, the distributions of Moran's I statistics for single alleles are well characterized, and now, because of the developments discussed in this chapter, they can be averaged over alleles as well as loci. Moreover, using Moran's I statistics maintains the single-locus information, and thus they may be used for detecting differences among loci that might be caused by natural selection. These problems of various multilocus estimators also apply to the study of genetic variation among populations.

F STATISTICS

F statistics, as they are related to Wright's measures of inbreeding at various hierarchical levels or spatial scales, can also be used to examine structure of genetic variation in a large continuous population. The most common method is to use F_{ST} as a measure of genetic differentiation among subgroups of the population. However, before discussing F_{ST}, response of another F statistic, the fixation index, to differences in structure should be noted. The fixation index in a two-allele system is usually estimated simply as $1 - h/2q(1 - q)$, where h is the total frequency of heterozygotes and q is the gene frequency. The situation is rather more complex for multiple alleles (e.g., Kirby 1975). Generally, as dispersal increases, all else being constant, the fixation index decreases (e.g., Rohlf and Schnell 1971; Turner et al. 1982). For example, in simulations with neighborhood size of 4.2 (dispersal model 1 in table 8.1) the fixation index averaged about 0.33 at the quasistationary phase. This model did allow a probability of self-fertilization (zero migration of both parents) of $s = 1/9$ (Epperson 1990a). However, the observed fixation index is far greater than that expected for the standard (nonspatial) mixed mating model with selfing and random outcrossing at equilibrium, $F_e = (1 - t)/(1 + t)$ (where $t = 1 - s$), or 0.06. The difference is caused by biparental inbreeding. Simulated populations with neighborhood size 12.6 had $F = 0.15$ at the quasi-stationary phase (Epperson 1990a), again much greater than F_e (0.02). However, generally, the fixation index has highly limited utility for spatial structure, because it is affected by any other form of inbreeding, for example, selfing, that may have little or nothing to do with spatial structure. In contrast, F_{ST} and similar measures of spatial differentiation are affected by selfing rate only to the degree that it restricts pollen movements, and thus acts through the neighborhood size (Wright 1946).

Whereas the theoretical values of inbreeding coefficients have been well described, initially by Wright (1943), most aspects of the statistical properties of F-statistic measures of spatial differentiation in continuous populations are not well studied. The usual and simplest estimator of F_{ST} for an allele is

$$F_{ST} = \frac{\sum_{i=1}^{n}(q_i - q)^2/(n-1)}{q(1-q)} \qquad (8.27)$$

where q_i is the gene frequency in a quadrat or subpopulation (population in the case of a system of discrete populations) i, n is the number of quadrats or subpopulations, and q is the average gene frequency. The numerator of equation 8.27 is the usual estimate of the variance, in this case among gene frequencies in quadrats or subpopulations (e.g., Weir and Cockerham 1984; Slatkin and Barton 1989). As noted, this F_{ST} is not the same as the inbreeding coefficients of Wright (1943), nor is it an unbiased estimate of it. The "less biased" estimator θ of Weir and Cockerham (1984) has a more complicated expression and is not displayed here.

We can examine F_{ST} in the types of Monte Carlo simulations described earlier. A variety of sampling schemes could be envisioned, but must involve grouping sampled individuals into subsamples or quadrats, to distinguish local inbreeding, F_{IS}, from global inbreeding (F_{IT}). One way to simulate sampling is analogous to those studied above for Moran's I statistics. Thus we may include total censusing of the entire population ($N = 10{,}000$, $n = 10{,}000$, "full sampling"), and other sampling whereby individuals are initially sampled on a grid (or, nearly equivalently, some average percentage of total individuals) or "porosity," perhaps over some subarea of the population. When all individuals in the sample area are sampled, porosity equals 1.0. Note that porosity still affects the spatial scale of sampling as well as the size of the sample, and in general quadrat size can also affect the scale of sampling and the number of quadrats.

In simulation studies sampled individuals were grouped into quadrats of 25. Observed values of the standard statistic F_{ST} (equation 8.27), as well as the less biased estimator θ, were calculated (Epperson and Li 1996). Because little is known about the distribution of F_{ST} and θ for single-locus data, jackknife variances for θ (Weir 1990) were used. Values for θ for sets of 100 simulations of various dispersal models at quasistationarity are shown in table 8.13. The mean value of θ for full sampling for the random model was 0.0002, close to zero, whereas the mean of the traditional measure was 0.02, indicating that θ is less biased, at least under the null hypothesis. Because F_{ST} is more biased, only the results for θ are shown in table 8.13, but it is noted that F_{ST} was consistently larger than θ for the various studied dispersal and sampling schemes. Values of θ decrease steadily as the level of dispersal increases, as predicted by the theory. They also decrease as porosity increases, again as expected because as porosity increases, the area (and the number of individuals in the subpopulation area) covered by each quadrat increases (Wright 1943; Cavalli-Sforza and Feldman 1990). In addition, jackknifed variances decrease as dispersal increases.

TABLE 8.13. Mean values of the theta estimate of F_{ST} among quadrats of 25 individual genotypes (standard deviation in parentheses), following different initial sampling schemes, for various dispersal models

Sampling Scheme			*Wright's Neighborhood Sizes*				
Porosity	*Area*[a]	*Size*	*4.2*	*25.1*	*50.2*	*115.2*	*Random*
1	1	10,000	0.227 (0.022)	0.075 (0.015)	0.058 (0.018)	0.018 (0.008)	0.000 (0.002)
4	1	2,500	0.145 (0.022)	0.059 (0.016)	0.050 (0.019)	0.017 (0.008)	0.000 (0.003)
9	1	1,089	0.090 (0.028)	0.046 (0.018)	0.037 (0.019)	0.014 (0.010)	−0.001 (0.005)
25	1	400	0.040 (0.026)	0.032 (0.018)	0.029 (0.021)	0.012 (0.011)	−0.001 (0.007)
4	0.5	1,250	0.143 (0.028)	0.057 (0.021)	0.043 (0.023)	0.015 (0.011)	−0.001 (0.004)
9	0.5	544	0.089 (0.039)	0.041 (0.024)	0.034 (0.023)	0.010 (0.013)	−0.001 (0.007)
25	0.5	200	0.038 (0.033)	0.027 (0.026)	0.024 (0.026)	0.008 (0.014)	−0.003 (0.008)
4	0.25	625	0.137 (0.043)	0.051 (0.028)	0.034 (0.024)	0.010 (0.009)	−0.001 (0.006)
9	0.25	272	0.082 (0.055)	0.035 (0.030)	0.025 (0.025)	0.004 (0.013)	−0.003 (0.009)
25	0.25	100	0.027 (0.040)	0.017 (0.033)	0.014 (0.025)	0.003 (0.018)	−0.006 (0.012)

[a]The proportion of the total population area actually sampled under a given sample design.

Measures of the statistical significance of values of θ can be obtained by observing whether the estimated 95-percent confidence intervals overlap with zero. However, unlike the Moran's *I* statistics, where the observed type I error rate is not different from the expected 5-percent level, those for θ were generally much larger than 5 percent, with an average of 23.4 percent and a range from 4 to 47 percent (Epperson and Li 1996). Given such inflated errors, tests of significance are invalid. Similarly high rejection rates were observed for F_{ST} based on use of an approximately normal test statistic. Moreover, it is thus meaningless to measure the statistical power function under alternative hypotheses for limited dispersal models.

There is little to recommend the use of *F* statistics to measure spatial structure within populations. There is no obvious basis for choosing quadrat sizes, and although it is somewhat popular to set it equal to the neighborhood size N_e, the latter is not a spatial unit of genetic structure. The spatial scale of structure is not linearly related to N_e. Moreover, there is neither random mating nor random spatial distribution within neighborhoods under the process of isolation by distance (Wright 1946). As noted earlier, combining individuals into subpopulations or quadrats simply causes loss of information and statistical power, whether it be for Moran's I statistics or presumably for F statistics. It is to be generally avoided, unless experimental feasibility dictates otherwise. However, methods for F statistics are somewhat improved by employing a large number of (hierarchical) levels, similar to the approaches sometimes used for analyzing differentiation among populations.

MEASURES OF KINSHIP

There are various multiple locus and multilocus estimates of the relative genetic relatedness among pairs of individuals (or among subpopulations) that bear some indirect relationships to the mathematical models of isolation by distance of Malécot (1948) and similar models. The terms kinship, consanguinity, relatedness, genetic relatedness, and recently even "probability of identity by descent" are often used interchangeably and without reference to their original definitions. Moreover, usually they are casually compared to the mathematical models of genetic isolation by distance. The mathematical relationships of estimators of kinship to theoretical values of conditional kinship and probabilities of identity by descent are further examined in chapter 9.

The common feature of kinship estimators is the summation of alleles shared by pairs of individuals, further summed over a locus, and often still further summed over multiple loci. Many of these measures build upon the single-allele measure of genetic relatedness of two individuals i and j given by $(q_i - q)(q_j - q)/q(1 - q)$ (Morton 1975). An average may be found for a set of pairs of individuals:

$$R = \frac{\sum_{ij}(q_i - q)(q_j - q)}{kq(1-q)} \tag{8.28}$$

where q_i is the converted frequency of the allele in individual i. The summation is once over all k pairs of n sample genotypes in the set, i.e., distance class. Note that the sum is one-half of that in the numerator of Moran's I statistics for converted genotypes, for the special case of binary weights over all joins in a given distance class. The kinship measure in equation 8.28 is biased with respect to "conditional kinship" (sensu Morton 1975, see chapter 9), and "less biased" measures have been promoted (e.g., Loiselle et al. 1995). A bit of algebra shows that for groups of quadrats or populations

$$R = \frac{(n-1)}{n} \frac{I}{F_{\mathrm{ST}}} \tag{8.29}$$

and thus if n is large, then $R \approx I/F_{\mathrm{ST}}$, where F_{ST} is the usual estimator, given by equation 8.27 (Barbujani 1987). When R is defined separately for each pair of individuals i and j, it can be shown by formulating genotypic frequencies in terms of allele frequencies and the fixation index (F) that

$$R = \frac{I(1+F)}{2} \tag{8.30}$$

(Hardy and Vekemans 1999), and R is considered an estimate of conditional kinship (Hardy and Vekemans 1999). Thus R can be converted to Moran's I statistics and vice versa if F is known, but the problems remain in connecting R to the theoretical models, particularly if estimators of F are substituted for theoretical ones. Multiple-locus estimators of R have been obtained by using sums of single-locus specific numerators of equation 8.28, and dividing by the sum of single-locus denominators (Rousset 2000). However, little has been documented about the properties of such estimators, either through probability theory or simulations. Recently, Smouse and Peakall (1999) have further generalized this basic approach in a way that allows for weighting of multiple alleles, and it can account for the covariance that exists among alleles for the same locus, although only with the results of this chapter do we now know what the covariances are. They further revealed the relationship of their estimator to R and to the AIDA measure of Bertorelli and Barbujani (1995).

Multiple-locus measures of R are first averaged over alleles and/or loci and then normalized. Alternatively, it is possible to normalize first and then average, as in the case of Moran's I statistics for individual alleles, now that we have found the covariances and variances based on the join counts (Epperson, n.d.(a)). Importantly, now the distributions of averages of Moran's I statistics are relatively easily obtained, because the distributions of the individual values are known, as are the covariances among alleles within a locus. It appears that expected value and variance (under the null hypothesis that the existing genotypes are randomly distributed) of the measure of Smouse and Peakall (1999) could also be characterized, because it can be expressed in terms of join counts. Again, this reveals the inherent advantage of starting with sample-based measures such as spatial autocorrelations and utilizing sampling without replacement probability theory, as was used to derive the variances and covariances of Moran's I statistics. Sample-based estimates of the local mean, together with non-free sampling theory are required to find the distributions (without knowing the mean and assuming normal distributions of the data [Cliff and Ord 1981]). The problem of not knowing the mean (gene frequency) of the underlying (and unobservable) stochastic process remains the main problem in directly utilizing Malécot's (and others') mathematical theory.

There has been considerable debate as to the role of Moran's I and these other measures, yet these methods are all measuring essentially the same thing. The very features that distinguish Moran's I statistics appear to be the same ones that allow the distribution to be known under the null hypothesis. Characterization of the distribution under the alternate hypothesis, that of genetic isolation by distance, requires space–time computer simulations, and the differences among the statistics are primarily unresolved issues of the relative statistical powers for the different ways of averaging. While Moran's I has been extensively characterized in many simulation studies of isolation by distance (and some other processes), the other measures have not. It has only recently been realized that simulation results for Moran's I statistics or join counts could be converted to these other measures (e.g., Hardy and Vekemans 1999). Finally, Smouse and Peakall (1999) applied their approach to a data set and found that how alleles were weighted made little difference. This is consistent with the simulation results discussed above, which showed (under the alternate hypothesis where there is spatial genetic structure) that the variance is nearly invariant with respect to allele frequencies, unless the frequencies of alleles are less than about 0.02.

The earlier results on correlations among Moran's I statistics for different alleles of a locus, under the null hypothesis of "non-free sampling," can be directly extended to some of the multiallele, multilocus estimators of conditional kinship, under the same null hypothesis, because then the allele-specific fixation index F is fixed in the denominator of Moran's I (Epperson n.d.[b]). In the simplest case, consider multiple locus estimators that are formed by first calculating allele-specific values of R, as defined in equation 8.28, and then taking the average. It is debatable whether the allele-specific F values may be considered constant over alleles and/or loci. If not, then such an average of R is simply equal to one-half the weighted (by $1 + F$, where F is allele-specific) average of I. The correlations for Moran's I, together with an equation analogous to the earlier equation by Feller (1957), but allowing for weights, can be used to find the variance of the average R. If F can be considered to be constant among alleles and/or loci, and hence averaged, then the average R is equal to $(1 + F)/2$ times the average I, and its variance equals $[(1 + F)^2/4]$ times the variance of the average I. Another multilocus conditional kinship estimator averages first over alleles and loci for each pair of multilocus genotypes, and then over pairs of individuals (Rousset 2000). If there is linkage disequilibrium then the loci are not independent, and this will affect the distribution of this estimator, including its variance. The distribution must be developed in terms of pairs of multilocus genotypes. In practice this would be inordinately complex, because the number of genotypes would be large and the number of products of join counts needed increases with the fourth power of the number of genotypes. If linkage disequilibrium is negligible, as would often be expected, then this measure is also a weighted average of the I statistics, and its variance can be calculated using the present methods.

CHAPTER 9

Theory of Spatial Structure within Populations

A priori means knowing only an initial generation. . . . It is unnecessary to subtract off the grand mean [to formulate covariances, correlations or kinship measures in theoretical models of genetic isolation by distance]. I subtracted off the grand mean in some of my early work because I wanted to model closely the models of Sewall Wright.

—Gustave Malécot (1998, personal communication)

Sewall Wright's theory of genetic isolation by distance centered on the levels of inbreeding within groups of individuals. Gustave Malécot's mathematical models, inspired by Wright's work, were more general (Epperson 1999b). To understand the theory of spatial structure within populations, it is best to trace its development from the very beginning. To connect spatial statistical measures to mathematical process models, it is critical to follow precise and consistent definitions of terms and coefficients. Recent use of multiple meanings of the same terms, casual comparisons and approximations, and biased estimators can lead to confusion of estimators with theoretical coefficients. All of the process models discussed in this chapter are stochastic. This is usually appropriate even when natural selection is acting, because it is to be expected that the role of chance will still be important in the development of spatial distributions of individuals or genotypes contiguous over a landscape within a population.

GENEALOGIES AND IDENTITY BY DESCENT

The original concept of inbreeding focused on covariances and variances of traits among relatives in known pedigrees. Wright (1921) developed and interpreted the inbreeding coefficient in terms of path coefficients and partial regression coefficients. In part because these interpretations assume linearity of genetic effects (Nagylaki 1989), Wright (1922) had difficulty in providing a mathematical definition that was completely generalized. Malécot defined the inbreeding coefficient based on what at the time were the relatively new ideas of prior and posterior probability distributions, and this became widely accepted as the proper basis of the inbreeding coefficient as well as the chain counting method of calculating it. Malécot defined probabilities, originally termed "*les coefficients de parente*," that two genes are descended from various genes of ancestors, using Mendel's laws (Malécot 1941, 1942). The studies of both Malécot and Wright were initially on pedigrees, and later moved toward the population level. This probability interpretation also facilitated the transition. It was Crow who first used the English term "identity by descent" (Crow 1954, 1989; see also Gillois 1996 and Epperson 1999b). Distinctions of genetic identity versus identity by descent are particularly critical when mutation is considered in population genetic models. Direct definition of the models in terms of probabilities turned out to be a very powerful approach, even when attention is limited to sets of pairs of genes.

The probability of identity by descent in a process where there is no mutation is the probability that two genes have a common ancestor, i.e., that they are descended from a common gene in the past, *relative* to the state of some ancestral generation. Naturally, in biology all genes are considered to ultimately share a common ancestor, and in any population or species after enough time has passed all genes will share a common ancestor. Sans mutation then, for this approach to be useful, we must refer to a particular ancestral population. A typical use then is to refer to the probability that two genes are descended from the same gene in the ancestral population. Note that this implies, looking now backward in time, that two genes coalesce either in the ancestral population or more likely in the generations intervening up to the present. Coalescence is an important and useful concept. Malécot called coalescence "*les chaines des kinship gametique*" or gametic kinship chains, a term that reflects the complete chain or circuit in the descendance of two genes back to a common ancestral gene. In this context, a probability of identity by descent is the sum of the coalescence probabilities over all appropriate generations. This definition is distinctly different from the term "identity by descent" as used in some recent literature, which means simply that two genes have the same allelic state (e.g., Loiselle et al. 1995).

Probabilities of identity by descent can refer to the two genes within a (diploid) genotype, in which case it reduces to the inbreeding coefficient of Wright (1921). Note that probabilities cannot be negative. The other important pairwise measure is the probability that two genes are identical by descent, where one gene is randomly chosen from one (diploid) individual and the other is randomly chosen from a second distinct individual. This is the original definition of the coefficients of consanguinity or kinship coefficients. Reference to an ancestral population is obviated if mutation or some other process, such as immigration from outside the population, continually brings newly defined lineages into the system, and when the population has reached equilibrium. For example, under the infinite alleles (Malécot 1942; Crow 1989) mutation process, where all new mutations result in unique alleles, the probability of identity by descent is the probability of shared gene ancestry and no mutation.

To proceed further, the assumption is made that a spatial distribution is of individual diploid genotypes, which is uniquely suited to the within-population case, and it is simply noted that some of the following developments can be adapted to problems of genetic variation among populations (chapter 5). We begin with models that make minimal assumptions and are exact, yet have sweeping generality. We assume that generations are discrete and we develop recursion equations for the incremental increase in probabilities of identity by descent, caused by inbreeding over a single generation. Consideration of the coalescences (i.e., gametic kinship chains) and/or zygotic kinship chain, "*les chaines de kinship zygotique*" (Malécot 1942, 1973a), over a single generation facilitates this formulation, as well as the tracing of parentage. The simplest way to deal with density is to invoke the lattice model, where all lattice points are occupied by one and only one individual every generation. It is assumed that the probability of dispersal as a function of distance is constant over generations, but this does not mean that dispersal is deterministic. This assumption is perhaps the major limitation of this model, although it is also made in almost all other models as well. Given these assumptions, the a priori probability that the two genes within a diploid individual at generation n are identical by descent ($\varphi_n(0)$), i.e., the inbreeding coefficient, is the same for all individuals.

To complete the most basic formulation we define probabilities, denoted by $\varphi_n(x, w)$, that pairs of genes present at time period n are identical by descent. The variables x and w represent samples, each of size one gene, respectively from each of two individuals with locations in space defined by x and w. The number of spatial dimensions in which the population exists is general. One dimension may sometimes be more appropriate, for example, for some riparian or river species; two dimensions are more appropriate for most species, including most terrestrial species; and three dimensions are best for some aquatic species. When x is equivalent to w, then $\varphi_n(w, w)$ is denoted $\varphi_n(0)$ (i.e., the inbreeding coefficient), else $\varphi_n(x, w)$ is the kinship coefficient (i.e., the coefficient of consanguinity, Malécot's "*coefficient de parenté*" [1973b]). Generally, it is necessary to set the dispersal rate as a probability function of some measure of the spatial difference between locations x and w. The function should depend only on the relative difference and, generally, not on the absolute locations. Any model must have some spatial regularity, otherwise it would be impossible to obtain theoretical results. This is presumably a much less serious limitation compared to the assumption of fixed probabilities of dispersal over repeated generations. However, the latter is an easy problem to fix if the variation of dispersal probabilities over time follows a simple periodic function, such as sinusoidal (Malécot 1973b). Thus, we set l_{xz} to be the rate of dispersal from z to x. The basic recursion equation (e.g., Malécot 1973b) is

$$\varphi_n(x, w) = \sum_z \sum_u l_{xz} l_{wu}\, \varphi_{n-1}(z, u) + \sum_z l_{xz} l_{wz} \frac{1 - \varphi_{n-1}(z, z)}{2} \quad (9.1)$$

The first term double sums over all individuals in the previous generation. The second term represents the added probability of identity by descent that arises from two genes descending from the same (diploid) individual, and it is critical to the model.

Equation 9.1 can be derived in various ways, including the use of coalescence probabilities or gametic kinship chains. Let $\pi_p(x, w)$ be the probability that the coalescence of two genes, randomly selected from diploid individuals at x and w, respectively, first occurs (now counting backward) p generations in the past. It is obvious that

$$\pi_p(x,\, w) = \sum_z \sum_u l_{xz} l_{wu} \pi_{p-1}(z,\, u) - \sum_z l_{xz} l_{wz} \frac{\pi_{p-1}(z,\, z)}{2} \tag{9.2}$$

because two genes must be in the same individual to coalesce. Equation 9.1 is generated by using the fundamental identity

$$\varphi_n(x,\, w) = \sum_{p=1}^{p=n} \pi_p(x,\, w) \tag{9.3}$$

If, in the infinite alleles process, mutation occurs with probability k, making two genes nonidentical even when descended from the same ancestor gene, then equation 9.1 is modified:

$$\varphi_n(x,\, w) = (1-k)^2 \left[\sum_z \sum_u l_{xz} l_{wu}\, \varphi_{n-1}(z,\, u) + \sum_z l_{xz} l_{wz} \frac{1 - \varphi_{n-1}(z,\, z)}{2} \right] \tag{9.4}$$

and equation 9.3 is modified:

$$\varphi_n(x,\, w) = \sum_{p=1}^{p=n} (1-k)^{2p}\, \pi_p(x,\, w) \tag{9.5}$$

Note that k can be related to recurrent mutation rates in finite alleles models (Malécot 1975), as well as to a parameter for some forms of selection. It is also sometimes termed the linearized systematic pressure and the "recall coefficient." It is the inclusion of k that allows stationary distributions to be obtained and analyzed. Otherwise, the system eventually goes to fixation and all genes become identical by descent. In that situation we can analyze only the dynamics, e.g., the rate of loss of heterozygosity, which is an inverse measure of the inbreeding coefficient (e.g., Malécot 1975).

General solutions to equation 9.4 appear to be difficult, and analytical results typically rely on an added assumption, that the dispersal system is "homogeneous" (Malécot 1973a). Homogeneity means that the dispersal probabilities depend only on the spatial distance, or, more specifically, the spatial lags in each spatial dimension. The most serious violation, or perhaps the one of most interest and importance in biological systems, is directionality of dispersal rates within a dimension. For example, prevailing winds may be from the west, causing seed and pollen to disperse further eastward than westward. In contrast, it is possible to include "dimensional directionality" in a two- or three-dimensional process (Epperson 1993b), which are differences in the dispersal distances and rates among but not within spatial dimensions. For example, dispersal may be greater along the north–south axis compared to east–west, and such a process can be modeled by appropriate orientation of the two spatial axes in the model. The importance of the homogeneity assumption is that it favors the utility of the Fourier transform. The homogeneity assumption allows us to develop recursions for probabilities of identity by descent solely in terms of spatial lags contained in a vector y:

$$\varphi_n(y) = (1-k)^2 \left[\sum_z \sum_x l_z l_x \ \varphi_{n-1}(y+z-x) + \sum_z l_z l_{y+z} \frac{1-\varphi_{n-1}(0)}{2} \right] \tag{9.6}$$

and

$$\varphi_n(y) = \sum_{p=1}^{p=n} (1-k)^{2p} \ \pi_p(y) \tag{9.7}$$

(Malécot 1973a, 1975). Equations completely analogous to equation 9.6 can be found for the "panmictic index," $P_n(y) = 1 - \varphi_n(y)$. Note that no other assumptions have been made about the distribution of l_z (the dispersal rate as a function of spatial lags). The nearest-neighbor dispersal model of Kimura and Weiss (1964) and Maruyama (1969) is a special case; for example, in the one-dimension case, $l_1 = l_{-1} = m$ and $l_0 = 1 - 2m$, and dispersal probabilities are zero else.

With mutation ($k > 0$), in a finite circular population of r locations (individuals) at equilibrium (i.e., $\varphi_n(y) = \varphi(y)$ as n goes to infinity) we have

$$\varphi(0) = \frac{1 + [1 - (2k)^{1/2}/\sigma]^r}{4\sigma(2k)^{1/2}\{1 - [1 - (2k)^{1/2}/\sigma]^r\} + 1 + [1 - (2k)^{1/2}/\sigma]^r} \quad (9.8)$$

and

$$\frac{\varphi(y)}{\varphi(0)} = \frac{[1 - (2k)^{1/2}/\sigma]^y + [1 - (2k)^{1/2}/\sigma]^{r-y}}{1 + [1 - (2k)^{1/2}/\sigma]^r} \quad (9.9)$$

where σ^2 is the variance in distances of dispersal (Malécot 1975). As r goes to infinity, equations 9.8 and 9.9 become

$$\varphi(0) = \frac{1}{1 + 4\sigma\sqrt{2k}} \quad (9.10)$$

and

$$\frac{\varphi(y)}{\varphi(0)} = e^{-y\sqrt{2k/\sigma^2}} \quad (9.11)$$

(Malécot 1975). Thus the probability of identity by descent decreases exponentially with distance, at equilibrium in an (infinitely) large population with one spatial dimension.

The case of populations in two spatial dimensions is more difficult, in several regards. Originally, Malécot (1948) attempted to examine a truly continuous space model, using probability density functions, and probabilities (here represented by l) of dispersal from one small area centered on an individual location to another such area, in continuous space. Although he gave a more general integral expression, he also developed a result specifically for isotropic, i.e., homogeneous, dispersal. Differences in dispersal among the two spatial axes could be analyzed using Fourier transforms (Malécot 1973a), but the system is considerably simplified if there is none, and then we can consider only the spatial distance (in two dimensions). Recall that to utilize Fourier transforms, there must be "spatial regularity" to the system. Thus, not only is the dispersal distance probability (technically in Malécot [1948], the probability that a gene at location area C at generation n came from an individual at a location area D in the previous generation, $n - 1$), assumed to be independent of absolute location, dependent only on relative location, and in all senses "homogeneous," it is also useful to assume that density is the same throughout the system, which creates some problems (see below). Technically, Malécot (1948) also assumed that the dispersal distances follow an "isotropic normal law" (Malécot 1969). However, without spatial stationarity there would seem to be little or no hope for analytic solutions. With the above added assumptions, the continuous model at equilibrium has

$$\varphi(0) = \frac{1/(1 - 8\pi\sigma^2\delta)}{\log(2k - 2k^2)} \tag{9.12}$$

where δ is the density per unit area, and σ^2 is the variance in distance of dispersal (Malécot 1948). Note this is quite different from that obtained in the one-dimension case. The function on distance is

$$\frac{\varphi(y)}{\varphi(0)} = \frac{\sum_{p=1}^{\infty}(1-k)^{2p}(1/4\pi p\sigma^2)e^{-y^2/4p\sigma^2}}{\sum_{p=1}^{\infty}(1-k)^{2p}(1/4\pi p\sigma^2)} \tag{9.13}$$

Although equation 9.12 appears to be easily calculable it is only in principle that equation 9.13 can be used to approximate the isolation by distance function for any k and σ^2, by using fixed limits on the summations. Application in statistical computing is made difficult by the fact that the (large number of) terms to be summed involve very large powers of a number $(1 - k)$ that is usually very close to one. Computing this formula appears to be beyond normal computational precision levels when k is on the order of typical mutation rates. Nonetheless, distance enters equation 9.13 only as the ratio y/σ, hence equation 9.13 does not appear to be especially close to an exponential form of decrease with distance. For long distances, specifically where $y\sqrt{k}$ is much larger than σ (thus where the absolute value of correlation is of less interest than the form of its decrease with distance),

$$\varphi(y) \sim \frac{a}{\sqrt{y}}e^{-y\sqrt{2k}/\sigma} \tag{9.14}$$

where a is a constant (Malécot 1948). This is not an exponential decrease.

Felsenstein (1975) has shown how difficult it is to obtain a spatially homogenous distribution of individuals based solely on densities and probabilities of dispersal. The precise assumptions of Malécot's (1948) continuous model lead to a contradiction. Felsenstein (1975) used the assumption of a Poisson distribution of the number of offspring produced by an individual, which seems appropriate if reproduction and movements are independent. If it is further assumed that the distribution of distances of dispersal is isotropic normal, then homogeneous density *cannot* obtain (Felsenstein 1975). Instead, the spatial distribution of individuals becomes clumped. Other examined dispersal models also tend to produce clumps (Felsenstein 1975; Sawyer and Felsenstein 1983). It remains an open question precisely what conditions of mating, reproduction, and dispersal probabilities may or may not lead to spatial homogeneity of density. It is not clear how seriously if at all clumping changes the predicted genetic isolation by distance form. Moreover, the same problem exists and was recognized in Wright's models (1943). It appears that clumping would act primarily through slight changes in the dispersal variance, and therefore it changes only the ratio of k/σ. This does appear to be a less serious problem than that which occurs in the application of the stepping stone migration model of Kimura and Weiss (1964), which further ignores the consanguinity produced each generation by pairs of genes derived from the same diploid ancestor in the previous generation. Malécot (1948) did not actually specify the mating system, but his model is explicitly diploid.

Many populations of plants and animals have highly clumped spatial distributions of individuals. The conditions that control clumping may be quite different in plants versus mobile animals. Among plant populations, newly founded ones (of sizes well below carrying capacities) may be more likely to be highly clumped. Spatial clumping may also be particularly pronounced in some social animals. However, as is pointed out in chapter 7, stable populations of plants may commonly feature local density dependency of survival and spatial competition. Indeed, many plant populations, such as lodgepole pine (Epperson and Allard 1989), are very uniform in density, and in some populations distributions could be even more uniform than random distributions (Pielou 1977). Highly territorial animals may also exhibit hyperuniformity in densities. Whatever the combinations of mating, reproduction, dispersal distances (either physical distance or combination of physical distance plus the operation of density regulation—i.e., effective dispersal distances), it is critical for the above that dispersal of each propagule be independent. Thus, for example, there can be no equivalent to "shared" stochastic migration (chapter 5) or "correlated seed or pollen dispersal" in the case of plants (chapter 7). These situations call for a closer look at lattice models of individuals within populations existing in two spatial dimensions.

The lattice-type models for two spatial dimensions examined in systems of discrete populations (chapter 5) also can be developed for lattices of individuals, but there are some differences that must be given attention. Because in effect there is only one individual per population, the probability of identity by descent "within a population" becomes the inbreeding coefficient. There can be no further parameterization regarding effective size (e.g., from added inbreeding) "within populations," and the mating system cannot be fully separated from the dispersal probabilities. Lattice models with homogeneous dispersal, at equilibrium, ensure spatial stationarity and the utility of the Fourier transform. For the lattice models in two spatial dimensions, we can use equations 9.1–9.7 above directly, except recognizing that x and y either are indices in multidimensional space or, better, are vectors of coordinates specifying locations. These equations are also identical in form to the discrete population lattice cases (chapter 5). The lattice may alternatively be infinite (Malécot 1973a), bounded (Malécot 1973a), or a torus (Malécot 1975). Large bounded systems may be approximated by infinite systems (Malécot 1973a). The Fourier transform requires indexing over both spatial dimensions, represented by α_1 and α_2. For any dispersal function in the homogeneous dispersal case we have

$$\varphi(y_1 y_2) = \frac{1-\varphi(0)}{2}\left(\frac{1}{2\pi i}\right)^2 \int_{C_1}\int_{C_2} \alpha^{y_1}{}_1 \alpha^{y_2}{}_2$$

$$\left[-1 + \frac{1}{1-(1-k)^2 L(\alpha_1\alpha_2) L\left(\dfrac{1}{\alpha_1\alpha_2}\right)}\right] d\alpha_1 d\alpha_2 \qquad (9.15)$$

where the integrals are contour integrals, $L(\alpha_1\alpha_2)$ is the Fourier transform of the dispersal function, and $L(1/\alpha_1\alpha_2)$ is the transform for negative spatial lags (because of the assumption of homogeneity or isotropy within and between dimensions). Thus solving equation 9.15 requires inversion of the Fourier transform (Malécot 1973a). Malécot (e.g., 1973a) developed various approximations and techniques (e.g., elliptic integrals). For example, if k (the recall coefficient or rate of mutation) is small, then the rate of decrease of probability of identity by descent along one (i) of the two spatial axes (keeping the other fixed) is proportional to

$$\frac{(\sqrt{2k}/\sigma_i)^{y_i}}{\sqrt{y_i}} \tag{9.16}$$

where y_i is the distance of separation measured along spatial axis i, and σ_i is the variance of dispersal distance along the same axis (Malécot 1973a). Note that this is similar to equation 9.14 and is not exponential. Indeed, it can be shown that when k is small,

$$\varphi(y) = \frac{a}{\sqrt{y}} e^{-y\sqrt{2k}/\sigma} \tag{9.17}$$

the same as equation 9.14, for distances y, much greater than the dispersal variance (Malécot 1969, 1973a). Note that an error by Malécot (1950) mistakenly gave an exponential decrease for this case, and this caused considerable confusion (for a discussion see Nagylaki 1989).

Finally, the probabilities of identity by descent are based either on an initial ancestral population, or, in the case of stationary distributions, the asymptotic (as time goes to infinity) probabilities, and the dispersal probability function on distance and the probability of mutation. If the probabilities were conditioned on realizations at any step of the process (i.e., any two genes in space at a particular time), where, in fact, genes must either be identical by descent or not, the subsequent probabilities would change. This is apparently why Malécot termed these a priori probabilities of identity by descent. They are very important for understanding the general theoretical effects of dispersal levels and systematic forces on genetic isolation by distance, but they are difficult to estimate in biological systems.

KINSHIP AND A PRIORI COVARIANCE

Probabilities of identity by descent can be equated to genetic correlations or covariances (e.g., Malécot 1955, 1972), but one must be careful to properly interpret the correlations. The basic approach is to substitute indicator variables X and X' for the states of the two randomly chosen genes (either from one individual, in the case of the inbreeding coefficient, or from two individuals), and take the expected values, which gives the probabilities of identity by descent. Note that these expected values are taken over the entire stochastic process, and hence may be called a priori expected values (Malécot 1972). In the simplest case, X and X' may imply inheritance traced back to the same gene in the duly noted ancestral generation. Alternatively, X and X' may represent allelic states if the process is stationary or the initial population can be considered to have unique genes and enough time has passed. Under the infinite alleles mutation process, once enough time has passed *and we can assume the process has become stationary*, the sharing of allelic state of two genes corresponds to their being identical by descent. If X is considered to be an indicator of allelic state then at any generation and location its expected value is a gene frequency q.

The a priori expected values at the stationary state are the same for all spatial locations. We may take the a priori expected value of $(X - C)(X' - C)$, where C is the a priori expected value of X. This expected value is a covariance, and by dividing it by the square root of the product of the variances of X and X' (or dividing by just the variance of X, because the *a priori* variance of X' is the same as that of X), we obtain a correlation coefficient (Malécot 1972). Malécot (1972) termed this the "*corrélation gamétique a priori*." The coefficient of coancestry (or probability of identity by descent) multiplied by the variance is the same as the a priori covariance at equilibrium. However, the allele frequency C is unknown, which causes problems. In the case of recurrent mutation with only two alleles, A and a, with mutation rates u and v, the asymptotic expected value is the usual mutation equilibrium $u/(u + v)$. In other cases, C might be the frequency of an allele in the initial ancestral population or the expected equilibrium frequency under a selection model. Such probabilities represent the expectation of many realizations, either over very long time periods, over many population realizations, or over many loci (Malécot 1972). In addition, and similar to the method of converting diploid genotypes into gene frequencies to calculate Moran's I statistics (chapters 7 and 8), we may consider the gene frequency in a diploid individual (or population—chapter 5) at a generation n and location x, $q_{x,n}$, as the stochastic variable and the correlations of the gene frequencies are also the probabilities of identity by descent (Malécot 1969).

The recursion equation corresponding to the lattice model of equation 9.1, but in terms of the a priori expected values of the covariances in gene frequencies between two sites x and w at generation n, $\sigma_n(x, w)$ (Malécot 1971), is

$$\sigma_n(x, w) = (1-k)^2 \left\{ \left[1 - \frac{\delta(w-x)}{2} \right] \sum_z \sum_u l_{xz} l_{wu} \, \sigma_{n-1}(z, u) + \delta(w-x) \frac{C - C^2}{2} \right\} \tag{9.18}$$

where C is the equilibrium gene frequency and δ is Kronecker's delta. Fourier and Laplace transforms can again be used (Malécot 1972) to extract exact and approximate analytic solutions. Similar equations also allow analysis of heterozygosity (Malécot 1972). In addition, such equations can be formulated for the continuous space models (Malécot 1973b), although they suffer from the same clumping problem mentioned above.

The second term of equation 9.18 is neglected in Kimura's approach (Kimura and Weiss 1964), and this is unacceptable in the individual-based model of structure within a population, because it ignores the additional consanguinity that arises from offspring sharing genes from the same individual in the previous generation. As noted in chapter 5, this is a serious problem also for the distributions among discrete populations especially if there is inbreeding within populations.

Various attempts have been made to examine diffusion approximations for the covariances and correlations in models of continuous space processes, (e.g., Malécot 1955, 1969; Maruyama 1971; Nagylaki 1974, 1986; Fleming and Su 1974). In 1955 Malécot presented (without giving the derivation, but noting its similarity to the Kolmogorov diffusion equations) the basic equation

$$\varphi_n(y) = (1-k)^2\left[\varphi_{n-1}(y) + \sigma^2 \frac{\delta^2 \varphi_{n-1}(y)}{\delta y^2} + \frac{1-\varphi(0)}{2\delta}\int l(\alpha,\ \mu) l(\beta,\ \mu) d\mu\right] \tag{9.19}$$

where $y = \beta - \alpha$, and $l(\alpha, \mu)$ is the probability of dispersal from location μ to location α. For the case of one spatial dimension, the diffusion approximation gives results consistent with the previous methods, for homogeneous dispersal for a system at equilibrium. After a long history of attempts, some mistakes and confusion in the field, it was finally and convincingly shown by Nagylaki (1978) that the diffusion approximation fails for systems with two dimensions. The scalings required are inconsistent.

RELATIONSHIP OF VARIOUS MEASURES OF OBSERVABLE SPATIAL COVARIANCES TO THE THEORETIC VALUES

Various methods for estimating the values of theoretic measures of isolation by distance have not carried consistent definitions. As shown above, the theoretical probabilities of identity by descent for two different individuals are the same as theoretical coefficients of kinship or consanguinity, and further they can be related to genetic covariances or correlations, at least for single loci. The original distinction between conditional kinships and covariances versus a priori kinships and covariances was clear in Malécot's (e.g., 1971) models. Conditional values are based on the conditional expectations of probabilities of genes, conditioned on knowing the genes in the previous generation(s). The conditional or a posteriori expected values are used as in Bayes theorem, except applied to a stochastic process.

Expressions among conditional expected values are analogous to the recursion equations (equations 9.1 and 9.18), with the addition that the higher moments involve stochastic inputs (stochastic and statistical error terms). As in the discrete population cases (chapter 5), the kinship conditional on the previous generations, like the conditional covariance, bears little relationship to the a priori kinships. In any case, we are unlikely to know the previous genotypic distributions (Malécot 1971), unless we have space–time data. Various attempts to measure the covariances (particularly for geographic analysis of genetic variation, e.g., in humans) have resulted in considerable confusion in the literature.

In 1955, Malécot suggested that the coefficient of correlation at distance y could be estimated by the empirical correlation $r(y)$:

$$r(y) = \frac{\sum_{x} (q_x - q)(q_{x+y} - q)}{\sum_{x} (q_x - q)^2} \tag{9.20}$$

where q is the mean for the system. This formulation is implicitly a discrete location model since it assumes that numbers of pairs of locations are separated by the same distance (class) y, and it would appear to be appropriate whether for discrete populations or individuals. In fact, if one assumes that the summation in the numerator is averaged over all pairs of points in the class, *equation 9.20 is identical to Moran's I statistics*. Note that the appropriate estimator of spatial variance appears in the denominator. Later estimators of kinship, or correlations, instead normalized over $q(1 - q)$ for reasons that are not clearly justified.

One problem with applying the a priori genetic correlations to real data is that we do not know what the expected frequencies of alleles are, and they are generally not the same as the allele frequencies in the population, let alone those in samples. Malécot (1973b) noted that the a priori expected values of gene frequencies may

> vary from place to place in some region R, perhaps in some unknown way: so, we often obtain only the "pseudocovariances" evaluated from some "regional mean" (Morton) or some "working mean" (Malécot, C. Smith). These may be considerably different

from the a priori covariances and will be negative [which the probabilities of identity by descent cannot be] at large distances if there is a cline.

Morton and colleagues (Morton et al. 1968; Imaizumi et al. 1970) developed a multiallelic estimator of kinship among locations i and j as $(H_0 - H_{ij})/H_0$, where H_0 is the estimated (using maximum likelihood or gene counting methods) total effective heterozygosity, and H_{ij} is the estimated heterozygosity of an F_1 population expected from the crossing of i and j. Note that for the diallelic case, this estimator is normalized by $q(1 - q)$. This approach appears to have been an attempt to adjust the variance (or heterozygosity) by the maximum possible value, given the gene frequency that might be equal to the initial or a priori expected gene frequency. However, this approach seems misguided because isolation by distance is a stochastic process, and the a priori expected values are usually unknown. The denominator of equation 9.20 is a better estimate of the variance among the locations, because variances among individuals (or among populations), under the process of genetic isolation by distance, depend on the parameters (dispersal rates and "recall" coefficient factors), in other words, the outcome of the stochastic process. Morton considered using the Wahlund principle (Workman and Niswander 1970), but, in contrast to Malécot (1973a), considered it disadvantageous because then correlations could *not* be negative. Morton also modified his measures to more efficiently estimate the gene frequencies in various scenarios (Morton et al. 1971). Morton (e.g., 1982) attempted to model his estimated gene frequency functions from migration matrices, and such attempts involved various approximations. More to the point is that Morton's (1975) most widely used estimator was the one discussed in chapter 8, utilized in Barbujani's (1987) demonstration of the relationship to Moran's I statistics (see equations 8.28–8.30). If the sample size is fairly large, this estimator is simply equal to Moran's I divided by the usual estimate of F_{ST}. Indeed, Morton (1973a,b) motivated the denominator by appealing to Wahlund's theory, but this too appears to ignore the distinction between the a priori and observable covariances or correlations. In sum, there is no formal justification for normalizing estimators such as that in equation 9.20 by F_{ST}, apart from an effort to adjust for differences in variance among alleles or loci because of differences in allele frequency. However, we have seen that allele frequency has little effect on correlations (chapter 8), it has no effect on the probabilities of identity by descent in theoretical models, and its relationship to the equilibrium assumptions may be misleading. Moreover, it is not helpful to divide by $q(1 - q)$ because the observed q is not the a priori expected value, and a better estimate of variance is given by Malécot (equation 9.20). To normalize by $q(1 - q)$ is unnecessary and it potentially destabilizes estimators, even under the null hypothesis of no spatial structure (effectively unlimited dispersal, with respect to spatial genetic isolation by distance). In contrast, more standard estimates of spatial variance, such as Malécot's, allows calculation of more standard spatial correlations such as Moran's I statistics, which have known distributions.

Malécot (1950) mistakenly obtained an exponential decline of genetic similarity with distance in the two-dimensional model, and he later corrected this error (Malécot 1973b). Malécot also suggested that the empirical estimates could be adjusted by a linear function correction using the "pseudocovariance" (say L) at long distances, and Morton (e.g., Imaizumi et al. 1970; Morton 1982) proposed the following formula

$$\varphi_n(y) = ae^{-by}(1 - L) + L \tag{9.21}$$

We now know that isolation by distance is not exponential for two-dimensional systems. Moreover, Harpending (1973) argued that this form of correction using L must be incorrect.

More confusing still, Morton (1973a) maintained that the a priori kinship $\varphi_n(y)$ could be related to what he termed the conditional coefficient (let us denote it $R_n(y)$), "conditional kinship relative to the contemporary gene pool," by the formula: $\varphi_n(y) = R_n(y) + [1 - R_n(y)]\varphi_R$, where φ_R is "the kinship of random individuals from the region relative to founders just sufficiently remote that their descendants are randomly distributed over the region." This definition appears to be quite different from Malécot's use of the Bayesian term a posteriori and the conditional probability arguments that he sometimes used to build recursion equations. To my knowledge, equation 9.21 as an interpretation of φ_R was never formally justified. Morton suggested that all of the kinship information is contained in products of gene frequencies among pairs of points, and he advocated averaging the values among loci. This appears to fit with the arguments presented in chapter 8, that there is little to be gained by true multi-locus measures over multiple locus ones (chapter 8) unless there is substantial linkage disequilibrium. On a final point, Morton (e.g., 1982) also tied the estimated values to the parameters of the process, but because those usually involved the sizes of populations, their discussion is presented in chapter 5.

Malécot (personal communication, 1998) later claimed that it might be better to entirely leave q out of the expectations, although up to now no published work has pursued this course. Nonetheless, the problem of not knowing the mean (allele frequency) showed up in the analogous way in chapter 8. As has been noted several times the a priori mean is the mean of a stochastic process that generally cannot be observed and often cannot even be posited. The advantage of the autocorrelation approach is that it is sample based and does not require estimation of the mean of the stochastic process. Note also that some of the various popular "kinship coefficients" mentioned above can be transformed into autocorrelation statistics only if the mean or other theoretical values are known, which they usually are not. It is better to use those that can be so transformed, in order to have known sampling distributions, or to use the spatial autocorrelation statistics directly.

WRIGHT'S VIEW

Wright's (1943) original theory was based on a spatially continuous population, and thus it presumed that there was uniform density over an infinitely large area (or line in the case of one spatial dimension). Various specific mechanisms of reproduction (uniparental and biparental) and dispersal were posited, and dispersal (parent–offspring) distance was assumed to be homogeneous and normally distributed in the one-dimension case and normally distributed along each axis in the two-dimension case (hence the term "axial variance" of dispersal). The displacement by dispersal (in the negative and positive directions along an axis within a dimension) was assumed to be normally distributed with mean zero and variance σ^2. Here we will discuss only the two-dimension case. In a second paper, Wright (1946) made it clear that the neighborhood size N_e is generally *not* a *panmictic unit* (i.e., there is not random mating within the neighborhood). This is distinct from Wright's reasonable proposal to relate the inbreeding in a structured population to that in another idealized population (of size N_e) that is unstructured in the sense that it has randomly mating throughout. Wright (1946) recognized that, although it was easier to derive the F values if there were random mating within neighborhoods, it is unlikely to be the case when dispersal is limited. He gave an alternative derivation that uses an inbreeding coefficient (his E value) that would be the inbreeding coefficient (F) "if there were random mating in the last generation" (Wright 1946). Wright (1943, 1946) standardized the actual density d by $n = 4\sigma^2 d$. Thus neighborhood size is most simply thought of as a standardized measure of dispersal (Epperson 1993a). To capture a high percentage of possible parents, Wright set the limits of dispersal to two standard deviations, and hence calculated the number of likely parents as $N_e = \pi n = 4\pi\sigma^2 d$. Thus N_e is the number of (reproducing) individuals in a circle of radius 2σ, and this circle includes "86.5 percent of parents of individuals at the center" (Wright 1946). Wright was interested in the fact that such populations would be spatially differentiated genetically despite being continuously distributed. He described structure in terms of relative inbreeding values at different hierarchical spatial scales, starting with the neighborhood size area and going up to scales just under that of the entire population. Neighborhood area may be defined as the area that contains the number of individuals of the neighborhood size. Let us be clear in defining the spatial grouping units of scale as "blocks," which might contain any number of neighborhood areas. Wright (1943) developed approximations for calculating theoretical values of F_{IS} (the inbreeding coefficient relative to the block), F_{IT} (the inbreeding coefficient relative to the entire population), and F_{ST} (the inbreeding within block relative to that in total). The theoretical F_{ST} is also a measure of the spatial differentiation or degree of genetic isolation by distance among blocks. Naturally, populations with larger neighborhood size (greater dispersal) have smaller F_{IS}, F_{IT}, and F_{ST}. For a given population with a certain neighborhood size, as the block size increases, F_{IS} increases and F_{ST} decreases (F_{IT} is constant for a fixed N_e) (Wright 1965).

There have been some attempts to reconcile Wright's theory with others. For example, Malécot (1972) showed that *if we may reasonably* consider an average consanguinity among genes within a block, defined for present purposes as $\varphi(0)$, to be approximately constant within blocks and that the (average) consanguinity of two genes from different blocks are also approximately constant, then the a priori expected variance of gene frequencies among blocks is approximately (as long as the block size is fairly large) equal to $\varphi(0)C(1 - C)$ (note that C is independent of block size). Here C is the a priori expected gene frequency, and the covariance among blocks is given by the average consanguinity multiplied by $C(1 - C)$. Note that the correlation, a ratio of the two, does not have C in the terms, and thus might provide a better estimator than the standard estimator of F_{ST}, which has $q(1 - q)$ in the denominator. Wright (e.g., 1965) duly noted that the F statistics were not the same as the F measures in theoretical models. Malécot (1955) also gave results for the overall inbreeding coefficient for the one- and two-dimensional models, in equilibrium populations where mutation is included.

As noted by Nagylaki (1989), there is every reason to believe that Wright's (1943) continuous model suffers the same inconsistency as Malécot's (1950) model, namely the clumping that occurs in violation of the uniform density assumption. Sawyer and Felsenstein (1983) made some progress on this problem. In explicit recognition of clumping, they sought to describe the hierarchical structure in terms of clumps. Here it is assumed that discrete clustering occurs at hierarchical levels, and, importantly, that the probabilities of dispersing depends on the hierarchical level of clustering. This is a nice bridge between Malécot's and Wright's theories, and it avoids the clumping problem, yet allows the Fourier transform, which would then be based on the regularity of the hierarchy rather than spatial lags (or spatial distance) of separation. It may be particularly relevant for animals that exist in groups whereby the groups might interact (exchange migrants) according to regular "rules" of spatial hierarchies.

Slatkin and Barton (1989) simplified the difficulties of finding F_{ST} for different block sizes, by instead focusing on groups of locations. They used approximations of Malécot's (1969) results for lattice models of isolation by distance, and defined F_{ST} as

$$F_{ST} = \frac{f_0 - f}{1 - f} \tag{9.22}$$

where f_0 is (for lattice models of individuals) the probability of identity by descent within the individual and f is "the probability of identity by descent of two alleles chosen at random from the entire population." Note that f is generally not observable, because it is a priori and it would require sampling over the entire system. This presents difficulties similar to those discussed above for estimating L in pairwise kinship formulations. Slatkin and Barton (1989) further simplified things by focusing at the two extreme spatial scales. First, one can consider a group of locations that are mutually separated by relatively large distances, separated by more than the "characteristic length," $l_c = \sigma/\sqrt{2\mu}$, where μ is the mutation rate of an infinite alleles mutation model, and, second, one can consider cases where the separation distances are smaller than l_c. They found that for both cases

$$F_{ST} \approx \frac{1}{1 + cN_e} \tag{9.23}$$

where N_e is Wright's neighborhood size and c is constant of order one that depends weakly on the mutation rate and distances of separation. Note that not only are there problems in defining and estimating F_{ST}, there are additional problems if the mutation rate is unknown, although the effects of mutation were characterized in simulations (Slatkin and Barton 1989).

Finally, Wright's theory has been extended to additional mating system models and considerations of sex specificity of dispersal. One critical adaptation is to recognize that seed and pollen dispersal are not equivalent. Crawford (1984) has shown that we may effectively retain Wright's isolation by distance formulation in terms of neighborhood size by equating the total dispersal (of both parents in Wright's original theory) σ^2 to $\sigma_s^2 + \sigma_p^2/2$, where σ_s^2 is the axial variance distance of seed dispersal and σ_p^2 is the axial variance distance for pollen, for example, in Wright's (1943) formula for neighborhood area $N_a = 4\pi\sigma^2$. Various special cases in animal behavior have been studied by Chesser and colleagues (e.g., Chesser and Baker 1996).

MECHANISTIC MODELS OF DISPERSAL

Genetic isolation by distance could be further and in some respects better characterized if there were a mechanistic understanding of the biological factors that determine movements of individual plant propagules or animals and how these determine the shape of the distribution of dispersal distances. This section presents some available mechanistic models. Most of these were developed by atmospheric scientists studying the dispersion of pollutants and some by ecologists, but rarely by geneticists, and more work in this area seems warranted in population genetics.

The standard curves of dispersal fitted in experiments are leptokurtotic, in particular the negative exponential curve $l(x) = ae^{-bx}$, and inverse power law curve $l(x) = ax^{-b}$, where x is distance, and a and b are parameters (e.g., reviews by Gregory 1968; Kiyosawa and Shiyomi 1972). There are several limitations to both of these models. Experimental curves often have a maximum observation distance that is fairly short, and neither of the above theoretical distributions have cutoffs. Nonetheless, both are often closely fitted by experimental data. Usually both models are not mechanistic and observed distances are not related to the determining factors, which may include, for example, wind speeds and settling velocity for passive transport of seed in plant species. Thus the models generally cannot extend the results of one experiment to predict dispersal distances in another setting where factors such as average wind speed may differ.

One set of sophisticated mechanistic models of dispersal may be particularly applicable to the passive transport of pollen or seed propagules by wind currents. These models can take into account vertical downward movements caused by gravity, and horizontal movements caused by various winds (advection), as well as diffusion (e.g., Fitt and McCartney 1986). One promising model of passive transport is the tilted Gaussian plume model with three spatial dimensions, where the rate of deposition D (which can be related to the probability density function of dispersal over an area) at a location (x, y) on the plane is given by $Q(x, y) = S(x, y, 0)W_s$, and where W_s is the settling velocity determined by air-resistance properties of the particles and gravity (Chamberlain 1975; Okubo and Levin 1989). This model usually makes the assumption that the vertical level (0) at which particles are stopped is an absorbing level, which often may be a reasonable approximation. This leads to the general diffusion equation, after aligning dimension x with the prevailing wind and y with the perpendicular coordinate:

$$Q(x, y) = \frac{n(x)W_s}{2\pi u \sigma_y \sigma_z} \exp\left[\frac{y^2}{2\sigma_y^2} - \frac{1}{2\sigma_z^2}\left(H - \frac{W_s x}{u}\right)^2\right] \tag{9.24}$$

(Okubo and Levin 1989). Here u is the wind speed in the direction of increasing x, $n(x)$ is the depleted plume concentration at distance x, σ_z^2 is the vertical diffusion, and σ_y^2 is the horizontal diffusion in the y dimension perpendicular to x (Okubo and Levin 1989; Pasquill and Smith 1983). The three terms $n(x)$, σ_z^2, and σ_y^2 generally are functions of x, hence these models are quite complex, and explicit general distributions have not been obtained, although it has been shown that the distribution has a nonzero mode (Okubo and Levin 1989). For very light particles, such as many kinds of pollen, even greater complexities may enter. If light enough, particles may have basically the same movements as air movements, but these may also include various currents, eddies, and updrafts around landscape structures (Di-Giovanni and Kevan 1991).

Tilted plume models compare favorably with the "ballistic models" for point sources, and particularly for passive transport of seed or other heavy particles. The simplest ballistic model specifies the function on distance, $x = Hu/F$, where H is the release height, u is the mean wind speed, and F is the descent velocity. An unrealistic feature of this model is that once the parameters are set all particles will land at the same distance. Greene and Johnson (1989) modify the ballistic models by incorporating variations in wind speed, crosswinds, and downward descent velocity. There is evidence that the "terminal velocity" of seed is either "Gaussian distributed," normal, or lognormal. For example, if it is log normal then

$$\frac{dQ}{dx} = \frac{Q}{x\sigma\sqrt{2\pi}} \exp - \left[\frac{\ln(xF/Hu)}{\sqrt{2\sigma}}\right]^2 \tag{9.25}$$

where dQ/dx can be interpreted as the frequency distribution of seed dispersal as a function of distance in the direction of the horizontal wind. The function of relative rates of gene dispersal on distance would be proportional to the expression in equation 9.25.

Active animal dispersal models traditionally have been motivated by quite different considerations, and typically are based on random walk models, where individuals make a series of steps. The simplest case is where each step has equal unit distance and random direction, and is independent of previous ones. Consider, for example, a system with only one spatial dimension. In random walk models, as the number of steps n becomes large, and if the size λ of the step is such that the limit of λ^2/t exists (say equal to $2D$ [Okubo 1980]) as λ and t (time) go to zero then the discrete (random walk) space and time model can be approximated by a diffusion process with coefficient D.

The appropriateness of using diffusion and diffusion-like models for biological dispersal processes depends on many factors (Okubo 1980). It appears that any passive transport system will have some sort of diffusion or randomness in the paths of travel. The most important aspect is to determine when the spatial and/or temporal scales of variable factors may be treated as diffusion. For example, for lightweight particles it may be reasonable to proceed by assuming negligible settling velocity and lack of effects of absorption by intervening areas. Further, air transport may consist of many changes in direction corresponding to small-scale temporal or spatial changes in prevailing winds, eddies and turbulence, etc., and the distance traveled between each change in direction may also be small. If all wind factors average out, this process is analogous to molecular diffusion. The role of small temporal or spatial scale changes in velocity and direction is a key consideration to dispersal, and so is the degree to which these may be treated as stochastic effects. It is worthwhile to make a distinction between the Lagrangian approach of modeling the movement of individual particles and the Eulerian models in terms of expected densities of particles at various distances or locations from the source (Aylor 1990; Andersen 1991; Tufto et al. 1997; Klein et al. 2003). Ideally, models are Lagrangian, but usually observations are Eulerian, because in most experiments the movements of individual particles are not traced.

In further examining diffusion models, it is useful to introduce some equations and notation first for specific types of systems with one spatial dimension. First, consider that there is a time period during which all migration occurs (that is, a relatively short period, or at least all dispersal occurs prior to growth and survival), and each source emits a continuous release of windborne propagules throughout this period. Then, the relative fluxes at distances from the sources should be proportional to dispersal rates. If a single, limited time period source is imagined then the variance of the concentration of propagules is paramount. The key parameter is

$$A = \frac{1}{2}\frac{\delta\sigma^2}{\delta t}$$

where A is the diffusivity constant and σ^2 is the time rate of change of the concentration and also a measure of the width of the diffusion at some point in time.

In the case of no directionality (i.e., mean movement direction is zero), A is constant and is analogous to the constant D of molecular diffusion, and extension to systems with two spatial dimensions is straightforward. The relative concentrations at sites located $\pm x$ and $\pm y$ from the source $C(x, y)$ have a two-dimensional Gaussian distribution:

$$C(x, y) = \frac{M}{2\pi\sigma_x\sigma_y}\exp\left[-\frac{1}{2}\left(\frac{x^2}{\sigma_x^2} + \frac{y^2}{\sigma_y^2}\right)\right] \tag{9.26}$$

Note that M is a constant that refers to the overall amount of propagules (it vanishes when we consider the relative rates), and that there may be "dimensional anisotropy" (i.e., σ_x might not equal σ_y), even though there are no directional trends within each dimension. Such a system could occur, for example, where there are stronger daily winds upslope and downslope on a mountainside compared to perpendicular winds.

Before proceeding further, it should be noted that many experimental studies in fact observe a Gaussian distribution (e.g., the classic experiments of Dobzhansky and Wright 1943; Wright 1968). It should also be noted that if there is a great inherent difference in ability to disperse (e.g., one group of propagules have much larger σ than other groups), then the mixed-Gaussian distribution obtained is *always leptokurtotic*. However, it might be expected that in most cases there should be few inherent differences among propagules in their ability to disperse by wind. On the other hand, if general conditions for dispersal, e.g., wind speed, varies, then this, too, could cause a mixed-Gaussian, leptokurtotic distribution.

In a second approach, when heavy particles are dispersing from some height H (at the source) and falling with significant settling velocity W_s and there is significant directionality to the average wind velocity u, a somewhat different dispersal pattern or distribution is obtained. Again, first consider the one-dimensional case. Okubo and Levin (Okubo 1980; Okubo and Levin 1989) found that the concentration $C(x)$, at a distance x in the same direction as the prevailing wind has values given by

$$C(x) = \frac{nW_s}{\sqrt{2\pi u \sigma_z^2}} \exp\left[-\frac{1}{2\sigma_z^2} \left(H - \frac{W_s x}{u} \right)^2 \right] \qquad (9.27)$$

Here we follow Okubo and Levin (1989), who assume that the concentration is not affected by deposition in a way inconsistent with the other parameters. We note that σ_z is a function of x, taken to be $\sigma_z = 2Ax/u$, where A is the diffusivity coefficient in the vertical dimension, and thus σ_z basically represents a transformation of x as a function of time. Note that again in terms of relative concentrations (to be rates of dispersal relative among emigrants from a source) the numerator nW_s is unimportant. Okubo and Levin (1989) ignore the probability of dispersal in the direction opposite to that of the prevailing wind u (i.e., for $x < 0$) in their examples and figures. However, if u is small relative to W_s this probability can be substantially greater than zero, almost mirroring migration in the primary direction.

The above result can be directly extended to a two-dimensional model if strong winds operate only in one dimension, aligned with x. Then equation 9.27 is simply the cross-wind integration deposition (CWID) of the two-dimensional case. It appears that this can also be extended to the case of directionality in each of the two dimensions by having a mean wind direction for each dimension. It should be pointed out that these distributions typically have modes that are nonzero. In addition, equation 9.27 can be modified to avoid the assumption of no loss of plume by replacing n with $n(x)$, effectively by multiplying the right-hand side by $(1 + a)/B$, where

$$a = \frac{1}{2}\left(\frac{uH}{W_s x} - 1\right)$$

and

$$B^2 = 1 + \frac{W_s^2 \sigma_x^2}{u^2 \sigma_z^2}$$

and replacing σ_z^2 with $B^2\sigma_z^2$ in the denominator of the exponential term (Csanady 1973; Pasquill and Smith 1983). The relative deposition rates become even more complex functions of x.

Animals show a very wide variety of behaviors in movement during dispersal, which necessitates a range of models. One most extreme cases is that of passive transport (i.e., no or "irrelevant" movement behavior), and the same models as above may be used. Another simple scenario is where behavior mimics diffusion and related models, as based on random walk arguments and taking the number of steps to infinity. However, for active movement, in many cases there may be a correlation between sequential steps in the walk (which still contains elements of stochasticity or randomness). Again, if we can assume that the distances between individuals is long relative to incremental (walk) movements, we can develop continuous time analogs over the time period during dispersal events. Many insects, for example, show high correlations among subsequent steps. Some bees, while foraging in a food-scarce region, tend to fly in relatively straight lines; whereas if food is abundant they tend to make sharp turns between succeeding flights between plants visited, in essence tending to fly in circles (Zimmerman 1981). In many cases, the association is positive and mostly contained in pairwise correlations. In other words, the number T of correlated flight or ground movements is generally around 2 for most insects. It could be quite larger for higher vertebrates. Flight patterns may be very complex, especially if both active behavior and passive transport occur, for example the flight of moths in a strong prevailing wind (see simulations studies of McCulloch and Cain [1989]).

A one-dimensional system was formulated by Goldstein (1951). Let $p = (x, t)$ be the probability of reaching x at time t; then

$$\frac{\delta^2 p}{\delta t^2} + \frac{1-\gamma}{\tau}\frac{\delta p}{\delta t} = \frac{\lambda^2}{\tau^2}\frac{\delta^2 p}{\delta x^2} \tag{9.28}$$

Here γ is a parameter representing the degree of correlation among successive steps, τ is a small period of time allotted for each step, and λ is the distance traveled during one step. Naturally, if γ is zero, the diffusion model is recovered. Moreover, for small γ relative to τ the same is approximately true. In contrast, if $\gamma \rightarrow 1.0$ as $\tau \rightarrow 0$, then we have

$$\frac{\delta^2 p}{\delta t^2} + \frac{2}{T}\frac{\delta p}{\delta t} = v^2 \frac{\delta^2 p}{\delta x^2} \tag{9.29}$$

where T is the "characteristic length" of time during the correlation of steps. As noted, in many insect cases $T = 2$ (two time lengths of steps, i.e., twice the length of time between steps). In many cases T is much smaller than the scaled values of t at which distances are reached and then the process follows a simplified diffusion equation:

$$\frac{\delta p}{\delta t} = v^2 \frac{T}{2}\frac{\delta^2 p}{\delta x^2} \tag{9.30}$$

On the other hand, if the time of correlation T is much larger than t, i.e., the correlation extends over times during which an animal is very likely to "pass" by many locations, then we have a wave equation:

$$\frac{\delta^2 p}{\delta t^2} = v^2 \frac{\delta^2 p}{\delta x^2} \tag{9.31}$$

Extensions to two-dimensional systems again appear to be straightforward.

Although correlated random walks may incorporate searching behaviors, we do not know enough about long-distance travel of animals dispersing for purpose of mating. We can consider a somewhat extreme case where animals are so adept at finding a mate that they always locate the nearest potential ones. Thus, this may be considered a (possibly correlated) random walk where the steps are travels to other individuals themselves. Additional considerations may involve the decision to stay, or keep moving without mating; mating and then moving on to other locations; and losses from lack of survival during movements. Plant propagules that are actively dispersed by animals may follow similar patterns of dispersal as the animals themselves. Added factors may include multiple "migrations" (i.e., a sequence of individual locations reached by the animal vector) per individual vector, "pollen carryover" or lack of deposition (seed), and a loss of propagule function with time and other factors.

The models discussed above are physical in nature, i.e., involve physical movements, yet they may not necessarily equate with gene flow. For example, pollen may move but not land on receptive stigmas. Plants may be absent, thus the distribution of receptive plants is also of issue. This may also change some of the other parameters, such as laminar air flow and flow near the settling surface. Seed may fall in inappropriate places, and thus distribution of proper habitat must be integrated with physical movements.

There is anecdotal evidence of very long-distance gene flow (e.g., Koski 1970). For example, seed from some pine species were able to travel 10–11 miles (Bannister 1965). Spores are found in high levels of the atmosphere, and some propagules are found in jet streams. Aircraft and marine vessels may also carry invasive species. There are even more complicated ecological models of dispersal that incorporate variations in population density. Genetical models involving the same features are rare and it is even rarer to find relevant experimental data. Such features would not fit with most genetical models and, as mentioned earlier, the problems of clumping potentially produced by dispersal models over multiple generations generally have not been characterized. Another popular idea is dispersal in the context of "metapopulation dynamics," where populations undergo extinction and recolonization. These events would also be difficult to fit into existing models of genetic isolation by distance. Most metapopulation models assume simple structure, usually Wright's island model.

CHAPTER 10

Emerging Study

> The regions for which data are available may be . . . irregularly spaced. In this kind of situation we recommend the use of interpolation methods to obtain values on a regular lattice . . . without the regular lattice assumption, it is extremely difficult to define a spatial lag structure so that . . . models are stationary and correlation functions well defined. . . . If hierarchical influences are assumed to operate, further complications arise.
>
> —Peter Hooper and Geoffrey Hewings (1981).

For both within and among populations, the spatial patterns observed in natural populations, statistical considerations, and space–time models all point to an intimate connection of spatial patterns to the underlying space–time process. A number of important considerations trace through the steps by which spatial patterns are used for making inferences about population genetics, from the framing of evolutionary, conservation, and ecological genetic questions, through the design of sample surveys and choice of markers and statistical methods, to the choice or construction of appropriate models of the processes. If the processes have any substantial stochastic components, then usually fairly sophisticated modeling is required. This chapter summarizes some of the key conclusions that emerge from the monograph and points out several outstanding problems for the continued development of the linkage of spatial patterns to space–time processes.

EMERGED CONCLUSIONS

Data considerations and choices of statistical models and space–time process models depend on whether the genetic questions addressed involve an averaged process versus a temporary effect of a temporally nonstationary process. Examples of averaged processes include the use of spatial patterns of genetic differentiation among populations for the estimation of migration routes and rates (chapter 2), and the use of spatial structure of genetic variation within populations to infer dispersal distances (chapter 7). If it is reasonable to expect that things have "averaged out," then a further consideration is spatial stationarity. Spatial stationarity is not expected in most cases where evolutionary questions center on spatial effects of an ancient event (chapter 3). If spatial stationarity is a reasonable expectation, then statistical power may be gained by sampling repeated spatial features of the system. Weak stationarity for a space–time process requires that the expected values and correlations depend only on definable spatial lags. The exact conditions under which spatial stationarity (chapters 4, 5, 8, and 9) obtains in a stochastic space–time process are fairly complex. They usually would require that the locations are fairly regularly distributed and that the migration or dispersal rates depend only on the spatial lag order representing spatial proximity, for example, Euclidean distance or lags along both axes in systems with two spatial dimensions. A key exception is the use of spatial autocorrelation measures, such as Moran's I statistic, for tests of significance of the null hypothesis of random distributions (i.e., no pattern) do not rely on spatial stationarity.

In experimental studies of spatial patterns per se, as opposed to simple measures of the average degree of genetic differentiation, especially for the purposes of inferring isolation by distance, there appears to be little to recommend for using F_{ST} estimators. The F_{ST} approach involves the formation of hierarchical groups of locations, and requires approximations arising from averaging within such groups as if all elements are equal, which generally they are not. In other words, it ignores structures within groups, unless these are specified as subgroups at a lower hierarchical level, for which the same problem remains. Moreover, spatial stationarity is more difficult to define, and migration generally does not act in hierarchical fashion. There seems to be no advantage to using hierarchical groups or hierarchical statistics.

Although as a theoretical parameter in genetic isolation by distance models F_{ST} can be formulated as an averaged pairwise value of probabilities of identity by descent, or similar pairwise measures, statistical estimators supposedly of theoretical F_{ST} in fact are not. What exactly they do measure is unknown, and in simulations the estimated values are biased with respect to input values, although the θ estimator of Weir and Cockerham (1984) appears to be less biased. F_{ST} cannot be used as an exact test of null hypotheses, because its distribution is unknown even under simple null hypotheses. In the few cases where it has been studied such tests yield unacceptably high rates of false rejection of null hypotheses of no spatial structure (Epperson and Li 1996). The problems of using F_{ST} for studies of genetic variation *within* populations are further compounded by the fact that there is no basis for choosing a minimum size subgroup. Moreover, small groups are likely to be highly structured internally, and random sampling within groups cannot be assumed. In this context, it is worth noting that Wright's neighborhood size is not a spatial unit of spatial structure, even though as a standardized measure of dispersal it determines the spatial structure in standard processes of isolation by distance.

Most estimates of pairwise kinship also do not correspond to theoretical values and have unknown distributions and biases. Many of the problems arise from the fact that the mean values of genetic traits are not the same as the a priori expected values in any stochastic process, and the a priori expected values are almost always unknown, if not unknowable. This applies to both single-locus and multilocus measures, typically of a form suggested by Cockerham (1969). Moreover, while, generally, greater information is expected in data for multiple loci than for single-locus data, it is unclear whether true multilocus measures, based on samples of multilocus genotypes, are manageable. It appears that in most cases there is little linkage disequilibrium, at least at many spatial scales, and if there is none then there is no additional information in multilocus data compared to multiple locus data. If there is linkage disequilibrium, then none of the theoretical models of isolation by distance can be used.

Multiple-locus or multilocus measures of pairwise kinship, such as those of Bertorelli and Barbujani (1995) and Smouse and Peakall (1999), are apparently attractive because, in principle, they should yield a better estimate of kinship between a specified pair of individuals or populations. This has some appeal. On the other hand, as noted, they are not true measures of kinship, they cannot be related to existing analytical results in mathematical models, and for some their distributions are unknown. Moreover, the appeal is less so for spatial pattern analyses, because one ends up averaging over many pairs of individuals or populations, for example, by distance classes. In many cases, the process of genetic isolation by distance is inherently highly stochastic. This means that the degree of relatedness among pairs is also highly stochastic, and the value for any given pair not necessarily highly predictable solely from the distance separating them. In most respects the difference between Moran's I values and multilocus measures is that Moran's I now can be used to average over pairs of locations first and then over alleles and loci, whereas multilocus or some multiple locus measures average over alleles and loci first and then over pairs of locations. If there is no linkage disequilibrium, then both types of measures use precisely the same information.

Some recently developed methods avoid many of the problems mentioned above. Perhaps the single most important development is the derivation of the distribution of Moran's I statistics under the null hypothesis that pairs of sample elements are randomized (Cliff and Ord 1973). Malécot (1955) had suggested this form of statistic. Cliff and Ord (1973) were able to find the distribution because the denominator of I is fixed under this condition. It is unnecessary to find the expected mean of a stochastic process. However, until recently it was possible to find the precise standard errors for Moran's I statistics only separately for each allele at a locus, and this limited its utility for multiple-locus data. Data for multiple loci are generally required to precisely analyze the controlling factors of the underlying space–time process. In chapter 8 the covariance and correlation were derived for Moran's I for different alleles of the same locus, under the randomization null hypothesis. The correlations depend on allele frequencies, but are robust with respect to variations in number of alleles, sample array and distance class, and actual amount of spatial autocorrelation. Together with the standard errors, the covariances can be used to find the standard errors of averages of I over multiple alleles, and hence data for all alleles can be used simultaneously to test the null hypothesis with greater statistical power. It is also generally straightforward to average among loci, because the spatial patterns for different loci should be nearly stochastically independent.

Thus it is now possible to obtain average estimates of I for data for multiple loci with multiple alleles, and to use this as a test of the null hypothesis for all data.

In addition, in simulations studies reported in chapter 8, the perhaps surprising result was that the stochastic correlations among values of Moran's I for different alleles of the same locus appear not to depend much on the amount of isolation by distance. This is helpful because it means that we do not need to posit the spatial structure to calculate standard errors and the distribution of I averaged over alleles within a locus, and we can simply use the covariances expected under the null hypothesis. However, perhaps not surprisingly, the stochastic correlations themselves appear to be highly stochastic, and will vary considerably for any actual set of alleles. The results in chapter 8 also show how the correlations among I for different alleles depend on allele frequencies, and these results provide guidelines for dealing with the correlations. Also, values of Moran's I, like pairwise kinship measures, cannot be directly compared to a priori expected ones in theoretical models. Nonetheless, they can be compared to values in simulation models. Simulations are also useful because they give the stochastic variation in values of correlations, not just the average values. Stochastic variances of spatial correlations are generally not studied in mathematical models of isolation by distance.

Another particularly useful advance in statistical methods is the Mantel (1967) statistic for associations or correlations of two different matrices of pairwise measures of distance. Mantel statistics of genetic distances and physical distances have similar properties as do spatial autocorrelation statistics. Moreover, modified forms of physical distance matrices can be used to construct specific spatial models of genetic variation (e.g., Sokal and Thomson 1987; Sokal et al. 1992). In addition, the multivariate extension of Mantel statistics of Smouse et al. (1986), essentially a modified form of multiple regression, allows simultaneous examination of several factors. An impressive display of this is the application to the Yanomama reviewed in chapter 2. Other important advances in spatial statistics include trend surface analysis, wombling, local indicators of spatial association, and directional autocorrelation analysis.

Observed geographical patterns of genetic variation among populations often fit standardizable forms of genetic isolation by distance, thus being characterized by the magnitudes of spatial autocorrelations and repeated spatial units, generically termed genetic patches, of predictable spatial scale. The structure then depends only on the amount of dispersal and the effective number of spatial dimensions in which the system exists. Within populations, the intensity and scale of spatial genetic units also fit expectations based on the amounts and distances of dispersal and the effective number of spatial dimensions. Regarding spatial dimensionality, if a system effectively exists in one spatial dimension, then the form of spatial autocorrelation function on distance is exponential, but not if it exists in two or more spatial dimensions. In two-dimensional systems genetic patches appear to be a main form of spatial genetic structure, caused by stochasticity in realizations of a process and not necessarily predicted by the mathematical theory of isolation by distance. Altered forms of genetic structure can be caused by the combination of localized stochastic effects, such as genetic drift, and limited migration or dispersal distances. One of the most prominent features is directionality. In addition, the spatial patterns of genetic variation among populations may be altered by various forms of stochastic migration. Within populations, some details of demographic processes may also affect spatial patterns. Finally, at both levels, how fast events average out and how close the system is to equilibrium or quasi-equilibrium are important factors for the types of spatial patterns expected.

It appears that in many cases the process of genetic isolation by distance is highly stochastic. Thus the degree of relatedness among pairs of locations separated by a given distance is itself highly stochastic, especially within populations and perhaps less so among populations. There are two important implications. First, this implies that the order of averaging, first over pairs of locations or first over alleles and loci, should not cause much difference in statistical power. Second, it points out that multiple locus data are generally needed to precisely infer important aspects of space–time processes. This is evident from both characterizations of statistical properties in simulation studies and from review of genetic survey studies, for example, those using wombling for identification of barriers to gene flow among populations or using spatial correlations to estimate dispersal within populations.

The spatial scales of the distances between sample locations and choices of distance classes for spatial autocorrelation analyses both have strong influences on the results. The smallest distance among sampled locations sets the lower limit on the distances at which correlations may be measured. If locations are grouped, for example, as individuals grouped into quadrats for analyses of spatial structure within populations, then the minimum distance of resolution is correspondingly larger. Moreover, grouping of locations mixes different spatial scales and can change the theoretical and statistical properties of spatial correlation measures, for example, their response to increases in dispersal within populations (e.g., Epperson and Li 1997).

Finally, studies of the temporary effects of ancient events on spatial distributions observed contemporarily, in particular among populations, can make use of haplotype or similar data, which may carry temporal information in the form of multiply mutated types. A key consideration is how long it takes before the effects are spread or averaged out spatially or erased by time. Another set of important considerations centers around the conditions under which particularly interesting effects may be validly "parsed out" from the generative space–time process. Also important are the conditions under which reasonable approximations may be made to simplify analyses by ignoring either space or time during various steps of the inferential process. These issues are just beginning to be characterized.

FUTURE STUDIES

There are several areas in which future work may produce especially important contributions to the characterization of the connection of observed spatial patterns to space–time processes through spatial or spatial–temporal statistical analyses. All of these issues involve one or more of the following: (1) the framing of evolutionary questions; (2) spatial or spatial–temporal context; (3) methods of statistical analysis; and (4) the construction of appropriate models of space–time processes. Some of these issues arise from the novelty of DNA sequence data and various forms of haplotype data. Others are general and have importance for all types of spatially distributed genetic data.

In part because of the power of modern computers, many complex space–time processes can be simulated in large numbers of sets with large numbers of replications. Hence, much of the complexity can be broken down into contrasting combinations or permutations of the causal factors, and many combinations of factors can be characterized. Other needed results may be derived analytically. One particularly important set of theoretical models that need better characterization is true multilocus genetic models of structured populations. None of Malécot's models, for example, were truly multilocus, in the sense that linkage disequilibrium is accounted for. It is also particularly important to examine spatially structured linkage disequilibrium between genes that respond to variable environmental or microenvironmental selection and genes that are selectively neutral markers. Further, this is relevant to statistical methods exemplified by the transmission disequilibrium test (Spielman et al. 1993). In addition, the variety of single-locus models of environmental or microenvironmental selection needs to be expanded, especially where migration among populations or dispersal within populations is great enough that it cannot be ignored. If possible, it would also be helpful to have analytical, theoretical models of the basic genetic isolation by distance process that do not involve the a priori expected values, since these are usually unobservable.

Further developments in metapopulation models that have spatially explicit structures would improve understanding of spatial patterns. They would further characterize the effects of fluctuating population size on spatial patterns. Also needed are models that better inform the conditions under which spatial structure may be ignored in constructing gene genealogies based on spatially located samples. Similarly, additional developments are needed in characterizing how spatial directionalities to migration rates cause directionalities in spatial correlations. It also appears that more could be learned by incorporating mechanistic models of dispersal and gene flow distances into models of genetic isolation by distance. Presently, most of the mechanistic modeling efforts for dispersal are concentrated in the field of ecology, and do not include a genetic component. In addition, empirical studies, probably for data collected for multiple life-stages, would yield direct results on the effects of stochastic migration on spatial patterns, complimentary to existing theoretical results. Additional theoretical results are also needed on (1) how deviations from the assumptions of the standard models of genetic isolation by distance may average out; (2) the signature effects of stochastic demographics; and (3) characterizations of nonequilibrium space–time processes. Finally, additional results on the effects of age structure and demographic shifts on spatial structures within populations are needed, as are results on the conditions under which systems with clumped distributions of individuals can be delineated into effectively discrete populations to better guide experimental studies.

Shifting balance theories (Wright 1931) are still controversial, particularly the third phase. The complexity of these theories in their full form involves multilocus selection that may also vary with the environment, a spatial context, and changes in the sizes or ranges of populations, as well as migration. Nonetheless, continued theoretical analyses, even if including only a subset of causal factors, may improve understanding of the importance of shifting balance selection in evolution. The importance, or not, of spatial context has yet to be well characterized, although there is some evidence that spatial structuring may make more likely the third phase of shifting balance processes (Peck et al. 1998).

This monograph reports considerable theoretical results on spatial–temporal measures, including space–time correlations, coalescences, and probabilities of identity by descent. These are useful for characterizing theoretical space–time processes and inferences about ancient events, but they are also needed for analysis of space–time data. Collection of space-time data is impossible or infeasible in many cases, but where feasible, it may often be well worth the effort. Experimental methods of isolating informative DNA samples from ancient materials will likely make increasingly important contributions to the study of geographic genetics, whether the data are treated as haplotypes or as allele frequencies. Methods of statistical computing for space–time data also need continued development and characterization.

Cross-correlations between spatial distributions of genetic variables and other variables, typically measures of environmental factors, need to take into account the spatial autocorrelations of each variable. The author is unaware of any theoretical models of environmental selection that are expressed in terms of such cross-correlations. Some statistical properties of cross-correlations have been developed (e.g., Cliff and Ord 1981).

The final set of issues is directly connected to new types of information that may be gained from modern molecular genotyping. First, there is the need for characterizing the conditions under which there *is* more information in such data for structured populations. It is often the case that methods designed to use any additional information also make additional assumptions, for example, about the mutational process, and these can sometimes be in error. Then, statistical methods only lose accuracy and do not gain precision, if there is no additional information to be gained in the data. In addition, the incorporation of the mutational state of genes tends to complicate theoretical models, which are already complicated by the inclusion of spatial context. For example, models that are fully explicit spatially have just recently begun to be developed for coalescence models (e.g., Nagylaki 2002), and further such developments would continue to constitute major advances.

Much of the added potential of modern molecular methods above allele frequency data can be attributed to a certain added temporal depth to the data, and this temporal depth can be applied in studies of ancient events, such as geographic origination of genetic variants. Nonequilibrium space–time models of probabilities of identity by descent are needed, as are further developments for nonequilibrium, spatially explicit coalescence models. Results from a full model would yield a firm basis for the probability theory needed to develop better statistical methods. In principle, spatially distributed data on haplotypes or DNA sequences could be analyzed spatially and temporally, in terms of overall likelihood. Even the study of theoretical processes would help inform as to the conditions under which it is valid to ignore either time or space during a stepwise procedure of analysis. It may also inform as to the conditions under which it is valid to "parse off" particularly interesting features of space–time processes.

There are two outstanding problems in studies of ancient events. Currently, the most appropriate statistical method is the nested cladistic analysis method of Templeton (1998). Steps of this procedure are perfectly valid, but most undoubtably lose some information. The latter steps of the procedure presume that a correct gene genealogy has been identified, although it should be noted that the methods can identify inconsistences and ambiguities in tree topology. Nonetheless, further development of statistical methods in this area is warranted. In addition, theoretical models of the distribution of haplotypes or sequences during the process of range expansion are particularly important. Presumably, such models are rather complex spatial–temporal processes of the distances at which populations are newly colonized and various population size fluctuations, as well as migrations among established populations.

Literature Cited

Allard, R. W. 1975. The mating system and micro-evolution. *Genetics* 79: 115–126.

Allee, W. C., Emerson, A. E., Park, O., Park, T., and Schmidt, K. P. 1949. *Principles of Animal Ecology*. Philadelphia: W. B. Saunders and Co.

Alvarez-Buylla, E. R. 1994. Density dependence and patch dynamics in tropical rain forest: matrix models and applications to a tree species. *Am. Nat.* 143: 155–191.

Alvarez-Buylla, E. R., and Garay, A. A. 1994. Population genetic structure of *Cecropia obtusifolia*, a neotropical pioneer tree species. *Evolution* 48: 437–453.

Ammerman, A. J., and Cavalli-Sforza, L. L. 1984. *Neolithic Transition and the Genetics of Populations in Europe*. Princeton, NJ: Princeton University Press.

Andersen, M. 1991. Mechanistic models for the seed shadows of wind-dispersed plants. *Am. Nat.* 137: 476–497.

Anselin, L. 1995. Local indicators of spatial associations—LISA. *Geogr. Anal.* 27: 93–115.

Argyres, A. Z., and Schmitt, J. 1991. Microgeographical genetic structure of morphological and life history traits in a natural population of *Impatiens capensis*. *Evolution* 45: 178–189.

Arnaud, J. F., Madec, L., Bellido, A., and Guiller, A. 1999. Microspatial genetic structure in the land snail *Helix aspersa* (Gastropoda: Helicidae). *Heredity* 83: 110–119.

Aroian, L. A. 1985. Time series in *m* dimensions: past, present and future. In O. D. Anderson, J. K. Ord, and E. A. Robertson, eds., *Time Series Analysis: Theory and Practice* 6, 241–261. Amsterdam: Elsevier/North Holland.

Avise, J. C. 1994. *Molecular Markers, Natural History and Evolution*. New York: Chapman and Hall.

Ayala, F. J. 1995. The myth of Eve: molecular biology and human origins. *Science* 270: 1930–1936.

Aylor, D. E. 1990. The role of intermittent wind in the dispersal of fungal pathogens. *Annu. Rev. Phytopathol.* 28: 73–92.

Bachmann, K. 2001. Evolution and the genetic analysis of populations: 1950–2000. *Taxon* 50: 7–45.

Bacilieri, R., Labbe, T., and Kremer, A. 1994. Intraspecific genetic structure in a mixed population of *Quercus petraea* (Matt.) Lelbl and *Q. robur* L. *Heredity* 73: 130–141.

Bannister, M. H. 1965. Variation in the breeding system of *Pinus radiata*. In H. G. Baker and G. L. Stebbins, eds., *The Genetics of Colonizing Species*, 353–372. New York: Academic Press.

Barbujani, G. 1987. Autocorrelation of gene frequencies under isolation by distance. *Genetics* 117: 777–782.

Barbujani, G., Oden, N. L., and Sokal, R. R. 1989. Detecting regions of abrupt change in maps of biological variables. *Syst. Zool.* 38: 376–389.

Barbujani, G., and Sokal, R. R. 1990. Zones of sharp genetic change in Europe are also linguistic boundaries. *Proc. Natl. Acad. Sci. USA* 87: 1816–1819.

———. 1991a. Genetic population structure of Italy, I: geographic patterns of gene frequencies. *Hum. Biol.* 63: 253–272.

———. 1991b. Genetic population structure of Italy, II: physical and cultural barriers to gene flow. *Am. J. Hum. Genet.* 48: 398–411.

Barker, J.S.F, and Mulley, J. C. 1976. Isozyme variation in natural populations of *Drosophila buzzatii*. *Evolution* 30: 213–233.

Bartlett, M. S. 1971. Physical nearest-neighbor models and non-linear time-series. *J. Appl. Prob.* 8: 222–232.

Barton, N. H., and Wilson, I. 1995. Genealogies and geography. *Philos. Trans. R. Soc. Lond. B* 349: 49–59.

Batzer, M. A., Arcot, S. S., Phinney, J. W., Alegria-Hartman, M., Kass, D. H., Milligan, S. M., Kimpton, C., Gill, P., Hochmeister, M., Iaonnou, P. A., Herrera, R. J., Boudreau, D. A., Scheer, W. D., Keats, B.J.B., Deininger, P. L., and Stoneking, M. 1996. Genetic variation of recent Alu insertions in human populations. *J. Mol. Evol.* 42: 22–29.

Batzer, M. A., Sherry, S. T., Deininger, P. L., and Stoneking, M. 1997. Alu repeats and human evolution- response. *J. Mol. Evol.* 45: 7–8.

Bennett, R. J. 1979. *Spatial Time Series*. London: Pion Ltd.

———. 1981. Statistical forecasting. *Concepts. Tech. Mod. Geogr.* 28: 1–43.

Bennett, R. J., Haining, R. P., and Wilson, A. G. 1985. Spatial structure, spatial interaction, and their integration: a review of alternative models. *Environ. Planning A* 17: 625–645.

Berg, E. E., and Hamrick, J. L. 1995. Fine-scale genetic structure of a turkey oak forest. *Evolution* 49: 110–120.

Bertorelle, G., and Barbujani, G. 1995. Analysis of DNA diversity by spatial autocorrelation. *Genetics* 140: 811–819.

Bird, J., Riska, B., and Sokal, R. R. 1981. Geographic variation in variability of *Pemphigus populicaulis*. *Syst. Zool.* 30: 58–70.

Bocquet-Appel, J. P., and Sokal, R. R. 1989. Spatial autocorrelation analysis of trend residuals in biological data. *Syst. Zool.* 38: 333–341.

Bodmer, W. F. 1960. Discrete stochastic processes in population genetics. *J. R. Stat. Soc. B Met.* 22: 218–236.

Bodmer, W. F., and Cavalli-Sforza, L. L. 1968. A migration matrix model for the study of random genetic drift. *Genetics* 59: 565–592.

Bohonak, A. J. 1999. Dispersal, gene flow, and population structure. *Q. Rev. Biol.* 74: 21–45.

Bowcock, A. M., Kidd, J. R., Mountain, J. L., Hebert, J. M., Carotenuto, L., Kidd, K. K., and Cavalli-Sforza, L. L. 1991. Drift, admixture, and selection in human evolution: a study with DNA polymorphism. *Proc. Natl. Acad. Sci. USA* 88: 839–843.

Box, G.E.P., and Jenkins, G. M. 1976. *Time Series Analysis: Forecasting and Control*. San Francisco: Holden-Day Inc.

Bradshaw, A. D. 1984. Ecological significance of genetic variation between populations. In R. Dirzo and J. Sarukhan, eds., *Perspectives on Plant Population Ecology*, 213–228. Sunderland, MA: Sinauer.

Brown, A.H.D. 1979. Enzyme polymorphism in plant populations. *Theor. Popul. Biol.* 15: 1–42.

Brown, B. A., and Clegg, M. T. 1984. Influence of flower color polymorphism on genetic transmission in a natural population of the common morning glory, *Ipomoea purpurea*. *Evolution* 38: 796–803.

Brown, W. M. 1985. The mitochondrial genome of animals. In R. J. MacIntyre, ed., *Molecular Evolutionary Genetics*, 95–130. New York: Plenum.

Caballero, A. and Hill, W. G. 1992. Effective size of nonrandom mating populations. *Genetics* 130: 909–916.

Campbell, D. R. 1991. Comparing pollen dispersal and gene flow in a natural plant population. *Evolution* 45: 1965–1968.

Campbell, D. R., and Dooley, J. L. 1992. The spatial scale of genetic differentiation in a hummingbird-pollinated plant: comparison with models of isolation by distance. *Am. Nat.* 139: 735–748.

Cann, R. L., Stoneking, M., and Wilson, A. C. 1987. Mitochondrial DNA and human evolution. *Nature* 325: 31–36.

Caujapé-Castells, J., and Pedrola-Monfort, J. 1997. Space-time patterns of genetic structure within a stand of *Androcymbium gramineum* (Cav.) McBride (Colchicaceae). *Heredity* 79: 341–349.

Caujapé-Castells, J., Pedrola-Monfort, J., and Membrives, N. 1999. Contrasting patterns of genetic structure in the south african species *Androcymbium bellum*, *A. guttatum*, and *A. pulchrum* (Colchicaceae). *Biochem. Syst. Ecol.* 27: 591–605.

Cavalli-Sforza, L. L., and Feldman, M. W. 1990. Spatial subdivision of populations and estimates of genetic variation. *Theor. Popul. Biol.* 37: 3–25.

Chamberlain, A. C. 1975. The movement of particles in plant communities. In J. L. Monteith, ed., *Vegetation and the Atmosphere*, Vol. 1, 155–203. London: Academic Press.

Chesser, R. K., and Baker, R. J. 1996. Effective sizes and dynamics of uniparentally and diparentally inherited genes. *Genetics* 144: 1225–1235.

Christiansen, F. B. 1975. Hard and soft selection in a subdivided population. *Am. Nat.* 109: 11–16.

Christiansen, F. B., and Feldman, M.W. 1975. Subdivided populations: a review of one- and two-locus deterministic theory. *Theor. Popul. Biol.* 7: 13–38.

Chung, M. G., Chung, J. M., Chung, M. Y., and Epperson, B. K. 2000. Spatial distribution of allozyme polymorphisms following clonal and sexual reproduction in populations of *Rhus javanica* (Anacardiaceae). *Heredity* 84: 178–185.

Chung, M. G., Chung, J. M., and Epperson, B. K. 1999. Spatial genetic structure of allozyme polymorphisms within populations of *Rhus trichocarpha* (Anacardiaceae). *Silvae Genet.* 48: 223–227.

Chung, M. G., and Epperson, B. K. 1999. Spatial distribution of clonal and sexual reproduction in populations of *Adenophora grandiflora* (Campanulaceae). *Evolution* 53: 1068–1078.

———. 2000. Clonal and spatial genetic structure in *Eurya emarginata* (Theaceae). *Heredity* 84: 170–177.

Chung, M. Y., Chung, G. M., Chung, M. G., and Epperson, B. K. 1998. Spatial genetic structure in populations of *Cymbidium goeringii* (Orchidaceae). *Genes Genet. Syst.* 73: 281–285.

Clarke, B. C. 1966. The evolution of morph-ratio clines. *Am. Nat.* 100: 389–402.

Cliff, A. D., and Ord, J. K. 1973. *Spatial Autocorrelation*. London: Pion.

———. 1981. *Spatial Processes*. London: Pion.

Cockerham, C. C. 1969. Variance of gene frequencies. *Evolution* 73: 72–84.

Conkle, M. T., and Critchfield, W. B. 1988. Genetic variation and hybridization in ponderosa pine. In D. M. Baumgartner and J. E. Lotan, eds., *Ponderosa Pine: The Species and Its Management*, 27–43. Pullman: Washington State University.

Coyne, J. A., Barton, N. H., and Turelli, M. 2000. Is Wright's shifting balance important in evolution? *Evolution* 54: 306–317.

Crawford, T. J. 1984. The estimation of neighborhood parameters for plant populations. *Heredity* 52: 273–283.

Crow, J. F. 1954. Breeding structure of populations, II: effective population number. In O. Kempthorne, T. A. Bancroft, J. W. Gowen, and J. L. Lush, eds., *Statistics and Mathematics in Biology*, 543–556. Ames: Iowa State University Press.

———. 1989. Twenty-five years ago in genetics: the infinite allele model. *Genetics* 121: 631–634.

Crow, J. F., and Denniston, C. 1988. Inbreeding and variance effective population numbers. *Evolution* 42: 482–495.

Csanady, G. T. 1973. *Turbulent Diffusion in the Environment*. Boston: D. Reidel.

Deininger, P. L., and Batzer, M. A. 1993. Evolution of retrotransposons. *Evol. Biol.* 27: 157–196.

Dewey, S. E., and Heywood, J. S. 1988. Spatial genetic structure in a population of *Psychotria nervosa*, I: distribution of genotypes. *Evolution* 42: 834–838.

Di-Giovanni, F., and Kevan, P. G. 1991. Factors affecting pollen dynamics and its importance to pollen contaminations: a review. *Can. J. For. Res.* 21: 1155–1170.

Di Rienzo, A., and Wilson, A. C. 1991. Branching pattern in the evolutionary tree for human mitochondrial DNA. *Proc. Natl. Acad. Sci. USA* 88: 1597–1601.

Dobzhansky, T., and Wright, S. 1943. Genetics of natural populations, X: dispersion rates in *Drosophila pseudoobscura*. *Genetics* 28: 304–340.

Doligez, A., Baril, C., and Joly, H. I. 1998. Fine-scale spatial genetic structure with nonuniform distribution of individuals. *Genetics* 148: 905–919.

Dow, M. M., and Cheverud, J. M. 1985. Comparison of distance matrices in studies of population structure and genetic microdifferentiation: quadratic assignment. *Am. J. Phys. Anthropol.* 68: 367–373.

Echt, C. S., DeVerno, L. L., Anzidei, M., and Vendramin, G. G. 1998. Chlororplast microsatellites reveal population genetic diversity in red pine, *Pinus resinosa*, Ait. *Mol. Ecol.* 7: 307–316.

Eller, E., and Harpending, H. C. 1996. Simulations show that neither population expansion nor population stationarity in a West African population can be rejected. *Mol. Biol. Evol.* 13: 1155–1157.

Ellstrand, N. C., and Roose, M. L. 1987. Patterns of genotypic diversity in clonal plant species. *Am. J. Bot.* 74: 123–131.

Ellstrand, N. C., Torres, A. M., and Levin, D. A. 1978. Density and the rate of apparent outcrossing in *Helianthus annus* (Asteraceae). *Syst. Bot.* 3: 403–407.

Endler, J. A. 1977. *Geographic Variation, Speciation, and Clines*. Princeton, NJ: Princeton University Press.

Ennos, R. A., and Clegg, M. T. 1982. Effect of population substructuring on estimates of outcrossing rate in plant populations. *Heredity* 48: 283–292.

Epling, C., Lewis, H., and Ball, F. M. 1960. The breeding group and seed storage: a study in population dynamics. *Evolution* 14: 238–255.

Epperson, B. K. 1990a. Spatial autocorrelation of genotypes under directional selection. *Genetics* 124: 757–771.

———. 1990b. Spatial patterns of genetic variation within plant populations. In A.H.D. Brown, M. T. Clegg, A. L. Kahler, and B. S. Weir, eds., *Plant Population Genetics, Breeding, and Genetic Resources*, 229–253. Sunderland, MA: Sinauer Associates.

———. 1992. Spatial structure of genetic variation within populations of forest trees. *New Forests* 6: 257–278.

———. 1993a. Recent advances in correlation studies of spatial patterns of genetic variation. *Evol. Biol.* 27: 95–155.

———. 1993b. Spatial and space–time correlations in systems of subpopulations with genetic drift and migration. *Genetics* 133: 711–727.

———. 1994. Spatial and space–time correlations in systems of subpopulations with stochastic migration. *Theor. Popul. Biol.* 46: 160–197.

———. 1995a. Spatial distribution of genotypes under isolation by distance. *Genetics* 140: 1431–1440.

———. 1995b. Spatial structure of two-locus genotypes under isolation by distance. *Genetics* 140: 365–375.

———. 1995c. Fine-scale spatial structure: correlations for individual genotypes differ from those for local gene frequencies. *Evolution* 49: 1022–1026.

———. 1999a. Gene genealogies in geographically structured populations. *Genetics* 152: 797–806.

———. 1999b. Gustave Malécot, 1911–1998: population genetics founding father. *Genetics* 152: 477–484.

———. 2000a. Spatial and space–time correlations in ecological models. *Ecol. Modell.* 132: 63–76.

———. 2000b. Spatial genetic structure and non-equilibrium demographics within plant populations. *Plant Species Biol.* 15: 269–279.

———. 2002. Spatial–temporal properties of gene genealogies in geographically structured populations. In M. Slatkin and M. Veuille, eds., *Modern Developments in Theoretical Population Genetics*, 165–182. Oxford, UK: Oxford University Press.

———. n.d.(a). Covariances among join count spatial autocorrelation measures.

———. n.d.(b). Multilocus estimation of genetic structure within populations.

Epperson, B. K., and Allard, R.W. 1984. Allozyme analysis of the mating system in lodgepole pine populations. *J. Hered.* 75: 212–214.

———. 1989. Spatial autocorrelation analysis of the distribution of genotypes within populations of lodgepole pine. *Genetics* 121: 369–377.

Epperson, B. K., and Alvarez-Buylla, E. 1997. Limited seed dispersal and genetic structure in life stages of *Cecropia obtusifolia*. *Evolution* 51: 275–282.

Epperson, B. K., and Chung, M. G. 2001. Spatial genetic structure of allozyme polymorphisms within populations of *Pinus strobus* (Pinaceae). *Am. J. Bot.* 88: 1006–1010.

Epperson, B. K., Chung, M. G., and Telewski, F. W. 2003. Spatial pattern of allozyme variation in a contact zone of *Pinus ponderosa* and *P. arizonica* (Pinaceae). *Am. J. Bot.* 90: 25–31.

Epperson, B. K., and Clegg, M. T. 1986. Spatial autocorrelation analysis of flower color polymorphisms within substructured populations of morning glory (*Ipomoea purpurea*). *Am. Nat.* 128: 840–858.

———. 1987. Frequency-dependent variation for outcrossing rate among flower color morphs of *Ipomoea purpurea*. *Evolution* 41: 1302–1311.

Epperson, B. K., Huang, Z., and Li, T.-Q. 1999. Spatial genetic structure of multiallelic loci. *Genet. Res. Cambridge*. 73: 251–261.

Epperson, B. K., and Li, T.-Q. 1996. Measurement of genetic structure within populations using Moran's spatial autocorrelation statistics. *Proc. Natl. Acad. Sci. USA* 93: 10528–10532.

———. 1997. Gene dispersal and spatial genetic structure. *Evolution* 51: 672–681.

Epperson, B. K., Telewski, F. W., Plovanich-Jones, A. E., and Grimes, J. E. 2001. Clinal differences and putative hybridization in a contact zone of *Pinus ponderosa* and *Pinus arizonica* (Pinaceae). *Am. J. Bot.* 88: 1052–1057.

Erlich, H. A., Bergstrom, T. F., Stoneking, M., and Gyllensten, U. 1996. HLA sequence polymorphism and the origin of humans. *Science* 274: 1552–1554.

Ewens, W. J. 1974. A note on the sampling theory for infinite alleles and infinite sites models. *Theor. Popul. Biol.* 6: 143–148.

Excoffier, L., Smouse, P. E., and Quattro, J. M. 1992. Analysis of molecular variance inferred from metric distances among DNA haplotypes: application to human mitochondrial DNA restriction data. *Genetics* 131: 479–491.

Feller, W. 1957. *An Introduction to Probability Theory and Its Applications*. New York: Wiley.

Felsenstein, J. 1975. A pain in the torus: some difficulties with models of isolation by distance. *Am. Nat.* 109: 359–368.

Fenster, C. B. 1991. Gene flow in *Chamaecrista fasciculata*, II: gene establishment. *Evolution* 45: 410–422.

Fisher, R. A. 1941. Average excess and average effect of gene substitution. *Ann. Eugen. (Lond.)* 11: 53–63.

Fisher, R. A., and Ford, E. B. 1947. The spread of a gene in natural conditions in a colony of the moth *Panaxia dominula* L. *Heredity* 1: 143–174.

Fitt, B.D.L., and McCartney, H. A. 1986. Spore dispersal in relation to epidemic models. In K. J. Leonard and W. E. Fry, eds., *Plant Disease Epidemiology*, Vol. I, 311–345. New York: Macmillan.

Fix, A. G. 1975. Fission–fusion and lineal effect: aspects of the population structure of the Semai Senoi of Malaysia. *Am. J. Phys. Anthropol.* 43: 295–302.

———. 1978. The role of kin-structured migration in genetic microdifferentiation. *Ann. Hum. Genet. Lond.* 41: 329–339.

———. 1993. Kin-structured migration and isolation by distance. *Hum. Biol.* 65: 193–210.

———. 1994. Detecting clinal and balanced selection using spatial autocorrelation analysis under kin-structured migration. *Am. J. Phys. Anthropol.* 95: 385–397.

Fleming, W. H., and Su, C.-H. 1974. Some one-dimensional migration models in population genetics theory. *Theor. Popul. Biol.* 5: 431–449.

Friedlaender, J. S. 1975. *Patterns of Human Variation*. Cambridge, MA: Harvard University Press.

Fu, Y.-X. 1997. Coalescent theory for a partially selfing population. *Genetics* 146: 1489–1499.

Furnier, G., Knowles, P., Clyde, M. A., and Dancik, B. P. 1987. Effects of avian seed dispersal on the genetic structure of whitebark pine populations. *Evolution* 41: 607–612.

Gabriel, K. R., and Sokal, R. R. 1969. A new statistical approach to geographic variation analysis. *Syst. Zool.* 18: 259–278.

Gehring, J. L., and Delph, L. F. 1999. Fine-scale genetic structure and clinal variation in *Silene acaulis* despite high gene flow. *Heredity* 82: 628–637.

Gerber, A. 1994. The semiotics of subdivision: an empirical study of the population structure of *Trimerotropis saxàtalis*. PhD thesis, Division of Biomedical and Biological Sciences, Washington University, St. Louis, MO.

Giles, B. E., and Goudet, J. 1997. Genetic differentiation in *Silene dioica* metapopulations: estimation of spatio-temporal effects in a successional plant species. *Am. Nat.* 149: 507–526.

Gillois, M. 1966. Le concept d'identité et son importance en génétique. *Ann. Génét.* 9: 58–65.

Gillois, M. 1996. Malécot, (Gustave), 1911–. In P. Tort, ed., *Dictionnaire du Darwinisme et de l'Evolution*, 2768–2785. Paris: Presses Universitaires de France.

Goldstein, D. B., and Harvey, P. H. 1999. Evolutionary inference from genomic data. *BioEssays* 21: 148–156.

Goldstein, S. 1951. On diffusion by discontinuous movements, and on the telegraph equation. *Q. J. Mech. Appl. Math.* 4: 129–156.

Govindaraju, D. R. 1988. Life histories, neighbourhood sizes, and variance structure in some North American conifers. *Biol. J. Linn. Soc.* 35: 69–78.

Gower J. C. 1971. Statistical methods of comparing different multivariate analyses of the same data. In F. R. Hodson, D. G. Kendall and P. Tauta, eds., *Mathematics in the Archaeological and Historical Sciences*, 138–149. Edinburgh: Edinburgh University Press.

Graven, L., Passarino, G., Semino, O., Boursot, P., Santachiara-Benerecetti, S., Langaney, A., and Excoffier, L. 1995. Evolutionary correlation between control region sequence and restriction polymorphisms in the mitochondrial genome of a large Sengalese Mandenka sample. *Mol. Biol. Evol.* 12: 334–345.

Greene, D. F., and Johnson, E. A. 1989. A model of wind dispersal of winged or plumed seeds. *Ecology* 70: 339–347.

Gregory, P. H. 1968. Interpreting plant disease dispersal gradients. *Annu. Rev. Phytopathol.* 6: 189–212.

Guttman, L. 1968. A general nonmetric technique for finding the smallest coordinate space for a configuration of points. *Psychometrika* 33: 469–506.

Haggett, P., Cliff, A., and Frey, A. 1977. *Locational Analysis in Human Geography*. London: Edward Arnold.

Haining, R. P. 1977. Model specification in stationary random fields. *Geogr. Anal.* 9: 107–109.

———. 1978. The moving average model for spatial interaction. *Trans. Inst. Br. Geogr.* 3: 202–225.

———. 1979. Statistical test and process generators for random field models. *Geogr. Anal.* 11: 45–64.

Hammer M. F., Karafet, T., Rasanayagam, A., Wood, E. T., Altheide, T. K., Jenkins, T., Griffiths, R. C., Templeton, A. R., and Zegura, S. L. 1998. Out of Africa and back again: nested cladistic analysis of human Y chromosome variation. *Mol. Biol. Evol.* 15: 427–441.

Hammer, M. F., Spurdle, A. B., Karafet, T., et al. (11 coauthors). 1997. The geographic distribution of human Y chromosome variation. *Genetics* 145: 787–805.

Hammer, M. F., and Zegura, S. L. 1996. The role of the Y chromosome in human evolutionary studies. *Evol. Anthropol.* 5: 116–134.

Hamrick, J. L., and Godt, M.J.W. 1990. Allozyme diversity in plant species. In A.H.D. Brown, M. T. Clegg, A. L. Kahler, and B. S. Weir, eds., *Plant Population Genetics, Breeding, and Genetic Resources*, 43–63. Sunderland, MA: Sinauer Associates.

Hardy, O. J., and Vekemans, X. 1999. Isolation by distance in a continuous population: reconciliation between spatial autocorrelation analysis and population genetics models. *Heredity* 83: 145–154.

Harpending, H. C. 1973. Discussion of relationship of conditional kinship to a priori kinship, following paper by N. E. Morton. In N. E. Morton, ed., *Genetic Structure of Populations*, 78–79. Honolulu: University of Hawaii Press.

———. 1994. Signature of ancient population growth in a low-resolution mitochondrial DNA mismatch distribution. *Hum. Biol.* 66: 591–600.

Harpending, H. C., Batzer, M. A., Gurven, M., Jorde, L. B., Rogers, A. R., and Sherry, S. T. 1998. Genetic traces of ancient demography. *Proc. Natl. Acad. Sci. USA* 95: 1961–1967.

Hedrick, P. W., and Gilpin, M. E. 1997. Genetic effective size of a metapopulation. In I. Hanski and M. E. Gilpin, eds., *Metapopulation Dynamics: Ecology, Genetics, and Evolution*, 166–182. New York: Academic Press.

Heywood, J. S. 1991. Spatial analysis of genetic variation in plant populations. *Annu. Rev. Ecol. Syst.* 22: 335–355.

Holsinger, K. E. 1988. Inbreeding depression doesn't matter: the genetic basis of mating system evolution. *Evolution* 42: 1235–1244.

Holsinger, K. E., and Mason-Gamer, R. J. 1996. Hierarchical analysis of nucleotide diversity in geographically structured populations. *Genetics* 142: 629–639.

Hooper, P. M., and Hewings, G.J.D. 1981. Some properties of space–time processes. *Geogr. Anal.* 13: 203–223.

Hubert, L. J. 1985. Combinatorial data analysis: association and partial association. *Psychometrika* 50: 449–467.

Hudson, R. R. 1990. Gene geneologies and the coalescent process. In D. J. Futuyma and J. Antonovics, eds., *Oxford Surveys in Evolutionary Biology*, Vol. 7, 1–44. Oxford, UK: Oxford University Press.

Hudson, R. R., Slatkin, M., and Maddison, W. P. 1992. Estimation of levels of gene flow from DNA sequence data. *Genetics* 132: 583–589.

Imaizumi, Y., Morton, N. E., and Harris, D. E. 1970. Isolation by distance in artificial populations. *Genetics* 66: 569–582.

Ingvarsson, P. K., and Giles, B. E. 1999. Kin-structured colonization and small-scale genetic differentiation in *Silene dioica*. *Evolution* 53: 605–611.

Jones, J. S., Selander, R. K., and Schnell, G. D. 1980. Patterns of morphological and molecular polymorphism in the land snail *Cepaea nemoralis*. *Biol. J. Linn. Soc.* 14: 359–387.

Jorde, L. B., Eriksson, A. W., Morgan, K., and Workman, P. L. 1982. The genetic structure of Iceland. *Hum. Hered.* 32: 1–7.

Jorde, L. B., Rogers, A. R., Bamshad, M., Watkins, W. S., Krakowiak, P., Sung, S., Kere, J., and Harpending, H. C. 1997. Microsatellite diversity and the demographic history of modern humans. *Proc. Natl. Acad. Sci. USA* 94: 3100–3103.

Jumars, P. A., Thistle, D., and Jones, M. L. 1977. Detecting two-dimensional spatial structure in biological data. *Oecologia* 28: 109–123.

Kawano, S., and Kitamura, K. 1997. Demographic genetics of the Japanese beech, *Fagus crenata*, in the Ogawa Forest Reserve, Ibaraki, central Honshu, Japan, III: population dynamics and genetic substructuring within a metapopulation. *Plant Species Biol.* 12: 157–177.

Kimura, M. 1953. "Stepping stone" model of population. *Annu. Rep. Natl. Inst. Genet., Jpn.* 3: 62–63.

Kimura, M., and Weiss, G. H. 1964. The stepping stone model of population structure and the decrease of genetic correlation with distance. *Genetics* 49: 561–576.

Kingman, J.F.C. 1982a. The coalescent. *Stochast. Proc. Appl.* 13: 235–248.

———. 1982b. On the genealogy of large populations. *J. Appl. Prob.* 19: 27–43.

Kirby, G. C. 1975. Heterozygote frequencies in small populations. *Theor. Popul. Biol.* 8: 31–48.

Kiyosawa, S. and Shiyomi, M. 1972. A theoretical evaluation of the effect of mixing resistant variety with susceptible variety for controlling plant diseases. *Ann. Phytopathol. Soc. Jpn.* 38: 41–51.

Klein, E. K., Lavigne, C., Foueillassar X., Gouyon, P. H., and Larédo, C. 2003. Corn pollen dispersal: quasi-mechanistic models and field experiments. *Ecological Monographs* 73: 131–150.

Klein, R. G. 1995. Anatomy, behavior, and modern human origins. *J. World Prehist.* 9: 167–198.

Knight, A., Batzer, M. A, Stoneking, M., Tiwari, H. K., Scheer, W. D., Herrera, R. J., and Deininger, P. L. 1996. DNA sequences of *Alu* elements indicate a recent replacement of the human autosomal genetic complement. *Proc. Natl. Acad. Sci. USA* 93: 4360–4364.

Knowles, P. 1991. Spatial genetic structure within two natural stands of black spruce [*Picea mariana* (Mill) B.S.P.]. *Silvae Genet.* 40: 13–19.

Koopmans, T. C. 1942. Serial correlation and quadratic forms in normal variables. *Ann. Math. Stat.* 13: 14–33.

Koski, V. 1970. A study of pollen dispersal as a mechanism of gene flow in conifers. *Commun. Inst. For. Fenn.* 70: 1–78.

Krings, M., Stone, A., Schmitz, R. W., Krainitzki, H., Stoneking, M., and Paabo, S. 1997. Neanderthal DNA-sequences and the origin of modern humans. *Cell* 90: 9–30.

Krishna-Iyer, P. V. 1949. The first and second moments of some probability distributions arising from points on a lattice and their applications. *Biometrika* 36: 135–141.

———. 1950. The theory of probability distributions of points on a lattice. *Ann. Math. Stat.* 21: 198–217.

Kudoh, H., and Whigham, D. F. 1997. Microgeographic genetic structure and gene flow in *Hibiscus moscheutos* (Malvaceae) populations. *Am. J. Bot.* 84: 1285–1293.

Lahr, M. M. 1996. *The Evolution of Modern Human Diversity: A Study of Cranial Variation.* Cambridge, UK: Cambridge University Press.

Lande, R. 1991. Isolation by distance in a quantitative trait. *Genetics* 128: 443–452.

Larrimore, W. E. 1977. Statistical inference on stationary random fields. *Proc. IEEE* 65: 961–970.

Latta, R. G., Linhart, Y. B., Fleck, D., and Elliot, M. 1998. Direct and indirect estimates of seed versus pollen movement within a population of ponderosa pine. *Evolution* 52: 61–67.

Latter, B.D.H., and Sved, J. A. 1981. Migration and mutation in stochastic models of gene frequency change, II: stochastic migration with a finite number of islands. *J. Math. Biol.* 13: 95–104.

Lee, C.-Y., and Epperson, B. K. n.d. Integrated environment for analyzing STARMA models.

Legendre, P., and Fortin, M.-J. 1989. Spatial pattern and ecological analysis. *Vegetatio* 80: 107–138.

Leonardi, S., and Menozzi, P. 1996. Spatial structure of genetic variability in natural stands of *Fagus sylvatica* L. (beech) in Italy. *Heredity* 77: 359–368.

Leonardi, S., Raddi, S., and Borghetti, M. 1996. Spatial autocorrelation of allozyme traits in a Norway spruce (*Picea abies*) population. *Can. J. For. Res.* 26: 63–71.

Levene, H. 1953. Genetic equilibrium when more than one ecological niche is available. *Am. Nat.* 87: 331–333.

Levin, D. A. 1981. Dispersal versus gene flow in plants. *Ann. MO Bot. Gard.* 68: 233–253.

Levin, D. A., and Fix, A. G. 1989. A model of kin-migration in plants. *Theor. Appl. Genet.* 77: 332–336.

Levins, R. 1969. Some demographic and genetic consequences of environmental heterogeneity for biological control. *Bull. Entomol. Soc. Am.* 15: 237–240.

Lewontin, R. C., and Krakauer, J. 1973. Distribution of gene frequency as a test of the theory of the selective neutrality of polymorphisms. *Genetics* 74: 175–195.

Li, W.-H. 1977. Distribution of nucleotide differences between two randomly chosen cistrons in a finite population. *Genetics* 85: 331–337.

Linhart, Y. B., Mitton, J. B., Sturgeon, K. B., and Davis, M. L. 1981. Genetic variation in space and time in a population of ponderosa pine. *Heredity* 46: 407–426.

Loiselle, B. A., Sork, V. L., Nason, J., and Graham, C. 1995. Spatial genetic structure of a tropical understory shrub, *Psychotria officinalis* (Rubiaceae). *Am. J. Bot.* 82: 1553–1564.

Lynch, M., and Gabriel, W. 1991. Mutation load and the survival of small populations. *Evolution* 44: 1725–1737.

Mahy, G., and Nève, G. 1997. The application of spatial autocorrelation methods to the study of *Calluna vulgaris* population genetics. *Belg. J. Bot.* 129: 131–139.

Maki, M., and Masuda, M. 1993. Spatial autocorrelation of genotypes in a gynodioecious population of *Chionographis japonica* var. *kurohimensis* (Liliaceae). *Int. J. Plant Sci.* 154: 467–472.

Malécot, G. 1941. Étude mathématique des populations "mendéliennes." *Ann. Univ. Lyon A* 4: 45–60.

———. 1942. Mendélisme et consanguinité. *C. R. Séances Acad. Sci. Paris* 215: 313–314.

———. 1948. *Les Mathématiques de l'Hérédité*. Paris: Masson.

———. 1950. Quelques schémas probabilistes sur la variabilité des populations naturelles. *Ann. Univ. Lyon A* 13: 37–60.

———. 1955. Remarks on the decrease of relationship with distance. Following paper by M. Kimura. *Cold Spring Harbor Symp. Quant. Biol.* 20: 52–53.

———. 1967. Identical loci and relationship. *Proc. 5th Berkeley Symp. Math. Stat. Prob.* 4: 317–332.

———. 1969. *The Mathematics of Heredity*. San Francisco: W. H. Freeman.

———. 1971. Génétique des populations diploïdes naturelles dans le cas d'un seul locus, I: evolution de la frequence d'un gene. Étude des variances et des covariances. *Ann. Génét. Sélection Anim.* 3: 255–280.

———. 1972. Génétique des populations naturelles dans le cas d'un seul locus, II: étude du coefficient de parenté. *Ann. Génét. Sélection Anim.* 4: 385–409.

———. 1973a. Génétique des populations diploïdes naturelles dans le cas d'un seul locus, III: parenté, mutations et migration. *Ann. Génét. Sélection Anim.* 5: 333–361.

———. 1973b. Isolation by distance. In N. E. Morton, ed., *Genetic Structure of Populations*, 72–75. Honolulu: University of Hawaii Press.

———. 1975. Heterozygosity and relationship in regularly subdivided populations. *Theor. Popul. Biol.* 8: 212–241.

Mantel, N. 1967. The detection of disease clustering and a generalized regression approach. *Cancer Res.* 27: 209–220.

Martin, R. L., and Oeppen, J. E. 1975. The identification of regional forecasting models using space-time correlation functions. *Trans. Inst. Br. Geogr.* 66: 95–118.

Maruyama, T. 1969. Genetic correlations in the stepping stone model with non-symmetrical migration rates. *J. Appl. Prob.* 6: 463–477.

———. 1970. On the rate of decrease of heterozygosity in circular stepping stone models of populations. *Theor. Popul. Biol.* 1: 101–119.

———. 1971. Analysis of population structure, II: two dimensional stepping stone models of finite length and other geographically structured populations. *Ann. Hum. Genet.* 35: 411–423.

Matula, D. W., and Sokal, R. R. 1980. Properties of Gabriel graphs relevant to geographic variation research and the clustering of points in the plane. *Geogr. Anal.* 12: 205–222.

McCulloch, C. E., and Cain, M. L. 1989. Analyzing discrete movement data as a correlated random walk. *Ecology* 70: 383–388.

McFadden, C. S., and Aydin, K. Y. 1996. Spatial autocorrelation analysis of small-scale genetic structure in a clonal soft coral with limited larval dispersal. *Mar. Biol.* 126: 215–224.

Merzeau, D., Comps, B., Thiébaut, B., Cuguen, J., and Letouzey, J. 1994. Genetic structure of natural stands of *Fagus sylvatica* L. (beech). *Heredity* 72: 269–277.

Michalakis, Y., and Excoffier, L. 1996. A generic estimation of population subdivision using distances between alleles with special reference for microsatellite loci. *Genetics* 142: 1061–1064.

Milgroom, M. G., and Lipari, S. E. 1995. Spatial analysis of nuclear and mitochondrial RFLP genotypes in populations of the chestnut blight fungus, *Cryphonectria parasitica*. *Mol. Ecol.* 4: 633–642.

Moran, P.A.P. 1950. Notes on continuous stochastic phenomena. *Biometrika* 37: 17–23.

Morton, N. E. 1969. Human population structure. *Annu. Rev. Genet.* 3: 53–74.

———. 1973a. Kinship and population structure. In N. E. Morton, ed., *Genetic Structure of Populations*, 66–69. Honolulu: University of Hawaii Press.

———. 1973b. Kinship bioassay. In N. E. Morton, ed., *Genetic Structure of Populations*, 158–163. Honolulu: University of Hawaii Press.

———. 1975. Kinship, fitness and evolution. In F. M. Salzano, ed., *The Role of Natural Selection in Human Evolution*, 133–154. Amsterdam: North-Holland.

———. 1982. Estimation of demographic parameters from isolation by distance. *Hum Hered.* 32: 37–41.

Morton, N. E., Miki, C., and Yee, S. 1968. Bioassay of population structure under isolation by distance. *Am. J. Hum. Genet.* 20: 411–419.

Morton, N. E., Yee, S., Harris, D. E., and Lew, R. 1971. Bioassay of kinship. *Theor. Popul. Biol.* 2: 507–524.

Murakami, N., Nishiyama, T., Satoh, H., and Suzuki, T. 1997. Marked spatial genetic structure in three populations of a weedy fern, *Pteris multifida* Poir., and reestimation of its selfing rate. *Plant Species Biol.* 12: 97–106.

Nagylaki, T. 1974. The decay of genetic variability in geographically structured populations. *Proc. Natl. Acad. Sci. USA* 71: 2932–2936.

———. 1978. A diffusion model for geographically structured populations. *J. Math. Biol.* 9: 101–114.

———. 1979. The island model with stochastic migration. *Genetics* 91: 163–176.

———. 1986. Neutral models of geographical variation. In P. Tauta, ed., *Lecture Notes in Mathematics*, No 1212: *Stochastic Spatial Processes*, 216–237. Berlin: Springer-Verlag.

———. 1989. Gustave Malécot and the transition from classical to modern population genetics. *Genetics* 122: 253–268.

———. 1994. Geographical variation in a quantitative character. *Genetics* 136: 361–381.

———. 2002. When and where was the most recent common ancestor? *J. Math. Biol.* 44: 253–275.

Neale, D. B., and Adams, W. T. 1985. The mating system in natural and shelterwood stands of Douglas-fir. *Theor. Appl. Genet.* 71: 201–207.

Neale, D. B., and Sederoff, R. R. 1989. Paternal inheritance of chloroplast DNA and maternal inheritance of mitochondrial DNA in loblolly pine. *Theor. Appl. Genet.* 77: 212–216.

Neel, J. V. 1967. The genetic structure of primitive human populations. *Jpn. J. Hum. Genet.* 12: 1–16.

———. 1978. The population structure of an Amerindian tribe, the Yanomama. *Annu. Rev. Gen.* 12: 365–413.

Neel, J. V., and Ward, R. H. 1972. The genetic structure of a tribal population, the Yanomama Indians, VI: analysis by F-statistics (including a comparison with the Makiritare). *Genetics* 72: 639–666.

Nei, M. 1972. Genetic distance between populations. *Am. Nat.* 106: 283–292.

———. 1973a. Analysis of gene diversity in subdivided populations. *Proc. Natl. Acad. Sci. USA* 70: 3321–3323.

———. 1973b. The theory and estimation of genetic distance. In N. E. Morton, ed., *Genetic Structure of Populations*, 53–54. Honolulu: University of Hawaii Press.

Nei, M., and Roychoudhury, A. K. 1993. Evolutionary relationships of human populations on a global scale. *Mol. Biol. Evol.* 10: 927–943.

Nevo, E. 1978. Genetic variation in natural populations: patterns and theory. *Theor. Popul. Biol.* 13: 121–177.

Nevo, E., Beiles, A., and Ben-Shlomi, R. 1984. The evolutionary significance of genetic diversity: ecological, demographic and life history correlates. In G. S. Mani, ed., *Evolutionary Dynamics of Genetic Diversity. Lecture Notes in Biomathematics*, Vol. 53, 13–213. New York: Springer-Verlag.

Nordborg, M. 1997. Structured coalescent processes on different time scales. *Genetics* 146: 1501–1514.

Nunney, L. 1999. The effective size of a hierarchically structured population. *Evolution* 53: 1–10.

Oden, N. L. 1984. Assessing the significance of a spatial correlogram. *Geogr. Anal.* 16: 1–16.

———. 1992. Spatial autocorrelation invalidates the Dow-Cheverud test. *Am. J. Phys. Anthropol.* 89: 257–264.

Oden, N. L., and Sokal, R. R. 1986. Directional autocorrelation: an extension of spatial correlograms to two dimensions. *Syst. Zool.* 35: 608–617.

Okubo, A. 1980. *Diffusion and Ecological Problems: Mathematical Models.* New York: Springer-Verlag.

Okubo, A., and Levin, S. A. 1989. A theoretical framework for data analysis of wind dispersal of seeds and pollen. *Ecology* 70: 329–338.

Olivieri, I., Michalakis, Y., and Gouyon, P. H. 1995. Metapopulation genetics and the evolution of dispersal. *Am. Nat.* 146: 202–228.

Oprian, C., Taneja, V., Voss, D., and Aroian, L. A. 1980. General considerations and interrelationships between MA and AR models, time series in m dimensions, the ARMA model. *Commun. Stat. B—Simul.* 9: 515–532.

Pasquill, F., and Smith, F. B. 1983. *Atmospheric Diffusion*, 3rd ed. Chichester, UK: Ellis Horwood.

Peakall, R., and Beattie, A. J. 1995. Does ant dispersal of seeds in *Sclerolaena diacantha* (Chenopodiaceae) generate local spatial genetic structure? *Heredity* 75: 351–361.

Peck, S. L., Ellner, S. P., and Gould, F. 1998. A spatially explicit stochastic model demonstrates the feasibility of Wright's shifting balance theory. *Evolution* 52: 1834–1839.

Perry, D. J., and Knowles, P. 1991. Spatial genetic structure within three sugar maple (*Acer saccharum* Marsh.) stands. *Heredity* 66: 137–142.

Pfeifer, P. E., and Deutsch, S. J. 1980a. A three-stage iterative procedure of space–time modelling. *Technometrics* 22: 35–47.

———. 1980b. Stationarity and invertibility regions for low order STARMA models. *Commun. Stat. B—Simul.* 9: 551–562.

———. 1980c. Identification and interpretation of first order space–time ARMA models. *Technometrics* 22: 397–408.

———. 1980d. A comparison of estimation procedures for the parameters of the STAR model. *Commun. Stat. B—Simul.* 9: 255–270.

———. 1980e. A STARIMA model-building procedure with application to description and regional forecasting. *Trans. Inst. Br. Geogr.* 5: 330–349.

Pielou, E. C. 1977. *Mathematical Ecology*, 2nd ed. New York: Wiley.

Pitman, E.J.G. 1937. The "closest estimates" of statistical parameters. *Proc. Cambridge Philos. Soc.* 33: 212–222.

Price, M. V., and Waser, N. M. 1979. Pollen dispersal and optimal outcrossing in *Delphinium nelsoni*. *Nature* 277: 294–297.

Prout, T. 1973. Appendix to J. B. Mitton and R. K. Koehn. Population genetics of marine pelecypods, III: epistasis between functionally related isozymes in *Mytilas edulis*. *Genetics* 73: 487–496.

Rausher, M. D., Augustine, D., and Vanderkooi, A. 1993. Absence of pollen discounting in a genotype of *Ipomoea purpurea* exhibiting increased selfing. *Evolution* 47: 1688–1695.

Rehfeldt, G. E. 1999. Systematics and genetic structure of *Ponderosae* taxa (Pinaceae) inhabiting the mountain islands of the southwest. *Am. J. Bot.* 86: 741–752.

Rehfeldt, G. E., Wilson, B. C., Wells, S. P., and Jeffers, R. M. 1996. Phytogeographic, taxonomic, and genetic implications of phenotypic variation in the *Ponderosae* of the southwest. *Southwest. Nat.* 41: 409–418.

Ricklefs, R. 1972. *Ecology*. New York: Chiron Press.

Ritland, K. 1985. The genetic mating structure of subdivided populations, I: open-mating model. *Theor. Popul. Biol.* 27: 51–74.

Rogers, A. R. 1987. A model of kin-structured migration. *Evolution* 41: 417–426.

———. 1988. Three components of genetic drift in subdivided populations. *Am. J. Phys. Anthropol.* 77: 435–449.

Rogers, A. R., and Eriksson, A. W. 1988. Statistical analysis of the migration component of genetic drift. *Am. J. Phys. Anthropol.* 77: 451–457.

Rogers, A. R., and Harpending, H. C. 1983. Population structure and quantitative characters. *Genetics* 105: 985–1002.

———. 1986. Migration and genetic drift in human populations. *Evolution* 40: 1312–1327.

———. 1992. Population-growth makes waves in the distribution of pairwise genetic-differences. *Mol. Biol. Evol.* 9: 552–569.

Rohlf, F. J., and Schnell, G. D. 1971. An investigation of the isolation-by-distance model. *Am. Nat.* 105: 295–324.

Rosenberg, M. S. 2002. *PASSAGE. Pattern Analysis, Spatial Statistics, and Geographic Exegesis, Version 1.0.* Tempe: Department of Biology, Arizona State University.

Rossi, M. S., Barrio, E., Latorre, A., Quezada-Diaz, J. E., Hasson, E., Moya, A., and Fontdevila, A. 1996. The evolutionary history of *Drosophila buzzatii*, XXX: mitochondrial DNA polymorphism in original and colonizing populations. *Mol. Biol. Evol.* 13: 314–323.

Rousset, F. 2000. Genetic differentiation between individuals. *J. Evol. Biol.* 13: 58–62.

Sakai, K. 1985. Studies on breeding structure in two tropical tree species. In: H. R. Gregorius, ed., *Lecture Notes In Biomathematics*, 60: *Population Genetics in Forestry*, 212–225. Berlin: Springer-Verlag.

Sawyer, S., and Felsenstein, J. 1983. Isolation by distance in a hierarchically clustered population. *J. Appl. Prob.* 20: 1–10.

Schaal, B. A. 1975. Population structure and local differentiation in *Liatris cylindracea*. *Am. Nat.* 109: 511–528.

Schmitt, J., and Gamble, S. E. 1990. The effect of distance from the parental site on offspring performance and inbreeding depression in *Impatiens capensis*: a test of the local adaptation hypothesis. *Evolution* 44: 2022–2030.

Schnabel, A., Lauschman, R. H., and Hamrick, J. L. 1991. Comparative genetic structure of two co-occurring tree species, *Maclura pomifera* (Moraceae) and *Gleditsia triacanthos* (Leguminosae). *Heredity* 67: 357–364.

Schoen, D. J., and Latta, R. G. 1989. Spatial autocorrelation of genotypes in populations of *Impatiens pallida* and *Impatiens capensis*. *Heredity* 63: 181–189.

Selander, R. K., and Kaufman, D. W. 1975. Genetic structure of populations of the brown snail (*Helix aspersa*), I: microgeographic variation. *Evolution* 29: 385–401.

Shapcott, A. 1995. The spatial genetic structure in natural populations of the Australian temperate rain forest tree *Atherosperma moschatum* (Labill.) (Monimiaceae). *Heredity* 74: 28–38.

Shaw, D. V., Kahler, A. L., and Allard, R. W. 1981. A multilocus estimator of mating system parameters in plant populations. *Proc. Natl. Acad. Sci. USA* 78: 1298–1302.

Sherry, S. T., Harpending, H. C., Batzer, M. A., and Stoneking, M. 1997. Alu evolution in human populations: using the coalescent to estimate effective population size. *Genetics* 147: 1977–1982.

Šidák, Z. 1967. Rectangular confidence regions for the means of multivariate normal distributions. *J. Am. Stat. Assoc.* 62: 626–633.

Simoni, L., Calafell, F., Pettener, D., Bertranpetit, J., and Barbujani, G. 2000. Geographic patterns of mtDNA diversity in Europe. *Am. J. Hum. Genet.* 66: 262–278.

Slatkin, M. 1995. A measure of population subdivision based on microsatellite allele frequency. *Genetics* 139: 457–462.

Slatkin, M., and Arter, H. E. 1991a. Spatial autocorrelation methods in population genetics. *Am. Nat.* 138: 499–517.

———. 1991b. Reply to Sokal and Oden. *Am. Nat.* 138: 522–523.

Slatkin, M., and Barton, N. H. 1989. A comparison of three indirect methods for estimating average levels of gene flow. *Evolution* 43: 1349–1368.

Slatkin, M., and Hudson, R. R. 1991. Pairwise comparisons of mitochondrial DNA sequences in stable and exponentially growing populations. *Genetics* 129: 555–562.

Slatkin, M., and Maddison, W. P. 1989. A cladistic measure of gene flow from the phylogenies of alleles. *Genetics* 123: 603–613.

Slatkin, M., and Voelm, L. 1991. F_{ST} in a hierarchical island model. *Genetics* 127: 627–629.

Smouse, P. E. 1998. To tree or not to tree. *Mol. Ecol.* 7: 399–412.

Smouse, P. E., and Peakall, R. 1999. Spatial autocorrelation analysis of individual multiallele and multilocus genetic structure. *Heredity* 82: 561–573.

Smouse, P. E., Long, J. C., and Sokal, R. R. 1986. Multiple regression and correlation extensions of the Mantel test of matrix correspondance. *Syst. Zool.* 35: 627–632.

Sneath, P.H.A., and Sokal, R. R. 1973. *Numerical Taxonomy*. San Francisco: WH Freeman.

Sokal, R. R. 1979a. Ecological parameters inferred from spatial correlograms. In G. P. Patil, and M. L. Rosenzweig, eds., *Contemporary Quantitative Ecology and Related Econometrics*, 167–196. Fairland, MD: International Cooperative Publishing House.

———. 1979b. Testing statistical significance of geographic variation patterns. *Syst. Zool.* 28: 227–231.

———. 1986. Spatial data analysis and historical processes. In E. Diday, Y. Escoufier, L. Lebart, J. Pages, Y. Schektman, and R. Tomassone, eds., *Data Analysis and Informatics*, IV, 29–43. Amsterdam: Elsevier Science Pub. B.V. (North Holland).

———. 1988. Genetic, geographic, and linguistic distances in Europe. *Proc. Natl. Acad. Sci. USA* 85: 1722–1726.

Sokal, R. R., Bird, J., and Riska, B. 1980. Geographic variation in *Pemphigus populicaulis* (Insecta: Aphididae) in eastern North America. *Biol. J. Linn. Soc.* 14: 163–200.

Sokal, R. R., and Friedlaender, J. 1982. Spatial autocorrelation analysis of biological variation on Bougainville Island. In M. H. Crawford and J. H. Mielke, eds., *Current Developments in Anthropological Genetics*, Vol. 2, 205–227. New York: Plenum.

Sokal, R. R., Harding, R. M., and Oden, N. L. 1989a. Spatial patterns of human gene frequencies in Europe. *Am. J. Phys. Anthropol.* 80: 267–294.

Sokal, R. R., and Jacquez, G. M. 1991. Testing inferences about microevolutionary processes by means of spatial autocorrelation analysis. *Evolution* 45: 152–168.

Sokal, R. R., Jacquez, G. M., and Wooten, M. C. 1989b. Spatial autocorrelation analysis of migration and selection. *Genetics* 121: 845–855.

Sokal, R. R., and Menozzi, P. 1982. Spatial autocorrelations of HLA frequencies in Europe support demic diffusion of early farmers. *Am. Nat.* 119: 1–17.

Sokal, R. R., and Oden, N. L. 1978a. Spatial autocorrelation in biology, 1: methodology. *Biol. J. Linn. Soc.* 10: 199–228.

———. 1978b. Spatial autocorrelation in biology, 2: some biological implications and four applications of evolutionary and ecological interest. *Biol. J. Linn. Soc.* 10: 229–249.

———. 1991. Spatial autocorrelation analysis as an inferential tool in population genetics. *Am. Nat.* 138: 518–521.

Sokal, R. R., Oden, N. L., and Barker, J.S.F. 1987. Spatial structure in *Drosophila buzzatii* populations: simple and directional spatial autocorrelation. *Am. Nat.* 129: 122–142.

Sokal, R. R., Oden, N. L., Rosenberg, M. S., and Thomson, B. A. 2000. Cancer incidences in Europe related to mortalities, and ethnohistoric, genetic, and geographic distances. *Proc. Natl. Acad. Sci. USA* 97: 6067–6072.

Sokal, R. R., Oden, N. L., and Thomson, B. A. 1992. Origins of the Indo-Europeans: genetic evidence. *Proc. Natl. Acad. Sci. USA* 89: 7669–7673.

———. 1997a. A simulation study of microevolutionary inferences by spatial autocorrelation analysis. *Biol. J. Linn. Soc.* 60: 73–93.

———. 1998a. Local spatial autocorrelation in biological variables. *Biol. J. Linn. Soc.* 45: 41–62.

———. 1998b. Local spatial autocorrelation in a biological model. *Geogr. Anal.* 30: 331–354.

Sokal, R. R., Oden, N. L., Walker J., and Waddle, D. M. 1997b. Using distance matrices to choose between competing theories and an application to the origin of modern humans. *J. Hum. Evol.* 32: 501–522.

Sokal, R. R., Oden, N. L., and Wilson, C. 1991. Genetic evidence for the spread of agriculture in Europe by demic diffusion. *Nature* 351: 143–145.

Sokal, R. R., Smouse, P. E., and Neel, J. V. 1986. The genetic structure of a tribal population, the Yanomama Indians, XV: patterns inferred by autocorrelation analysis. *Genetics* 114: 259–287.

Sokal, R. R., and Thomson, J. D. 1987. Applications of spatial autocorrelation in ecology. In P. Legendre and L. Legendre, eds., *Developments in Numerical Ecology* (NATO ASI Series, Vol. G14), 431–466. Berlin: Springer-Verlag.

Sokal, R. R., and Wartenberg, D. E. 1981. Space and population structure. In

D. Griffith and R. McKinnon, eds., *Dynamic Spatial Models*, 186–213. Alphen aan den Rijn, The Netherlands: Sijthoff & Noordhoff.

———. 1983. A test of spatial autocorrelation analysis using an isolation-by-distance model. *Genetics* 105: 219–237.

Sokal, R. R., and Winkler, E. M. 1987. Spatial variation among Kenyan tribes and subtribes. *Hum. Biol.* 59: 147–164.

Spielman, R. S., McGinnis, R. E., and Ewens, W. J. 1993. Transmission test for linkage disequilibrium: the insulin gene region and insulin-dependent diabetes mellitus (IDDM). *Am. J. Hum. Genet.* 52: 506–516.

Stanley, S. E. 1997. Alu repeats and human evolution. *J. Mol. Evol.* 45: 6–7.

Stoneking, M., and Soodyall, H. 1996. Human evolution and the mitochondrial genome. *Curr. Opin. Genet. Dev.* 6: 731–736.

Subramaniam, B., and Rausher, M. D. 2000. Balancing selection on a floral polymorphism. *Evolution* 54: 691–695.

Sykes, B., Leiboff, A., Low-Beer, J., Tetzner, S., Richards, M., et al. 1995. The origins of the Polynesians: an interpretation from mitochondrial lineage analysis. *Am. J. Hum. Genet.* 57: 1463–1475.

Takahata, N. 1991. Genealogy of neutral genes and spreading of selected mutations in a geographically structured population. *Genetics* 129: 585–595.

———. 1993. Allelic genealogy and human evolution. *Mol. Biol. Evol.* 10: 2–22.

Taneja, V. A., and Aroian, L. A. 1980. Time series in m dimensions, autoregressive models. *Commun. Stat. B—Simul.* 9: 491–513.

Templeton, A. R. 1993. The 'Eve' hypothesis: a genetic critique and reanalysis. *Am. Anthropol.* 95: 51–72.

———. 1997. Out of Africa? What do genes tell us? *Curr. Opin. Genet. Dev.* 7: 841–847.

———. 1998. Nested clade analyses of phylogeographic data: testing hypotheses about gene flow and population history. *Mol. Ecol.* 7: 381–397.

Templeton, A. R., Routman, E., and Phillips, C. A. 1995. Separating population structure from population history: a cladistic analysis of the geographical distribution of mitochondrial DNA haplotypes in the tiger salamander, *Ambystoma tigrinum*. *Genetics* 140: 767–782.

Tobler, W. R. 1970. A computer movie simulating urban growth in the Detroit region. *Econ. Geogr.* 46: 234–240.

Torroni, A., Schurr, T. G., Cabell, M. F., Brown, M. D., Neel, J. V., Larsen, M., Smith, D. G., Vullo, C. M., and Wallace, D. C. 1993a. Asian affinities and continental radiation of the four founding Native American mtDNAs. *Am. J. Hum. Genet.* 53: 563–590.

Torroni, A., Sukernik, R. I., Schurr, T. G., Starikovskaya, Y. B., Cabell, M. F., Crawford, M. H., Comuzzie, A. G., and Wallace, D. C. 1993b. mtDNA variation of aboriginal Siberians reveals distinct genetic affinities with Native Americans. *Am. J. Hum. Genet.* 53: 591–608.

Tufto, J., Engen, S., and Hindar, K. 1997. Stochastic dispersal processes in plant populations. *Theor. Popul. Biol.* 52: 16–26.

Turner, M. E., Stephens, J. C., and Anderson, W. W. 1982. Homozygosity and patch structure in plant populations as a result of nearest-neighbor pollination. *Proc. Natl. Acad. Sci. USA* 79: 203–207.

Underhill, P. A., Jin, L., Lin, A. A., Mehdi, S. Q., Jenkins, T., Vollrath, D., Davis, R. W., Cavalli-Sforza, L. L., and Oefner, J. 1997. Detection of numerous Y chromosome biallelic polymorphisms by denaturing high-performance liquid chromotography (DHPLC). *Genome Res.* 7: 996–1005.

Upton, G.J.G., and Fingleton, B. 1985. *Spatial Data Analysis by Example,* Vol. 1: *Point Pattern and Quantitative Data.* New York: Wiley.

Uyenoyama, M. K. 1986. Inbreeding and the cost of meiosis: the evolution of selfing in populations practicing partial biparental inbreeding. *Evolution* 40: 388–404.

Uyenoyama, M. K., and Waller, D. M. 1991. Coevolution of self-fertilization and inbreeding depression, I: mutation–selection balance at one and two loci. *Theor. Popul. Biol.* 40: 14–46.

VanStaaden, M. J., Michener, G. R., and Chesser, R. K. 1996. Spatial analysis of microgeographic genetic structure in Richardson's ground squirrels. *Can. J. Zool.* 74: 1187–1195.

Vigilant, L., Stoneking, M., Harpending, H., Hawkes, K., and Wilson, A. C. 1991. African populations and the evolution of human mitochondrial DNA. *Science* 253: 1503–1507.

Wade, M. J., and Goodnight, C. J. 1998. Perspective: the theories of Fisher and Wright in the context of metapopulations: when nature does many small experiments. *Evolution* 52: 1537–1553.

Wagner, D. B., Sun, Z.-X., Govindaraju, D. R., and Dancik, B. P. 1991. Spatial patterns of chloroplast DNA and cone morphology variation within populations of a *Pinus banksiana–Pinus contorta* sympatric region. *Am. Nat.* 138: 156–170.

Wakely, J. 2001. The coalescent in an island model of population subdivision with variation among demes. *Theor. Popul. Biol.* 59: 133–144.

Walter, R., and Epperson, B. K. 2001. Geographic pattern of genetic variation in *Pinus resinosa*: area of greatest diversity is not the origin of postglacial populations. *Mol. Ecol.* 10: 103–111.

Waser, N. M., and Price, M. V. 1983. Optimal and actual outcrossing, and the nature of plant–pollinator interaction. In C. E. Jones, and R. J. Little, eds., *Handbook of Experimental Pollination Biology*, 341–359. New York: Van Nostrand Reinhold.

———. 1989. Optimal outcrossing in *Ipomopsis aggregata*: seed set and offspring fitness. *Evolution* 43: 1097–1109.

Waser, P. M., and Elliott, L. F. 1991. Dispersal and genetic structure in kangaroo rats. *Evolution* 45: 935–943.

Watterson, G. A. 1975. On the number of segregating sites in genetical models without recombination. *Theor. Popul. Biol.* 7: 256–276.

Weir, B. S. 1990. *Genetic Data Analysis*. Sunderland, MA: Sinauer Assoc. Inc.

Weir, B. S., and Cockerham, C. C. 1984. Estimating *F*-statistics for the analysis of population structure. *Evolution* 38: 1358–1370.

Weiss, G. H., and Kimura, M. 1965. A mathematical analysis of the stepping stone model of genetic correlation. *J. Appl. Prob.* 2: 129–149.

Whitlock, M. C., and Barton, N. H. 1997. The effective size of a subdivided population. *Genetics* 146: 427–441.

Whitlock, M. C., and McCauley, D. E. 1990. Some population genetic consequences of colony formation and extinction: genetic correlations within founding groups. *Evolution* 44: 1717–1724.

Whitten, E.H.T. 1975. The practical use of trend-surface analysis in the geographical sciences. In J. C. Davis and M. J. McCullough, eds., *Display and Analysis of Spatial Data*, 282–297. New York: Wiley.

Whittle, P. 1954. On stationary processes in the plane. *Biometrika* 41: 434–449.

Williams, S. L., and Di Fiori, R. E. 1996. Genetic diversity and structure in *Pelvetia fastigiata* (Phaeophyta: Fucales): does a small effective neighborhood size explain fine-scale genetic structure? *Mar. Biol.* 126: 371–382.

Wilson, P. 1997. Mitochondrial DNA variation and biogeography among some etheostomid darters of the Central Highlands. PhD thesis, Division of Biomedical and Biological Sciences, Washington University, St. Louis.

Wolpoff, M. H. 1989. Multiregional evolution: the fossil alternative to Eden. In P. Mellars and C. Stringer, eds., *The Human Revolution: Behavioural and Biological Perspectives on the Origins of Modern Humans*, 62–108. Princeton, NJ: Princeton University Press.

Womble, W. H. 1951. Differential systematics. *Science* 114: 315–322.

Workman, P. L., and Niswander, J. D. 1970. Population studies on southwestern indian tribes, II: local genetic differentiation in the Papago. *Am. J. Hum. Gen.* 22: 24–49.

Wright, J. W. 1976. *Introduction to Forest Genetics*. New York: Academic Press.

Wright, S. 1921. Correlation and causation. *J. Agric. Res.* 20: 557–585.

———. 1922. Coefficients of inbreeding and relationship. *Am. Nat.* 56: 330–338.

———. 1931. Evolution in mendelian populations. *Genetics* 16: 97–159.

———. 1938. Size of population and breeding structure in relation to evolution. *Science* 87: 430–436.

———. 1940. Breeding structure of populations in relation to speciation. *Am. Nat.* 74: 232–248.

———. 1943. Isolation by distance. *Genetics* 28: 114–138.

———. 1946. Isolation by distance under diverse systems of mating. *Genetics* 31: 39–59.

———. 1951. The genetical structure of populations. *Ann. Eugen.* 15: 323–354.
———. 1965. The interpretation of population structure by *F*-statistics with special regard to systems of mating. *Evolution* 19: 395–420.
———. 1968. Dispersion of *Drosophila melanogaster*. *Am. Nat.* 102: 81–84.
———. 1978. *Evolution and the Genetics of Populations*, Vol. 4: *Variability within and among Populations*. Chicago: University of Chicago Press.
Zhivotovsky, L. A., Feldman, M. W., and Bergman, A. 1996. On the evolution of phenotypic plasticity in a spatially heterogeneous environment. *Evolution* 50: 547–558.
Zimmerman, M. 1981. Optimal foraging, plant-density and the marginal value theorem. *Oecologia* 49: 148–153.

Index

Page references to figures are in italics.

admixture, 30, 55, 59, 62, 66, 168–169
age structure, 223–231
Alu retroposable elements, 61–62
Analysis of Molecular Variance AMOVA, 112
ancient DNA samples, 44, 52, 62. *See also* space–time data
a posteriori expected value, 125–126
a priori expected value, 69, 75, 108, 114, 124–126, 128, 150, 174, 180, 244, 288, 291, 298–300, 302, 303, 305, 307
autocorrelation indices for DNA analysis AIDA, 64, 112–113
autoregression, spatial, 91, 92
autoregressive structure, space–time, 82, 96–101, 177

backshift operator, 71, 129, 140–141
barriers to gene flow, 31, 32, 33, 34, 63, 83, 89, 90
biparental inbreeding and mating by proximity, 185–187, 192, 220–223
bottleneck in population size, 37, 65
Bougainville Island, 21

Cecropia obtusifolia, 225–227
chloroplast DNA, 66, 111, 120, 162, 216, 242–243
cline, 21, 25, 26, 27–29, 37–40, 63, 64, 90, 92, 164–167, 238–239
clonal reproduction, 202–203, 225, 231–233
clumping, 223–231, 296–297, 307
coalescence, 42, 44, 46, 53, 111–112, 121–122, 160–161, 289–290. *See also* gametic kinship chain
colonization, 22, 43, 52, 53, 55, 65, 203, 231
connectivity, 31–32, 83–84, 176–177
correlated gene dispersal, 225, 297. *See also* stochastic migration
correlations among distance classes, 78–79, 254
covariance, a priori, 299–300
cross-correlations of genetic and environmental factors, 5, 13, 241

deme, 186, 187
demic diffusion, 63
demographic factors and spatial structure within populations, 223–233
density, standardizing by, 186. *See also* Wright's neighborhood size
density effects on spatial structure, 204
diffusion approximation, 2, 8, 123, 301
dimensional directionality, 87, 134, 293, 294
direct measure of dispersal, 205
directional autocorrelation, 25–31, 84–87, 176, 294
directional migration or dispersal, 26, 39, 86–87, 121, 128–129, 130, 133–134, 293, 312–313
dispersal mechanics, 170–171, 308–316
distance classes, configuration of, 15, *72–73*, 74, 82, 84, 106, 205–206, 245
distance/direction classes, configuration of, 26–27, 28–29, *28*, 86
DNA sequence data, information content of, 42, 44, 56–58, 64, 110–113, 161–163, 167–168, 181–182, 326–327
Dow and Cheverud test, 81
Drosophila buzzatii, 27–29, 33, 65–66

effective population size, 37, 46–47, 57, 60–61, 104, 110, 120
elliptical mound, *25*

environmental patches, 40, 87, 98, 105–106, 146–147, 166, 171, 238, 239–242, 316
Europe, population genetics of, 23, 34, 63–64
Eve hypothesis. *See* out-of-Africa theory
exponential decrease of genetic similarity with distance, 15, 82, 107, 123, 124, 126, 135, 144, 294, 296, 298, 304

finest (spatial) lag structure, 71, 100, 102, 103, 135, 176
forecasting, 101, 105
fossil DNA. *See* ancient DNA samples
Fourier transform, 122–123, 156, 293, 294, 297–298, 300, 307
fragmentation, 43, 55, 65, 66
F statistics, 34, 40, 41, 71, 106, 107–110, 139, 149–152, 163, 173–174, 176, 177, 187–188, 191–192, 208, 280–284, 305–308. *See also* hierarchical models
F_{ST}. *See F* statistics
fusion/fission of populations, 30, 35–36, 147–148

Gabriel connectivity, 32, *72–73*, 83–84, *85*, 97, 176
gametic kinship chain, 160–162, 289–292. *See also* coalescence
gene frequency surface, 29, 30, 33, 38, 39, 89–92
genetic differentiation among populations, per se, 33–34, 40–42, 107–110, 149–152, 163–164, 168–170
genetic disease, 5, 13, 114, 168–169
genetic distance, 24, 28, 30, 63, 79–81, 107
genetic diversity statistics, 40–42, 107–110, 112
genetic patches, 17–18, 27, 190–191, *191*, 195, 206–208, 211, 213, 215, 219, 261, 264, 266–267
geographic origin of anatomically modern humans, 45–62
glaciation, 65, 66

habitat patches, 95, 224, 230–231, 238, 239–242, 316. *See also* environmental patches
haplotypes, 42, 52–55, 56–57, 58–60, 64–66, 110–113, 161–162, 168, 182, 216, 243, 326–327
heterozygosity, per se, 40, 160
hierarchical models, 71, 108, 109, 112, 116, 149–152, 163, 176, 189, 280–289, 305–308. *See also F*-statistics
HLA, major histocompatibility complex, 60–61
homogeneity test of allele frequencies among populations, 108–109
hybridization zone, 242–243

immigration, 21–22, 23, 26, 29–30, 43, 63–64, 90, 92, 98, 120, 136, 290
inbreeding depression and spatial structure, 220–223
infinite alleles mutation model, 42, 110, 118, 120, 160–162, 290, 292, 308
infinite sites mutation model, 42, 110–113, 161–163, 167
invasive species, 29, 316
Ipomoea purpurea, 213–214, 237–238
isotropy, meaning of, 26, 294, 295. *See also* directional migration or dispersal

join counts, 245–267
join-count statistics, 209–217, 225–231, 251, 255–267; multilocus form of, 218, 219; SND correlogram of, 251

Kenya, population genetics of, 18–21
kinship, conditional, 24, 106–107, 180, 285, 287, 301–305. *See also* kinship measures
kinship measures, 24, 68–69, 76, 106–107, 180, 189, 218–219, 284–287, 299–305
kin-structured migration, 110, 120, 137, 144–146, 151–152, 165–166, 177, 178, 179, 225. *See also* correlated gene dispersal

Laplace transform, 160, 300
life history factors, effect of on differentiation among populations, 41–42

Linanthus parryae, 187–188
linguistic groups, 30, 34, 63, 90
linkage disequilibrium, 9–10, 30, 107, 112, 168–170, 218, 219, 280, 287
local autocorrelation, 87–89
local indicator of spatial association LISA, 87–89, 167, 240–241

Manhattan distance, 82, 94
Mantel statistic, 24, 26, 27, 30, 36, 63, 79–81, 94–95
mating system, 119, 197–204, 220–221, 296–297
metapopulation, 14, 37, 203, 316, 324
microsatellite, 42, 62, 163–164, 272, 279
migration, complex paths of, 21, 31–34, 81–84, 149–150
migration behavior, 31, 83, 170–171, 314–315
migration matrix models, 149–150, 174–175, 303
mitochondrial DNA, 46–58, 64, 65–66, 111, 120, 162, 181, 216–217
mitochondrial Eve, 46–49, 52–53. *See also* out-of-Africa theory
molecular genetic data, special attributes of, 44, 56–57, 110–113, 160–164, 181–182, 272, 278–279, 326–327
Moran's *I* statistic, 15, 68–69, 71–79, 81–82, 93, 94, 95, 106, 107, 166–167, 180, 192–209; construction of correlograms for, 193; for converted genotypes, 192–193, 197–207, 252–254, 268–280, 285–287; normality null hypothesis for, 75, 76–77; for quadrat samples, 192, 194–195, 208–209, 267–268; randomization null hypothesis for, 75–77; test of significance for correlograms of, 77–79; weighted, 71–77, 81–82; *X* intercept of correlograms of, 195, 206–208, 211
most recent common ancestor MRCA, 46, 181. *See also* coalescence
moving average, 99, 129–134, 141–148
multilocus genotypes, 9, 12, 92–95, 107, 168–170, 179, 181–182, 217–219, 235, 272, 279–280, 284–287, 305, 319–320, 322, 324
multilocus processes, 168–170
multiregional hypothesis, 45–46, 47, 61
mutation models, 160–164

nearest neighbor, 32, 83. *See also* connectivity
nested cladistic analysis, 53–56, 58–59, 64–66

origination, geographic, 4, 44, 48–52, 66, 157–158, 180–181
out-of-Africa theory 45–62
outside systematic force or pressure, 120, 134, 136, 137, 189, 292. *See also* recall coefficient
overlapping generations, 101, 135

patches, matrilineal, 224–231
Pemphigus populicaulis, 22–23
phylogenetic reconstruction, 44, 53, 55, 65, 66, 111, 161
pollen dispersal mechanisms, 197–204
population size: fluctuations in, 36–37; sudden increases in, 167–168
population size/density, localized regulation of, 223–231, 241–242, 296–297
porosity, 197, 206, 256–262, 268–269, 282–283
probabilities of identity by descent, general definition, 288–299
process identification, 101, 103

quadrat samples, 194–196, 267, 281–284
quantitative traits, 19, 37, 38, 39, 116, 196, 219–220
quasistationary state, 190, 255

random field, 67, 96, 97
random walk model of dispersal, 310–311, 314–316
range expansion, 43, 52–56, 58–59, 63, 64, 65, 66
recall coefficient, 136, 167, 298, 303. *See also* outside systematic force or pressure
refugia, 43, 65, 66

sample density. *See* porosity
sample design for within-populations, 255–262, 266–268
sampled populations vs. unsampled populations per unit area, 20
seasonal trends, 105
seed dispersal, correlated, 224, 225–227
seed dispersal mechanisms, 197–204
selection, 37–40, 94, 100, 120, 233–243; environmental, 37–40, 87, 90, 98, 106, 136, 146, 164–167, 174; gradients of, 37–40, 90, 165; microenvironmental, 235–236, 238–243; seasonal or cyclic, 147
Silene dioica, 227–231
space time autoregressive moving average STARMA models; statistical, 99–106, 178; of stochastic processes, 136–148
space time autoregressive STAR models; statistical, 99–106; of stochastic processes, 126–136
space–time coalescence 52, 56, 158–159
space–time correlation; statistics of, 101–103, 105; in theoretical models, 128–135, 139–148
space–time data, 10, 68, 96, 101, 152, 159, 178–179, 181–182, 302; analysis of, 99–106
space–time moving average STMA models, 129, 141
space–time partial correlations, 101, 103, 105, 135
space–time probabilities of identity by descent, 48–52, 54, 55, 56, 64, 116–124, 152–158, 159, 181
spatial autoregressive SAR model, 96–98
spatial correlations in theoretical models, 124–135, 139–140, 142–148, 178. *See also* Moran's *I* statistic
spatial lag operator, 71
spatial lag structure, 69–71, 83–84, 97, 116, 121
spatial models, 95–99
spatial probabilities of identity by descent, 116–124, 288–299
spatial processes. *See* spatial models
spatial regularity, 69–71, 74, 81–84, 116, 120, 147, 176, 294
spatial scale, 7, 12, 17–18, 26–27, 30, 38, 39–40, 90, 99, 135, 151, 166–167, 169, 186, 197, 206–207, 208, 211, 221, 235–236, 238–239, 256, 257, 266–267, 284, 308, 311
spatial stationarity, 69–71, 74, 81, 84, 87, 89, 90, 100, 103, 116, 147, 176–177, 240, 295, 297
spatial time series. *See* space time autoregressive moving average STARMA models
stepping stone migration model, 121, 125, 130–131, 132–133, 147–148, 170–171
stepwise mutation model, 42, 163–164
stochastic migration, 34–36, 40–41, 99, 103, 110, 136–148, 151–152, 177–178; negatively shared effects of, 35–36, 147–148, 178; positively shared effects of, 35, 140–147, 178, 297. *See also* correlated gene dispersal
substitution rates, 160

temporal stationarity, 69
time (back) to the most recent common ancestor TMRCA, 46–48
torus, 120, 297
transmission disequilibrium test TDT, 169, 324
trend surface analysis, 39–40, 63, 90–92, 105

weak stationarity, 70
weights, 71–73, 74–75
wombling, 33–34, 36, 89–90
Wright's island model, 14, 37, 40, 108, 109, 149–152, 163, 164, 173–174
Wright's neighborhood size, 183, 186–187, 189, 195, 305–306
Wright's shifting balance theory, 13–14, 168, 241–242

Yanomama, 30–31, 33–34, 35–36
Y-chromosome, 58–59, 64